NATURAL GAS – EXTRACTION TO END USE

Edited by **Sreenath Borra Gupta**

Natural Gas – Extraction to End Use
http://dx.doi.org/10.5772/2582
Edited by Sreenath Borra Gupta

Contributors

Michal Netušil, Pavel Ditl, Pramathesh R. Mhaskar, Arun S. Moharir, Mert Atilhan, Santiago Aparicio, M. F. Attallah, N. S. Awwad, H. F. Aly, V. Yildirim, T. Yomralioglu, R. Nisanci, Miguel Edgar Morales Udaeta, Jonathas Luiz de Oliveira Bernal, Luiz Claudio Ribeiro Galvão, José Aquiles Baesso Grimoni, Marcelo Ramos Martins, Adriana Miralles Schleder, Eui-Seob Park, So-Keul Chung, Dae-Hyuk Lee, Taek-Gon Kim, Liji Huang, Sreenath B. Gupta, Munidhar Biruduganti, Bipin Bihari, Raj Sekar, Tatiana Morosuk, George Tsatsaronis, Gaetano Iaquaniello, Elena Antonetti, Barbara Cucchiella, Emma Palo, Annarita Salladini, Alessandra Guarinoni, Andrea Lainati, Luca Basini, Fernando Juárez

Published by InTech

Janeza Trdine 9, 51000 Rijeka, Croatia

Notice

Publishing Process Manager Marina Jozipovic
Technical Editor InTech Prepress, Novi Sad
Cover InTech Design team

Additional hard copies can be obtained from orders@intechopen.com

Natural Gas – Extraction to End Use, Edited by Sreenath Borra Gupta
p. cm.
ISBN 978-953-51-0820-7

For EU product safety concerns:
IN TECH d.o.o., Prolaz Marije Krucifikse Kozulić 3, 51000 Rijeka, Croatia,
info@intechopen.com or visit our website at intechopen.com.

Meet the editor

Dr. Sreenath Borra Gupta obtained his B. Tech from IIT-Madras, and subsequently a M.S. and a Ph.D. from The Pennsylvania State University. After graduation, he held several combustion engineering positions at Abar Ipsen, Gas Technology Institute and Rolls-Royce. Dr. Gupta currently holds the position of a research engineer at Distributed Energy Research Center (DERC) belonging to Argonne National Laboratory, USA. In this capacity he conducts research to improve the performance and reduce the emissions footprint of stationary natural gas engines. His efforts include (i) development of a laser ignition system for natural gas engines, (ii) evaluation of polymeric membranes for nitrogen enrichment of combustion air, (iii) development of advanced combustion laser diagnostics, (iv) promoting the use of alternative gaseous fuels, and (v) improving the viability of micro-CHP systems. Dr. Gupta is happily married, has two lovely kids and a dog, who keep him busy on the home front.

Contents

Preface

With horizontal drilling, hydraulic fracturing and other such advances in NG extraction, vast amounts of shale gas have become available over the last 5 years. This has resulted in the NG reserves from increasing from a 63 year supply to over 100 years in the US alone. As this trend continues worldwide, the NG distribution landscape is likely to change, and various new technologies are likely to be developed.

For those of us deeply entrenched in natural gas research, it is mind boggling to recognize the ramifications that natural gas has from drilling and exploration, distribution and storage, end use, and finally in the financial markets. This book set out to document modeling practices in drilling, exploration, distribution and financial markets of natural gas, however, to a large extent has also captured the state-of-the-art in end use natural gas technologies. My personal favorites in the end use technologies are

1. Underground LNG storage
2. Advances in natural gas metering
3. LNG based cogeneration systems evaluation using energy analysis, and
4. Natural gas treatment using adsorptive separation.

With the help of excellent staff at InTech, we have collated articles submitted by 38 authors from 14 different countries spread throughout the world. Though the contents of each chapter are rich and valuable, they are as diverse as the culture and topology associated with each of these countries. For example, the population distribution patterns, and difficult terrain conditions have resulted in Bolivia and Western Brazil in developing Virtual Distribution Pipelines and gas compression and distribution systems. On the other hand, the diversity also proved to be a challenge as we had to overcome the language barrier and had to spend an extra effort to translate the author's ideas for clarity. For the occasional editorial mistakes, that are inevitable in this process, I wish to apologize to the reader ahead of time.

I would like to thank my manager, Mr. Raj Sekar for his support, and my colleagues Bipin Bihari and Munidhar Biruduganti for their help in reviewing the contents.

Finally, I would like to thank my family, who have endured my long absences over several weekends.

Sreenath Borra Gupta
Distributed Energy Research Center
Argonne National Laboratory
USA

Section 1

Exploration, Drilling and Processing

Chapter 1

Natural Gas Dehydration

Michal Netušil and Pavel Ditl

Additional information is available at the end of the chapter

http://dx.doi.org/10.5772/45802

1. Introduction

The theme of natural gas (NG) dehydration is closely linked with storage of natural gas. There are two basic reasons why NG storage is important. Firstly, it can reduce dependency on NG supply. With this in mind, national strategic reserves are created. Secondly, NG storage enables the maximum capacity of distribution lines to be exploited. NG is stored in summer periods, when there is lower demand for it, and is withdrawn in the winter periods, when significant amounts of NG are used for heating. Reserves smooth seasonal peaks and also short-term peaks of NG consumption. Underground Gas Storages (UGS) are the most advantageous option for storing large volumes of gas. Nowadays there are approximately 135 UGSs inside the European Union. Their total maximum technical storage capacity is around 10^9 m_s^3. According to the latest update, over $0{,}7 \cdot 10^9$ m_s^3 of additional storage capacity will come on stream in Europe by 2020 [1]. There are three types of UGSs: (1) Aquifers, (2) Depleted oil/gas fields, and (3) Cavern reservoirs (salt or hard rock). Each of these types possesses distinct physical characteristics. The important parameters describing the appropriateness of UGS use are storage capacity, maximum injection/withdrawal performance, and gas contamination during storage. Generally, the allowable pressure of stored gas inside a UGS is up to 20 MPa. The pressure inside increases as the gas is being injected, and decreases when gas is withdrawn. The output gas pressure depends on further distribution. Distribution sites from UGS normally begin at 7 MPa. The temperature of the gas usually ranges from 20 - 35°C. The exact temperature varies with the location of the UGS and with the time of year.

1.1. Water in the gas

A disadvantage of UGSs is that during storage the gas become saturated by water vapors. In the case of depleted oil field UGSs, vapors of higher hydrocarbons also contaminate the stored gas. The directive for gas distribution sets the allowable concentration of water and concentration of higher hydrocarbons. In the US and Canada, the amount of allowable water in the gas is specified in units: pounds of water vapor per million cubic feet (lbs/MMcft). This amount should be lower than 7 lbs/MMcft [2], which is equivalent to 0,112 g_{H2O}/m_s^3. In Europe, the concentration of water and higher hydrocarbons is specified by their dew point temperature (T_{dew}). T_{dew} for water is -7°C for NG at 4 MPa, and T_{dew} for hydrocarbons is 0°C for NG at the operating pressures [3]. This value for water is equivalent to roughly 0,131 g_{H2O}/m_s^3 of NG at 4 MPa. As was stated above, the distribution specifications depend on the geographic region in which they are applied. For example, in Nigeria water T_{dew} should be below 4°C for NG at 4 MPa, which means that the NG can contain more than twice as much water vapors as in Europe.

The water content of NG at saturation is dependent on temperature and pressure. With increasing pressure of the gas the water content decreases, and with increasing temperature the water content in the gas increases. This is well presented in Figure No. 20, Chapter 20, in the GPSA Data Book, 12th Edition. The water content of the gas can be calculated using the following equation [4, 5]:

$$w_{water} = 593{,}335 \cdot exp(0.05486 \cdot t_G) \cdot {P_G}^{-0.81462} \tag{1}$$

Where w_{water} is in kilograms of water per 10^6 m_s^3 of NG, t_G is temperature of NG in °C, and P_G is pressure of NG in MPa.

The average value of water in NG withdrawn from UGS is 2 - 5 times higher than required. An NG dehydration step is therefore essential before further distribution.

1.2. Problems with water in the gas

If the temperature of pipeline walls or storage tanks decreases below the T_{dew} of the water vapors present in the gas, the water starts to condense on those cold surfaces, and the following problems can appear.

- NG in combination with liquid water can form methane hydrate. Methane hydrate is a solid in which a large amount of methane is trapped within the crystal structure of water, forming a solid similar to ice. The methane hydrate production from a unit amount of water is higher than the ice formation. The methane hydrates formed by cooling may plug the valves, the fittings or even pipelines.
- NG dissolved in condensed water is corrosive, especially when it contains CO_2 or H_2S.
- Condensed water in the pipeline causes slug flow and erosion.
- Water vapor increases the volume and decreases the heating value of the gas.
- NG with the presence of water vapor cannot be operated on cryogenic plants.

2. Dehydration methods

2.1. Absorption

The most widely-used method for industrial dehydration of NG is absorption. Absorption is usually performed using triethyleneglycol sorbent (TEG). Absorption proceeds at low temperatures and the absorbed water is boiled out from TEG during regeneration in a reboiler at high temperatures. Some physical properties of pure TEG are given in the following text.

Viscosity data vs. temperature is shown in Table 1 and is shown in a graph in Figure 1 [6].

[°C]	4	10	16	21	27	32	38	43	49	54	60	66
[m^2/s 10^{-5}]	9,53	7,094	5,367	4,124	3,214	2,539	2,032	1,646	1,348	1,116	0,934	0,788
[°C]	71	77	82	88	93	99	104	110	116	121	127	132
[m^2/s 10^{-5}]	0,672	0,577	0,5	0,436	0,384	0,34	0,303	0,272	0,245	0,222	0,203	0,186
[°C]	138	143	149	154	160	166	171	177	182	188	193	199
[m^2/s 10^{-5}]	0,171	0,159	0,147	0,138	0,129	0,121	0,115	0,109	0,103	0,099	0,095	0,091

Table 1. Kinematic viscosity of TEG according to temperature

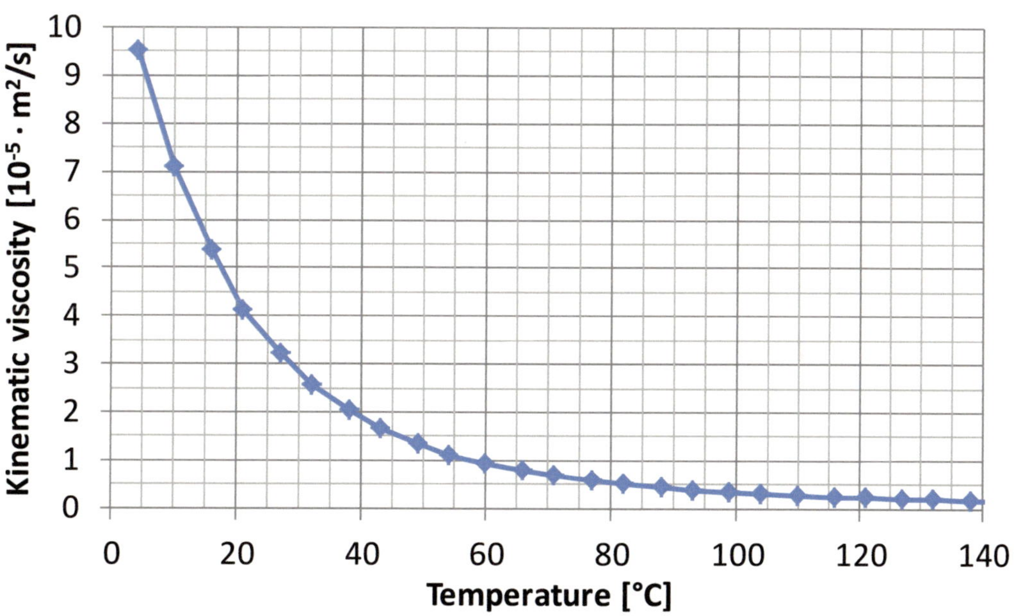

Figure 1. Kinematic viscosity of TEG as a function of temperature

It follows from Figure 1 that the kinematic viscosity of TEG increases dramatically with low temperatures. The temperature of TEG during a process should never decrease below 10°C. The reason is to prevent pump damage or even clogging of the flow. For temperatures above 100°C, the viscosity changes just slightly and the average kinetic viscosity value $5 \cdot 10^{-6}$ m^2/s can be used. For a description of the dependency of viscosity on temperature, the following polynomial interpolation coefficients have been calculated.

$$\nu = 1{,}159 \cdot 10^{-4} - 5{,}655 \cdot 10^{-6} t + 1{,}347 \cdot 10^{-7} t^2 + 1{,}872 \cdot 10^{-9} t^3 + 1{,}572 \cdot 10^{-11} t^4 - 7{,}820 \cdot 10^{-14} t^5 + 2{,}116 \cdot 10^{-16} t^6 - 2{,}393 \cdot 10^{-19} t^7 \quad (2)$$

The density of TEG at various temperatures is shown in the following table [7].

[°C]	10	16	21	27	32	38	43	49	54	60	66	71	77
[kg/m³]	1132	1128	1124	1119	1114	1111	1106	1101	1098	1093	1089	1084	1080
[°C]	82	88	93	99	104	110	116	121	127	132	138	143	149
[kg/m³]	1076	1071	1066	1063	1058	1053	1050	1045	1041	1036	1032	1028	1023
[°C]	154	160	166	171	177	182	188	193	199	204	210	216	221
[kg/m³]	1019	1015	1010	1007	1002	997	993	989	984	980	975	972	967

Table 2. Density of TEG according to temperature

Table 2 shows that in the range of working temperatures the density of TEG is a linear function of temperature. The following equation can be used for determining the density at a certain temperature.

$$\rho = -0{,}7831 \cdot t \; + \; 1140 \tag{3}$$

Finally, the thermal conductivity of TEG does not change in the range of working temperatures and has a value of 0,194 W/m²/°C.

For determining the physical properties of TEG solutions with water (concentrations c_{TEG} above 95 wt.%), the activity coefficient of water in TEG can be approximated by the following equation.

$$\gamma = -0{,}0585 \cdot c_{TEG} \; + 6{,}2443 \tag{4}$$

The industrial absorption dehydration process proceeds in a glycol contactor (a tray column or packet bed). In a contactor, a countercurrent flow of wet NG and TEG is arranged. During the contact, the TEG is enriched by water and flows out of the bottom part of the contactor. The enriched TEG then continues into the internal heat exchanger, which is incorporated at the top of the still column in the regeneration section of the absorption unit. It then flows into the flash drum, where the flash gases are released and separated from the stream. The TEG then runs to the cold side of the TEG/TEG heat exchanger. Just afterwards, the warmed TEG is filtered and then runs into the regeneration section, where is it sprayed in the still column. From there, the TEG runs into the reboiler. In the reboiler, water is boiled out of the TEG. The regeneration energy is around 282 kJ per liter of TEG. The temperature inside should not exceed 208°C, due to the decomposition temperature of TEG. Regenerated (lean) TEG is then pumped back through the hot side of the TEG/TEG and NG/TEG heat exchanger into the top of the contactor. The entire method is depicted in Figure 2 [8].

The circulation rate (l_{TEG}/kg_{H_2O}) and the purity of the regenerated TEG are the main limiting factors determining the output T_{dew} of NG. The amount of circulating TEG is around 40 times the amount of water to be removed. The minimal TEG concentration should be above 95 wt.%, but the recommended value is higher. However, to obtain TEG concentration above 99 wt.% enhanced TEG regeneration has to be implemented.

The simplest regeneration enhancing method is gas stripping. Proprietary designs DRIZO®, licensed by Poser-NAT, and COLDFINGER®, licensed by Gas Conditioners International,

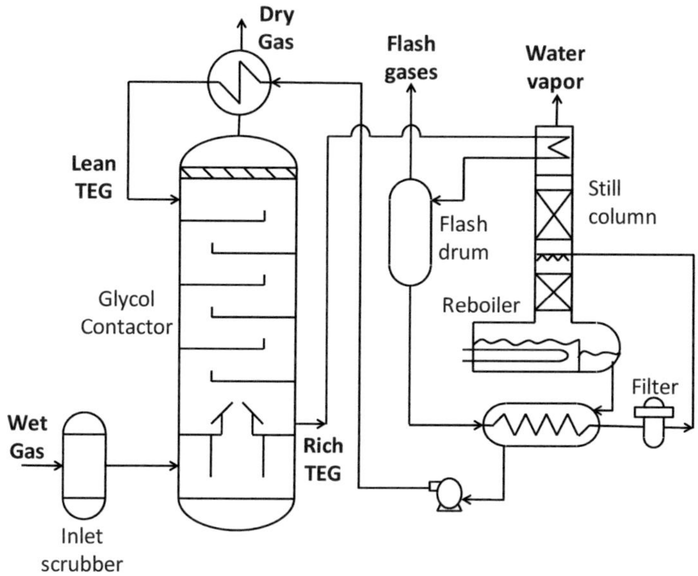

Figure 2. TEG absorption dehydration scheme

have been patented as an alternative to traditional stripping gas units. The Drizo regeneration system utilizes a recoverable solvent as the stripping medium. The patent operates with iso-octant solvent, but the typical composition of the stripping medium is about 60 wt.% aromatic hydrocarbons, 30 wt.% naphthenes and 10 wt.% paraffins. The three-phase solvent water separator is crucial for this method. The Coldfinger regeneration system employs a cooling coil (the "coldfinger") in the vapor space of the surge tank. The cooling that takes place there causes condensation of a high amount of vapors. The condensate is a water-rich TEG mixture, which is led to a further separation process [9]. Figure 3 depicts enhanced regeneration systems which replace the simple reboiler in the regeneration section shown in Figure 2.

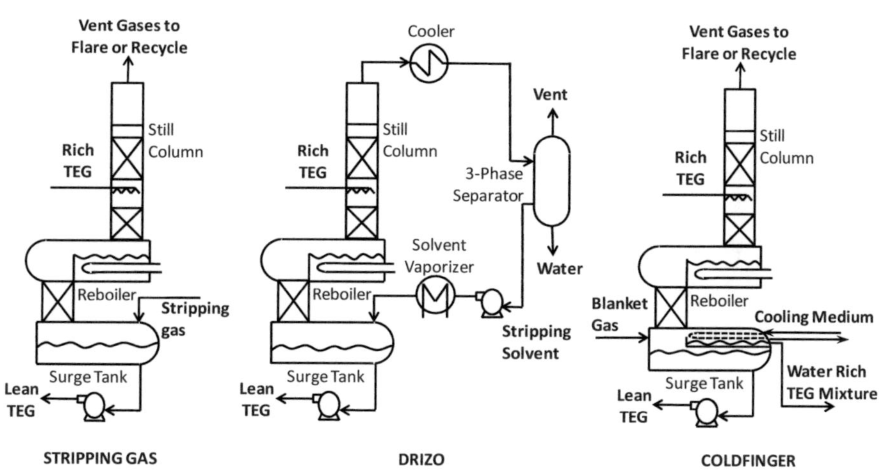

Figure 3. Enhanced TEG regeneration systems

2.2. Adsorption

The second dehydration method is adsorption of water by a solid desiccant. In this method, water is usually adsorbed on a mole sieve, on a silica gel or on alumina. A comparison of the physical properties of each desiccant is shown in Table 3 [4,10].

Properties	Silica gel	Alumina	Mol. sieves
Specific area [m²/g]	750 – 830	210	650 – 800
Pore volume [cm³/g]	0,4 – 0,45	0,21	0,27
Pore diameter [Å]	22	26	4-5
Design capacity [kg $_{H2O}$/100 kg $_{desiccant}$]	7-9	4-7	9-12
Density [kg/m³]	721	800 - 880	690 – 720
Heat capacity [J/kg/°C]	920	240	200
Regeneration temperature [°C]	230	240	290
Heat of desorption [J]	3256	4183	3718

Table 3. Comparison of the physical properties of desiccants used for dehydration of NG

The amount of adsorbed water molecules increases with the pressure of the gas and decreases with its temperature. These facts are taken into account when the process parameters are designed. Adsorption dehydration columns always work periodically. A minimum of two bed systems are used. Typically one bed dries the gas while the other is being regenerated. Regeneration is performed by preheated gas, or by part of the dehydrated NG, as depicted in Figure 4.

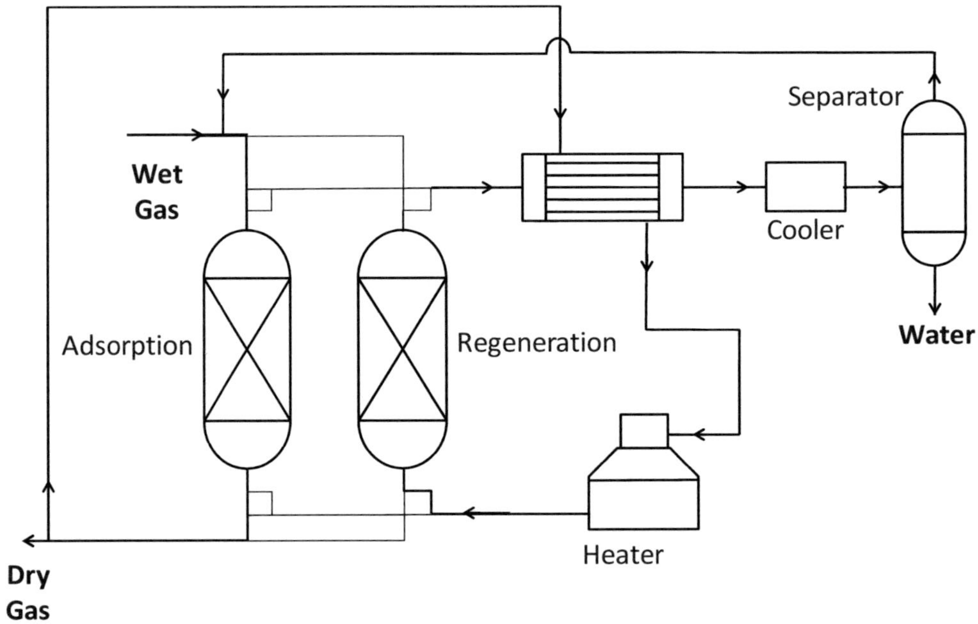

Figure 4. Scheme of the temperature swing adsorption dehydration process

This method is known as temperature swing adsorption (TSA). Regeneration can also be performed by change of pressure - pressure swing adsorption (PSA). However, PSA is not industrially applied for NG dehydration. Further details about PSA can be found in [11,12]. A combination of those two methods (PSA and TSA) seems to be a promising future option for adsorption dehydration of NG. This idea is still in the research process.

In classical applications, the TSA heater is realized as an ordinary burner or as a shell and tube heat exchanger warmed by steam or by hot oil. The regeneration gas warms in the heater and flows into the column. In the column passes through the adsorbent and the water desorbs into the regeneration gas. The water saturated regeneration gas then flows into the cooler. The cooler usually uses cold air to decrease the temperature of the regeneration gas. When the water saturated regeneration gas is cooled, partial condensation of the water occurs. The regeneration gas is led further into the separator, where the condensed water is removed.

A downstream flow of wet NG through the adsorption column is usually applied. In this way, floating and channeling of an adsorbent is avoided. Regeneration is performed by countercurrent flow in order to provide complete regeneration from the bottom of the column, where the last contact of the dried NG with the adsorbent proceeds. The typical temperature course for 12 h regeneration of molecular sieves is shown in Figure 5 [13].

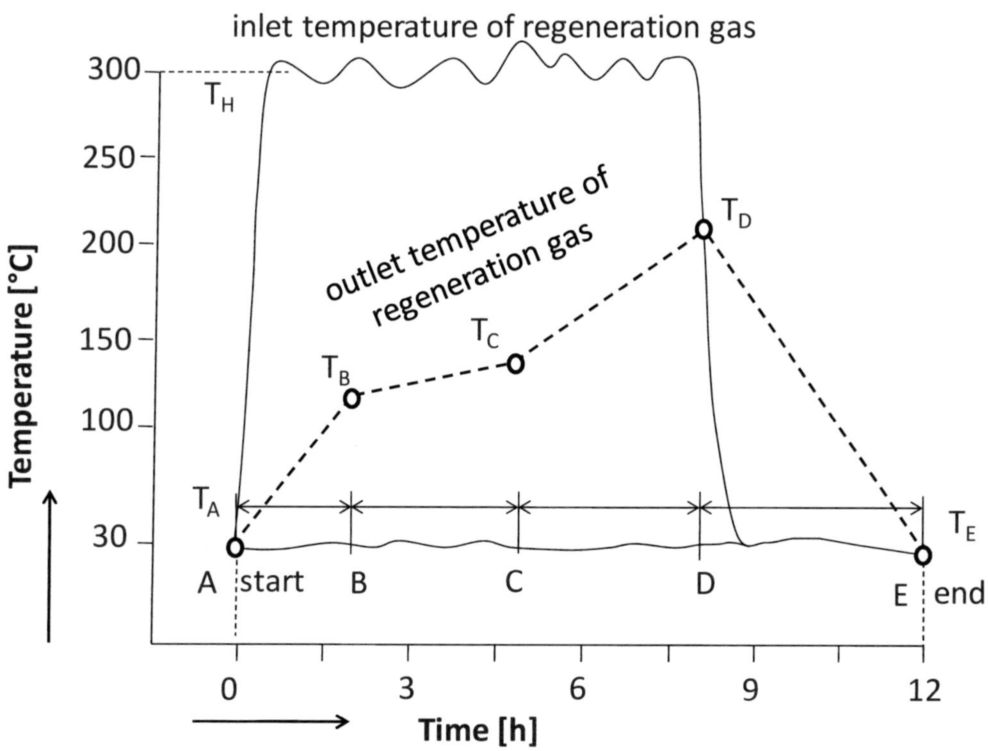

Figure 5. Typical temperature course for 12 h TSA regeneration of molecular sieves

The shape of the curve representing the course of the outlet regeneration gas temperature is typically composed of four regions. They are specified by time borders A, B, C and D with appropriate border temperatures T_A, T_B, T_C and T_D. Regeneration starts at point A. The inlet regeneration gas warms the column and the adsorbent. At a temperature around 120°C (T_B) the sorbed humidity starts to evaporate from the pores. The adsorbent continues warming more slowly, because a considerable part of the heat is consumed by water evaporation. From point C, it can be assumed that all water has been desorbed. The adsorbent is further heated to desorb C_{5+} and other contaminants. The regeneration is completed when the outlet temperature of the regeneration gas reaches 180 - 190°C (T_D). Finally, cooling proceeds from point D to point E. The temperature of the cooling gas should not decrease below 50°C, in order to prevent any water condensation from the cooling gas [13].

Part of the dehydrated NG is usually used as the regeneration gas. After regenerating the adsorbent the regeneration gas is cooled, and the water condensed from it is separated. After water separation, the regeneration gas is added back to the inlet stream or alternatively to the dehydrated stream.

The total energy used for regeneration is composed of heat to warm the load (30%), heat for desorption (50%) and heat going into the structure (20%). With proper internal insulation of the adsorption towers, the heat going to the structure can be minimized and around 20% of the invested energy can be saved.

So-called LBTSA (Layered Bed Temperature-Swing Adsorption) processes are an upgrade of the TSA method. Here, the adsorption column is composed of several layers of different adsorbents. Hence the properties of the separate adsorbents are combined in a single column. For example, in NG dehydration a combination of activated alumina with molecular sieve 4A is used. Alumina has better resistance to liquid water, so a thin layer is put in first place to contact the wet NG. This ordering supports the lifetime of the molecular sieve, which is placed below the alumina layer. The effect of adsorbent lifetime extension is shown for two cases in Figure 6. It can be seen that contact with liquid water dramatically decreases the lifetime of the molecular sieve [14].

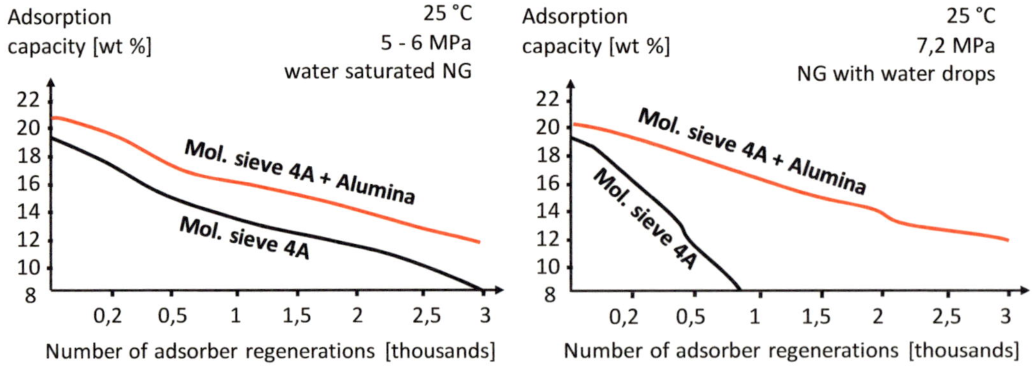

Figure 6. Effect of layered bed adsorption on the lifetime of the adsorbent

2.3. Condensation

The third conventional dehydration method employs gas cooling to turn water molecules into the liquid phase and then removes them from the stream. Natural gas liquids and condensed higher hydrocarbons can also be recovered from NG by cooling. The condensation method is therefore usually applied for simultaneous dehydration and recovery of natural gas liquids.

NG can be advantageously cooled using the Joule-Thompson effect (JT effect). The JT effect describes how the temperature of a gas changes with pressure adjustment. For NG, thanks to expansion, the average distance between its molecules increases, leading to an increase in their potential energy (Van der Waals forces). During expansion, there is no heat exchange with the environment, and no work creation. Therefore, due to the conservation law, the increase in potential energy leads to a decrease in kinetic energy and thus a temperature decrease of NG. However, there is another phenomenon connected with the cooling of wet NG. Attention should be paid to the formation of methane hydrate. Hydrates formed by cooling may plug the flow. This is usually prevented by injecting methanol or monoethylenglycol (MEG) hydrate inhibitors before each cooling. Figure 7 depicts an industrial application of dehydration method utilizing the JT effect and MEG hydrate inhibition.

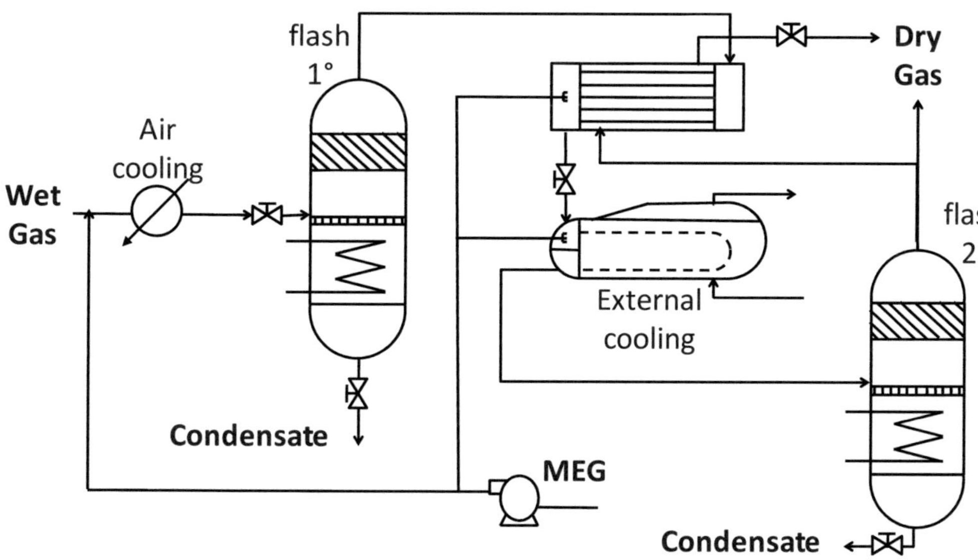

Figure 7. Dehydration method utilizing the JT effect and hydrate inhibition

The wet NG is throttled in two steps inside the flash tanks. The lower temperature (due to the JT effect) of the gas stream in the flash tanks leads to partial condensation of the water vapors. The droplets that are created are removed from the gas stream by a demister inside the flashes. In cases where cooling by the JT effect is insufficient (the usable pressure difference between the inlet and outlet of the gas is insufficient), the air pre-cooler and the external cooler are turned on. Since dehydration is normally applied to large volumes of NG, the external coolers need to have high performance, so this type of cooling is very energy expensive. For dehydration of low pressures NG the external coolers consumes up to 80% of total energy of dehydration unit. However, if the usable pressure difference is high, the JT effect inside the flashes is so strong that internal heating of the flashes is required to defreeze any methane hydrate or ice that may form. A condensation method is applied when suitable conditions for the JT effect are available.

2.4. Supersonic separation

The principle of this method lies in the use of the Laval Nozzle, in which the potential energy (pressure and temperature) transforms into kinetic energy (velocity) of the gas. The velocity of the gas reaches supersonic values. Thanks to gas acceleration, sufficient temperature drops are obtained. T_{dew} of water vapor in NG is reached, and nucleation of the droplets proceeds. Figure 8 depicts the basic design of a supersonic nozzle [15].

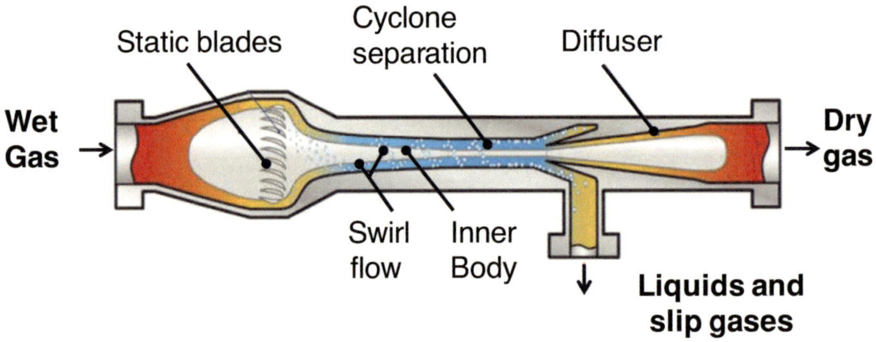

Figure 8. Design of a supersonic nozzle for NG dehydration

At the inlet to the nozzle there are static blades which induce a swirling flow of the gas. The water droplets that form are separated by the centrifugal force on the walls. The centrifugal force in the supersonic part of the nozzle can reach values up to 500 000 g [16]. The thin water film on the walls moves in the direction of flow into the separation channel. The separation channel leads into the heated degas separator. From here, the slip gas is returned back to the main stream and the water condensate is removed. After separation of the water it is important to recover the pressure of the gas from its kinetic energy. A shock wave is used to achieve this. Generally, shock waves form when the speed of a gas changes by more than the speed of sound. In supersonic nozzles, the shock wave is created by rapid enlargement of the nozzle diameter. This part of the nozzle is called the diffuser. Thanks to the diffuser 65 - 80% of the inlet pressure is recovered [17]. This section might also include another set of static devices to undo the swirling motion. The profile of pressure, temperature and velocity of a gas passing through the supersonic nozzle is depicted below.

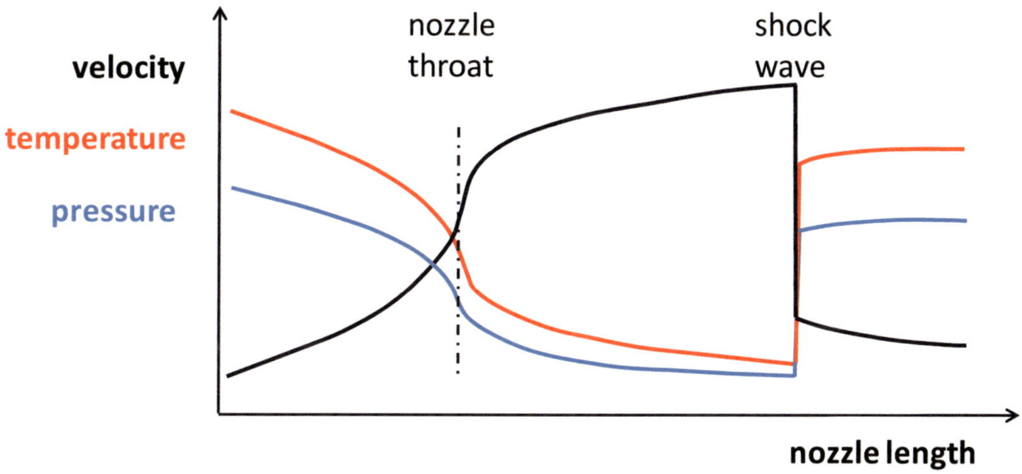

Figure 9. Profile of pressure, temperature and velocity of a gas passing through the supersonic nozzle

The scheme of a supersonic dehydration line working on the principle introduced here is depicted in Figure 10.

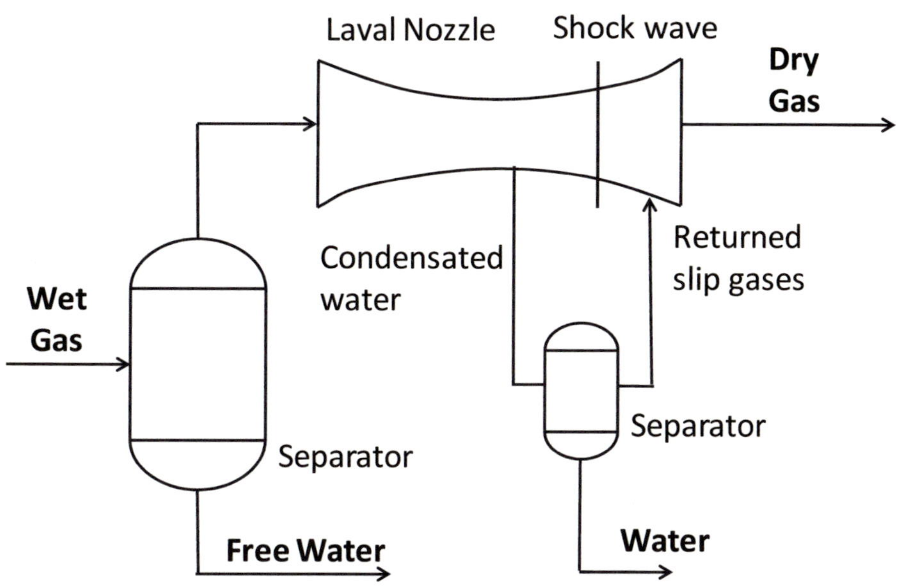

Figure 10. Scheme of a supersonic dehydration line

The gas residence time in the supersonic nozzle is below two milliseconds [18]. This time interval is too short for any methane hydrate formation, so no inhibitors are needed. To obtain supersonic velocity of the gas, the inlet diameter should be minimally $\sqrt{5}$ times higher than the nozzle throat. The geometry of the tapered section of the Laval nozzle is calculated by the following equations [16]

$$\frac{D-D_{cr}}{D_1-D_{cr}} = 1 - \frac{1}{X_m^2}\left(\frac{x}{L}\right)^3 \qquad \left(\frac{x}{L} \leq X_m\right) \tag{5}$$

$$\frac{D-D_{cr}}{D_1-D_{cr}} = \frac{1}{(1-X_m)^2}\left(1-\frac{x}{L}\right)^3 \qquad \left(\frac{x}{L} > X_m\right) \tag{6}$$

Where D_1, D_{cr}, L, X_m stand for the inlet diameter, the throat diameter, the length of the tapered section, and the relative coordinate of tapered curve, respectively; x is the distance between an arbitrary cross section and the inlet, and D is the convergent diameter at an arbitrary cross section of x.

A model of a supersonic dehydration unit was analyzed with the use of numerical simulation tools, and the separation efficiency in respect to lost pressure was evaluated. The simulations were performed on water saturated NG at 30 MPa and 20°C. The results are presented in Table 4 [19].

Pressure lost in nozzle (%)	17,3	20,0	27,6	49,0	51,5
Water separation efficiency (%)	40	50	90	94	96

Table 4. Supersonic water separation efficiency in respect to pressure lost in the nozzle

The supersonic separation is a promising new technology. The main advantage of the method is the small size of the supersonic nozzle. For example, a nozzle 1,8 m in length placed in a housing 0,22 m in diameter was used for dehydrating 42 000 m_s^3 per hour of water-saturated NG at 25°C compressed to 10 MPa to output water T_{dew} < -7°C [20]. The corresponding absorption contactor would be 5 m in height and 1,4 m in diameter, and the corresponding adsorption line would be composed of two adsorbers 3 m in height and 1 m in diameter. A further advantage is the simplicity of the supersonic dehydration unit. The supersonic nozzle contains no moving parts and requires no maintenance. The operating costs are much lower than for other methods. The only energy-consuming devices are the pumps for removing the condensate and the heater for the degas separator. However, during supersonic dehydration a pressure loss occurs. However, if the same pressure loss were used for the JT effect, the temperature drop would be 1,5 – 2,5 times lower [21]. Supersonic separation enables simultaneous removal of water and higher hydrocarbons from the treated gas, and can be used as pretreatment method before NG liquefaction. This method could also be usable for other applications of gas separation.

The application of supersonic separation has some disadvantages. The most limiting condition of use is the need for stationary process parameters. Fluctuations in temperature, pressure or flow rate influence the separation efficiency. However, it is in many cases impossible to achieve constant process parameters. For example, this is the case when withdrawing NG from UGS. However, supersonic dehydration can be used even in this case. The problem is solved by arranging several nozzles into a battery configuration with a single common degas separator. The battery configuration enables an optimal number of nozzles to be switched on, depending on the inlet parameters of the gas. The scheme of a possible arrangement is depicted in Figure 11 below [22].

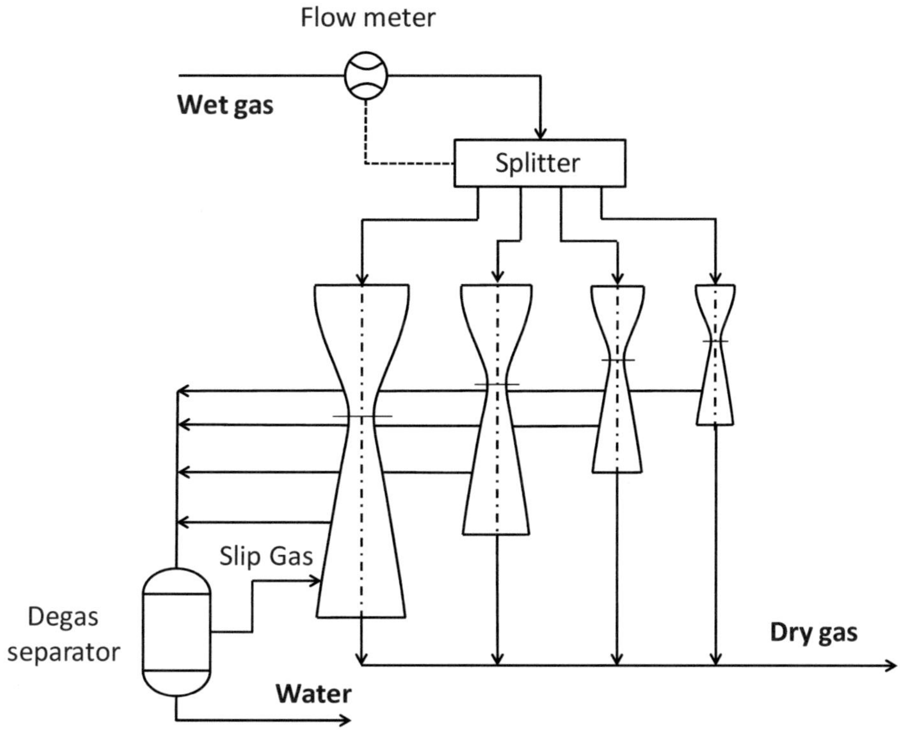

Figure 11. Arrangement of supersonic nozzles for unsteady inlet parameters of NG

Nozzle A is designed to process 80% of the nominal gas flow. Nozzles B, C, D are in the proportion 4:2:1 with respect to the processed gas flow. Nozzles B, C, D together can process 40% of the nominal gas flow. This arrangement therefore enables ± 20% deviation of the nominal flow to be covered. With appropriate switching of the nozzles, the maximal deviation between the real gas flow and the designed flow for the combination of nozzles is below 4%.

A further disadvantage of the supersonic dehydration is its novelty. The appropriate nozzle design is complicated, and "know how" is expensive. The geometry of the nozzle ranges in the order of micrometers. In addition, the construction material has to withstand abrasion and the impacts of a shock wave.

3. Comparison of conventional dehydration methods

3.1. General comparison

Each of the methods presented here has its advantages and disadvantages. Absorption by TEG is nowadays the most widely used method. Outlet T_{dew} around -10°C is usually reached and this water concentration is sufficient for pipeline distribution of NG. Indeed, with improved reboiler design (Vacuum Stripping, Drizo, Coldfinger), the outlet T_{dew} is even 2 - 3 times lower. However TEG has a problem with sulfur, and with gas contaminated with higher hydrocarbons. The TEG in the reboiler foams, and with time it degrades into a "black mud". BTEX emissions (the acronym stands for benzene, toluene, ethylbenzene and xylenes) in the flash gases and in the reboiler vent are a further disadvantage.

Adsorption dehydration can achieve very low outlet water concentration $T_{dew} < -50°C$, and contaminated gases are not a problem. Even corrosion of the equipment proceeds at a slower rate. However, adsorption requires high capital investment and has high space requirements. The adsorption process runs with at least two columns (some lines use three, four, or as many as six). Industrial experience indicates that the capital cost for an adsorption line is 2 - 3 times higher than when absorption is used [5]. In addition, the operating costs are higher for adsorption than for absorption.

Expansion dehydration is the most suitable method in cases where a high pressure difference is available between UGS and the distribution connection. However, the difference decreases during the withdrawal period and becomes insufficient, so that an external cooling cycle is needed. A cycle for regenerating hydrate inhibitor from the condensate separated inside the flashes is also required.

The general overview of areas suitable for application of target dehydration method is depicted on the following Figure 12.

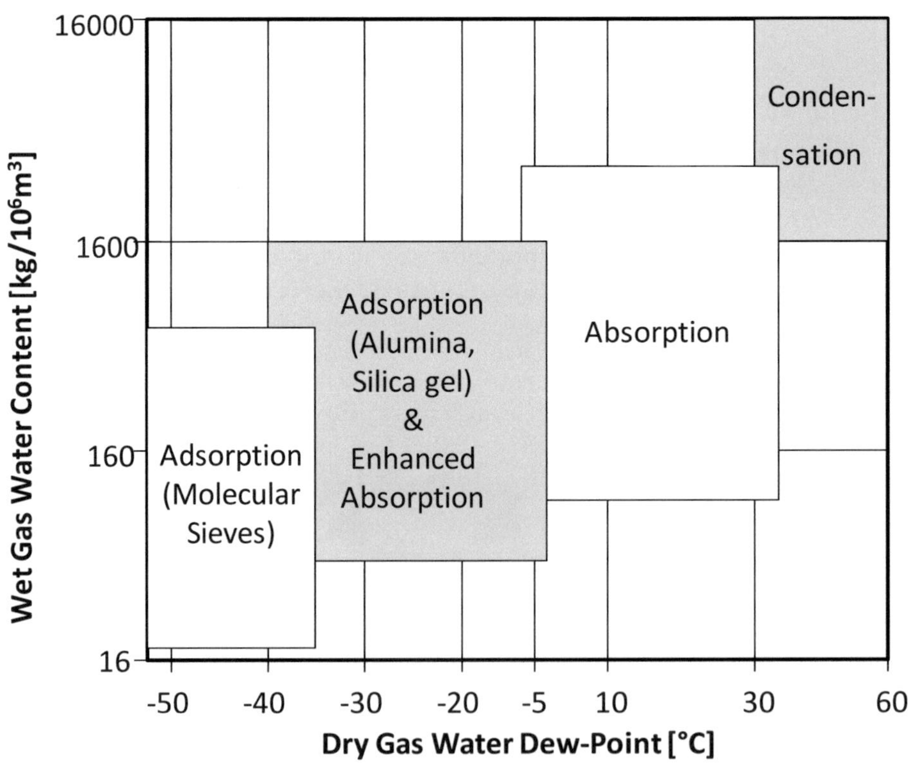

Figure 12. Overview of areas suitable for application of target dehydration method

3.2. Energy comparison based on own analyses and available data

The energy demand of the conventional methods presented here is compared on the basis of a model, where a volume of 10^5 m_s^3/h of NG from UGS is processed. The NG is water saturated at a temperature of 30°C. The pressure of the gas is varied from 7 to 20 MPa, but in the case of the condensation method the pressure range starts at 10 MPa. The required outlet water concentration of in NG is equivalent to dew point temperature -10°C at gas pressure 4 MPa [23].

The calculation of TEG absorption is based on GPSA (2004) [24]. The results are compared with the paper by Gandhidasan (2003) [5] and with industrial data provided by ATEKO a.s. The total energy demand is composed of heat for TEG regeneration in the reboiler, energy for the pumps, filtration and after-cooling the lean TEG before entering the contactor. Enhanced regeneration is not considered. The basic parameters for the calculation are: regeneration temperature 200°C, concentration of lean TEG 98,5 wt.%, and circulation ratio 35 l_{TEG}/kg_{H2O} [24].

For calculating adsorption dehydration, molecular sieve 5A is considered to be the most suitable adsorbent. The total energy demand is directly connected to the regeneration gas heater, and no other consumption is assumed. The calculations are again based on GPSA (2004) [24]. The results are compared with the paper by Gandhidasan (2001) [4] and also with the publication by Kumar (1987) [7]. The calculation procedure for GPSA and Gandhidasan arises from the summation of the particular heats, i.e. the heat for adsorbent warming, the heat for column warming (insulation of the adsorption towers is considered), and the heat for water desorption. Kumar's calculation procedure runs differently. The regeneration step is divided into four regions (as depicted in Figure 5). Afterwards, we determine what individual phenomena proceed in each region, what the border and average temperatures are, and how much energy is required to cover these phenomena. Finally, the demands for each region are added. The basic parameters for all procedures are: temperature of the regeneration gas 300°C, time of adsorption/regeneration 12 h, and two column designs.

The condensation method was calculated on the basis of industrial data provided by TEBODIN s.r.o. and supplementary calculations of the JT effect. The key parameter influencing energy demand is the pressure of NG. Because it is not feasible to apply this method for low pressures, and because the provided data starts at 10 MPa, the pressure range was adjusted. The total energy demand consists of the air pre-cooling unit, the external cooling, the pumps for MEG injection and condensate off take, the heat for MEG regeneration, and flash heating.

3.3. Results of the analyses

The results obtained for the TEG absorption method are the same for each of the calculation procedures, and good agreement with industrial data was also obtained. However, the calculation procedures for the adsorption method lead to different results. Hence the average energy demand value was taken as the reference. The maximum deviation from it is below 20% for all calculation procedures. The source of the deviation lies in the "loss factor and the non-steady state factor". In the case of the condensation method, the calculated values for the JT effect were in good agreement with the industrial data, but the amount of data was limited, resulting in limited representation of the condensation method.

The final energy consumption results for each dehydration method are summarized by the graph in Figure 13.

For low pressures (pressure of NG from UGS < 13 MPa), the condensation method is the most demanding. Its demand decreases linearly with pressure to a value of 145 kW for 13 MPa. At this point, the energy demand for the condensation method is roughly the same as for the adsorption method. When the NG pressure is further increased from 13 MPa to 16 MPa, the energy demand for the condensation method still decreases, but with a lowering tendency. For a high pressure of NG (> 16 MPa), the energy demand of the condensation method is at its lowest, and it remains nearly constant, with an average value around 36 kW.

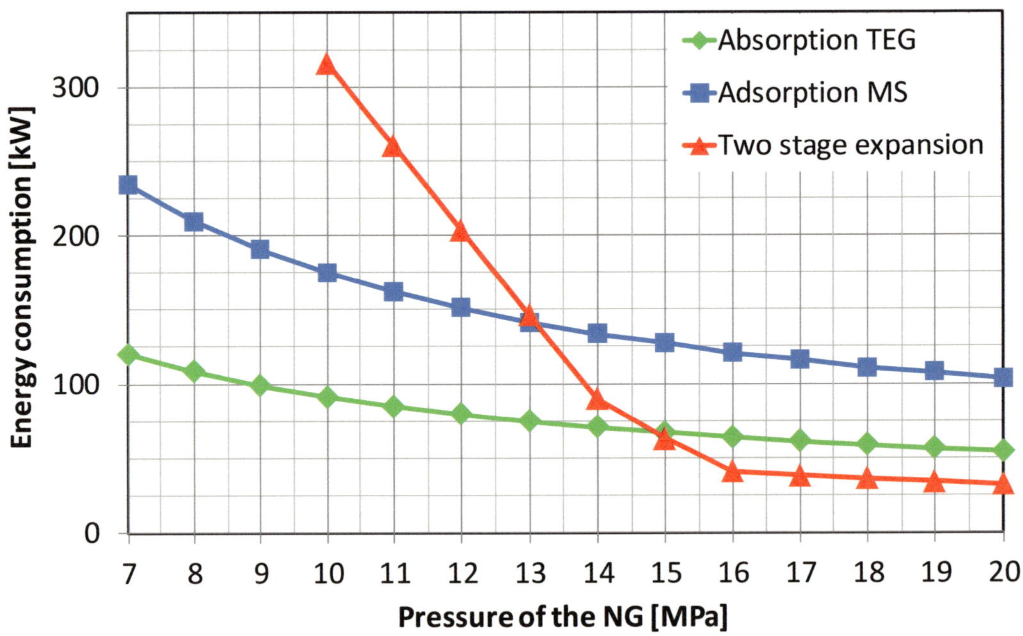

Figure 13. Energy consumption results for conventional dehydration methods

The course of the energy demand for the adsorption and absorption methods is quite similar: with increasing pressure of dehydrated NG the energy demand slowly decreases. The absorption method is less demanding on the whole pressure scale, and begins with consumption of 120 kW at 7 MPa. The adsorption method starts with 234 kW at 7 MPa, but the energy demand decreases slightly more as the pressure of NG in UGS rises. This leads to a gradual decrease in the difference between these methods, and the energy demand at the final pressure value of 20 MPa is equal to 54 kW for absorption and 103 kW for adsorption.

3.4. Conclusions of the analyses

By far the highest energy demand of the condensation method at low pressures of NG from UGS is due to the pressure being close to the distribution pressure, so that pressure cannot be used for the JT effect in flashes. Cooling is then compensated by the air pre-cooler and the external cooling device, which are unsuitable for large volumes of processed NG. However, as the pressure difference between UGS and the distribution site increases, the space for expansion rises and the JT effect proceeds with increasing impact. This is projected into a linear decrease in the energy demand of the air pre-cooler and the external cooling device. From the point where there is a pressure of NG > 14 MPa, flash heating is gradually turned on to prevent any freezing caused by the strong JT effect. The energy demand of flash heating is reflected in the total energy consumption. Finally, for pressures of NG > 16 MPa, total cooling and subsequent condensation is achieved by the JT effect. The total energy demand remains constant, and consists of flash heating and inhibitor injection and regeneration.

In case of the adsorption and absorption dehydration method, the similar falling course of the energy demand with increasing pressure of NG can be explained by the fact that with increasing pressure within a UDG the amount of water present in the NG decreases. The absorption method generally consumes less energy, because the regeneration of TEG is less demanding than adsorbent regeneration. The composition of the total energy demand of the adsorption method can be divided into three parts. The heat for water desorption is approximately 55%, for warming the adsorbent it is 31%, and for warming the column it is 14%. It also has to be assumed that just part of the heat in the regeneration gas transfers to the adsorbent, the column and heat loss leaves to the atmosphere, and the balance leaves with the hot gas.

In brief, in cases of high pressure the most appropriate dehydration method from the energy demand point of view is the stored NG condensation method. This holds for NG from UGS with pressure > 15 MPa and distribution pressure requirements 7 MPa. For lower pressures, the condensation method is used if the objective is to recover NGL and remove water simultaneously. However, this is usually not the case when storing NG in a UGS. In cases where insufficient pressure difference is available, the absorption method is therefore favored over the adsorption method in terms of energy demand. TEG absorption is almost twice less demanding. However, if a gas contaminated with sulfur or higher hydrocarbons is being processed, the TEG in the reboiler foams and degrades with time. This can occur when a depleted oil field is used as a UGS. Adsorption is preferred in cases where very low T_{dew} (water concentration lower than 1 ppm can be achieved) of NG is required, for example when NG is liquefied.

It is worth to note that the power comparison can be used as a measure of the technical excellence. From power data the specific energy consumption was calculated and its values indicate that the energy cost is much lower than the investment cost (depreciations). On the other hand the energy cost represents more than 60 % of the total operating cost.

4. Conclusions and recommendations

The chapter should help in the selection of a proper dehydration method and in calculating NG dehydration. The following methods are available as options: absorption, adsorption and condensation. Absorption is used in cases when emphasis is not placed on the water content of the output stream, and when low operating and capital investment are required. Adsorption is used in cases when bone dry NG is required. Low temperature separation employing the JT effect is used in cases where a sufficient pressure drop is available between the input and the output of the dehydration unit. Supersonic nozzles are a promising method that will in future displace these three conventional methods.

We have selected for citation here articles and procedures that we consider to provide reliable results.

Author details

Michal Netušil and Pavel Ditl
Czech Technical University in Prague, Faculty of Mechanical Engineering, Department of Process Engineering, Prague, Czech Republic

Acknowledgement

The authors are grateful for the financial support provided by Ministry of Industry and Trade of the Czech Republic (program TIP nr. FR-TI1/173).

Abbreviations

NG - natural gas
UGS - underground gas storage
lbs - pounds
MMcft - millions of cubic feet
TEG - triethyleneglycol
MEG - monoethylenglycol
TSA - temperature swing adsorption
LBTSA - layered bed temperature swing adsorption
JT effect - Joule-Thompson effect
BTEX - benzene, toluene, ethylbenzene and xylenes

Symbols

T_{dew} - dew point temperature
g_{H2O} - grams of water
m_s^3 - standard cubic meters of gas (293,15 K; 101,325 kPa)
w_{water} - kilograms of water per 10^6 m_s^3 of NG
t_G - temperature of NG in °C
C_{5+} - pentane and higher hydrocarbons
P_G - pressure of NG in MPa
ϱ - density of NG in kg/m^3
γ - activity coefficient dimensionless

c_{TEG} - weight concentration of TEG in TEG/water solution
l_{TEG} - liters of TEG
kg_{H2O} - kilograms of water
D_1 - inlet diameter of nozzle mm
D_{cr} - throat diameter of nozzle mm
L - length of tapered section of nozzle in mm
X_m - relative coordinate of tapered curve in mm
D - convergent diameter in mm
x - distance between arbitrary cross section and the inlet in mm
g - standard gravity

5. References

[1] Gas infrastructure Europe (2011) Map Dataset in Excel-format Storage map. Available: http://www.gie.eu/maps_data/storage.html. Accessed 2011 Mar 8.

[2] Foss M (2004) Interstate Natural Gas Quality Specifications and Interchangeability. Center for Energy Economics.

[3] NET4GAS (2011) Gas quality parameters. Available at: http://extranet.transgas.cz/caloricity_spec.aspx. Accessed 2011 Mar 8.

[4] Gandhidasan P, Al-Farayedhi A, Al-Mubarak A (2001) Dehydration of natural gas using solid desiccants. Energy 26: 855-868.

[5] Gandhidasan P (2003) Parametric Analysis of Natural Gas Dehydration by Triethylene Glycol Solution. Energy Sources 25: 189-201.

[6] CHEM Group, Inc. (2012) Triethylene Glycol - Liquid Density Data. Available at: http://www.chem-group.com/services/teg-density.tpl. Accesed 2012 Mar 6.

[7] CHEM Group, Inc. (2012) Triethylene Glycol - Kinematic Viscosity Data. Available at: http://www.chem-group.com/services/teg-viscosity.tpl. Accesed 2012 Mar 6.

[8] Bahadori A, Vuthaluru H.B (2009) Simple methodology for sizing of absorbers for TEG gas dehydration systems. Energy 34: 1910–1916

[9] Hubbard R.A, Campbell J.M (2000) An appraisal of gas dehydration processes. Hydrocarbon Engineering 5: 71-74.

[10] Tagliabue M, Farrusseng D, Valencia S, Aguado S, Ravon U, Rizzo C (2009) Natural gas treating by selective adsorption: Material science and chemical engineering interplay. Chemical Engineering Journal 155: 553-566

[11] Roušar I, Ditl P (1993) Pressure swing adsorption: analytical solution for optimum purge Original Research, Chemical Engineering Science 48: 723-734.

[12] Roušar I, Ditl P, Cekal M (1993) Pressure swing adsorption – the optimization of multiple bed units, Precision Process Technology: perspectives for pollution prevention: 483-492.

[13] Kumar S (1987) Gas Production Engineering. Houston: Gulf Professional Publishing 239 p.

[14] Jochem G (2002) Axens Multibed Systems for the Dehydration of Natural Gas, PETEM.

[15] Schinkelshoek P, Epsom H.D (2008) Supersonic gas conditioning – commercialization of twister technology. 87th Annual Convention. Grapevine, Texas.

[16] Wen C, Cao X, Zhang J, Wu L (2010) Three-dimensional Numerical Simulation of the Supersonic Swirling Separator. Twentieth International Offshore and Polar Engineering Conference. Beijing, China.

[17] Okimoto F, Brouwer J.M (2002) Supersonic gas conditioning, World Oil 34: 89-91

[18] Ma Q, Hu D, Jiang J, Qiu Z (2010) Numerical study of the spontaneous nucleation of self-rotational moist gas in a converging–diverging nozzle. International Journal of Computational Fluid Dynamics: 29–36.

[19] Karimi A, Abdi M.A (2006) Selective dehydration of high-pressure natural gas using supersonic nozzles, Chemical Engineering and Processing 48: 560–568

[20] Twister BV (2012) Twister supersonic separator – Experience. Available: http://twisterbv.com/products-services/twister-supersonic-separator/experience/. Accesed 2012 Mar 7.

[21] Betting M, Epsom H (2007) High velocities make a unique separator and dewpointer, World Oil, 197-200.

[22] Netušil M, Ditl P, González T (2012) Raw gas dehydration on supersonic swirling separator, 19th International Conference Process Engineering and Chemical Plant Design, Krakow.

[23] Netusil M, Ditl P (2011) Comparison of three methods for natural gas dehydration, Journal of Natural Gas Chemistry 20: 471 - 476

[24] GPSA (2004) Engineering Data Book. 12th ed. Tulsa: GPSA Press.

Chapter 2

Natural Gas Treatment Using Adsorptive Separation

Pramathesh R. Mhaskar and Arun S. Moharir

Additional information is available at the end of the chapter

http://dx.doi.org/10.5772/53269

1. Introduction

Natural Gas (NG) is used both as a process gas as well as a fuel. The two uses make contrary demands on its composition. Natural Gas is a mixture of Methane(C_1), Ethane (C_2), Propane (C_3), Butane isomers (C_4) and Pentanes, to name only the more significant components. It can have Carbon-Dioxide, Hexanes etc. as other components. Higher hydrocarbons are there as the gas comes off the well, but are removed in the gas treatment plants. Similarly, water, if any, is removed. Winterization and dehydration are two major aspects of gas treatment before it is transported to the user through pipelines [1].

When the use of Natural Gas is as a fuel replacing conventional sources such as coal, or furnace oil or petroleum fractions, the most important thing is its calorific value. Higher the content of higher hydrocarbons, higher is the calorific value of the feed gas. Natural Gas is predominantly Methane with second most abundant component being Ethane. C_{2+} alkanes (Propane and higher alkanes) together are about 4 or less %. Even this small percentage of higher hydrocarbons can change the calorific value significantly. For example, pure Methane has a lower calorific value of 850 kJ/mole, while a mixture of 97% Methane and 3% n-Butane has a lower calorific value of 904 kJ/mole. Users of Natural Gas as fuel thus would like more presence of the heavier hydrocarbons in the supply gas.

Gas based fertilizer plants, metallurgical units which need to carry out ore reduction using gases such as CO and H_2 would like to use Natural Gas as mainly a feedstock for reformers. Even a small presence of C_{2+} could be detrimental to the reformer catalyst and necessitate its regeneration frequently. Natural Gas is thus necessarily stripped of its C_{2+} content before use as reformer feedstock. One of the techniques used is cooling the gas below the boiling points of Propane so that C_{2+} hydrocarbons are condensed. Joule-Thompson expansion is often employed to achieve temperatures as low as 203 K for this purpose. This leaves a mixture predominantly of C_1 and C_2 for reforming. C_{2+} alkanes are recovered almost entirely and can contribute to Liquefied Petroleum Gas (LPG) pool. Turbo expansion is a costly technology, both in terms of capital and operating cost. An alternative separation technique is desired for large scale gas treatment for removal of C_{2+} from Natural Gas.

This paper deals with an adsorption based technique developed and implemented for this purpose. Pressure Swing Adsorption (PSA) was used as an alternative. Due the large volume that needed to be treated, we could conceive a significant modification in PSA technology. By suitable application of model based design and operation, we could convert the PSA to a surge-less PSA technology. This eliminated or minimized the need for surge vessels, so integral to any PSA process because of pressure swings. The job of surge tanks was sought to be done by appropriate pressure/flow control of feed to and product from the adsorption beds. This paper discusses the conceptual design process design and process plant engineering of a surge-less PSA plant designed to process a feed rate as high as 2000000 m^3 per day of Natural Gas at standard temperature and pressure conditions. This makes it one of the largest functional adsorptive separation units.

1.1. Design problem

The separation plant's objective was to reduce C_{2+} content of Natural Gas to below 0.6% (mole fraction) for all feed compositions with C_{2+} content not exceeding 2%. It was even this relatively low content of C_{2+} hydrocarbons which was perceived to be causing catalyst deactivation due to coking in downstream reactors.

The main components of the Natural Gas are given in Table 1 and a typical feed gas composition used in design was as given in Table 2. The operating temperature was the ambient value at site equal to 303.15 K. The viscosity of the feed gas at this temperature was 0.000012 Pa-s. This value was used to estimate the pressure drop across the bed.

i	Component Formula	b_i (m^3/mole)	q_i^{max} (moles adsorbed/ m^3 particles)
1	C_3H_8	0.008400	9091
2	n-C_4H_{10}	0.014970	12771
3	i-C_4H_{10}	0.019770	8019
4	CO_2	0.006400	9470
5	C_2H_6	0.002500	9009
6	CH_4	0.000550	7812

Table 1. Fitted values of pure component Langmuir adsorption isotherm parameters on Silica Gel at 303.15 K temperature

Component	y_i	Molecular Weight (g/mole)	w_i	b_i (m³/mole)	q_{max} (moles adsorbed/m³ particles)	q_i^* at total pressures of 1 bara and 34 bara respectively (moles adsorbed/m³ particles)
Propane	0.0065	44	0.0171	0.00840	7925.4	16.7 and 296.9
n-Butane	0.0011	58	0.0038	0.01497	7925.4	5.0 and 89.6
i-Butane	0.0010	58	0.0035	0.01977	7925.4	6.0 and 107.5
Carbon Dioxide	0.0074	44	0.0194	0.00640	7925.4	14.5 and 257.6
Ethane	0.0188	30	0.0337	0.00250	7925.4	14.4 and 255.6
Methane	0.9652	16	0.9225	0.00055	7925.4	162.3 and 2886.9

Table 2. Adsorption capacity of each component of the NG mixture as per Extended Langmuir isotherm at 303.15 K temperature (Case 1: no component lumping)

The hydrocarbons as well as CO_2 are adsorbed by several adsorbent materials. Most important among them are Activated Carbon and Silica Gel [1]. Silica Gel was the adsorbent material considered in this work. Limited data on adsorption thermodynamics is available in terms of adsorption isotherms [2]. The adsorption capacities increase significantly with increase in the number of Carbon atoms in the homologous series of alkanes of interest, namely C_1 to C_4. Typical pure-component isotherms at the operating temperature and pressures up to 35 bara and the parameters of the fitted curves for the five hydrocarbons are shown in Figure 1 and Table 1 respectively. Carbon Dioxide has also been reported for its adsorption behavior and its isotherm and isotherm expressions are also included in Figure 1 and Table 1. The isotherms show typical favorable isotherm shape and fit very well with Langmuir equation.

The isotherms and their mathematical expression forms show the adsorption equilibrium constant increasing with the number of Carbon atoms in the alkane molecule. The mathematical expressions can be used to appreciate the separation potential offered by Silica Gel for the desired separation task. If the Natural Gas composition given above is equilibrated with Silica Gel, the equilibrium adsorbed phase concentration of each of the six components is tabulated in Table 2 based on Extended Langmuir expression for multi-component adsorption. The pressure of gas was taken as 1 bara in this calculation. If the Natural Gas is available at higher pressure, the selectivities can be even better. For example, the feed gas pressure for the design case was 34 bara. If the feed composition is the same, but equilibration is done at 34 bara, the equilibrium adsorbed phase concentration of each component is also tabulated in Table 2. Higher pressures do not seem to help as the saturation capacities are reached. Physical adsorption being reasonably weak, multi-layer adsorption is less possible and monolayer saturation capacities are reached at the pressures considered. At both pressures, the solid phase has considerable amount of adsorbed CH_4 (74% on molar basis or 52% on mass basis of adsorbed phase is Methane). It is the major component of the adsorbed phase as its mole fraction in the feed itself exceeds 95%. So significant amounts of this adsorbed Methane will be lost during the depressurization step.

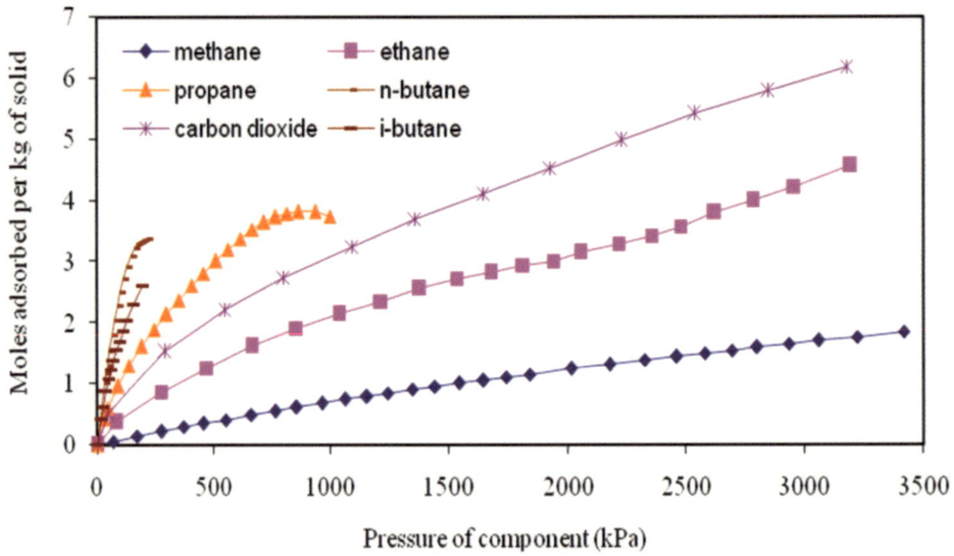

Figure 1. Pure component adsorption isotherms on Silica Gel at 303.15 K

These preliminary adsorbent evaluation studies indicated suitability of Silica Gel as a suitable adsorbent. If a bed of Silica Gel is subjected to a flow of Natural Gas, the simultaneous adsorption of the hydrocarbons and CO_2 will occur. The concentrations of higher hydrocarbons (due to their lower value and stronger preference by the adsorbent) would deplete fast and one can expect a raffinate gas leaner in the C_{2+} content.

The bed would get gradually saturated and C_{2+} gases would begin to break through into the raffinate. If the feed gas is then diverted to another identical bed and the saturated bed is regenerated, continuous raffinate production is possible. Regeneration can be carried out by depressurization of the bed. The desorption of C_{2+} can also be facilitated by passing through the bed a fraction of the lean gas (LG) produced earlier or purge-in gas (PG) which will reduce the partial pressure of C_{2+} components in the bulk phase surrounding the adsorbent particles. This would help effect further regeneration and carry away of the desorbed component. The gas collected during these blow down and purging steps will be richer in C_{2+} components than the feed gas and is termed as rich gas (RG). The adsorbed bed switched between alternate high and low pressure regimes thus offer a potential to separate the Natural Gas into two streams, namely a raffinate leaner in C_{2+} than the feed and an extract gas richer in C_{2+} components. We thus have a potential to employ a suitably designed PSA process for the desired separation task.

The design was carried out using a comprehensive simulation model for PSA processes developed by the authors [3]. These simulation studies and their findings in terms of an optimal PSA cycle is discussed in the next sections.

1.2. Cycle description

Refer to Figure 2 and Table 3. The description of the symbols in Figure 2 is given in the nomenclature section. In Natural Gas-Silica Gel system, the desired product is the weakly adsorbed component (Methane and Ethane). This is the characteristic feature of raffinate PSA systems. Conventional 4-step cycle of the Skarstrom type is used for dealing with raffinate PSA systems [4]. The steps of the Skarstrom cycle in sequence are: pressurization of the bed with feed gas to the feed tank pressure which is the adsorption pressure, adsorption of feed gas at the adsorption pressure to give raffinate product, depressurization of the bed to the desorption pressure in a direction counter-current to that of the feed gas flow direction and counter-current purge with fraction of raffinate product at the desorption pressure.

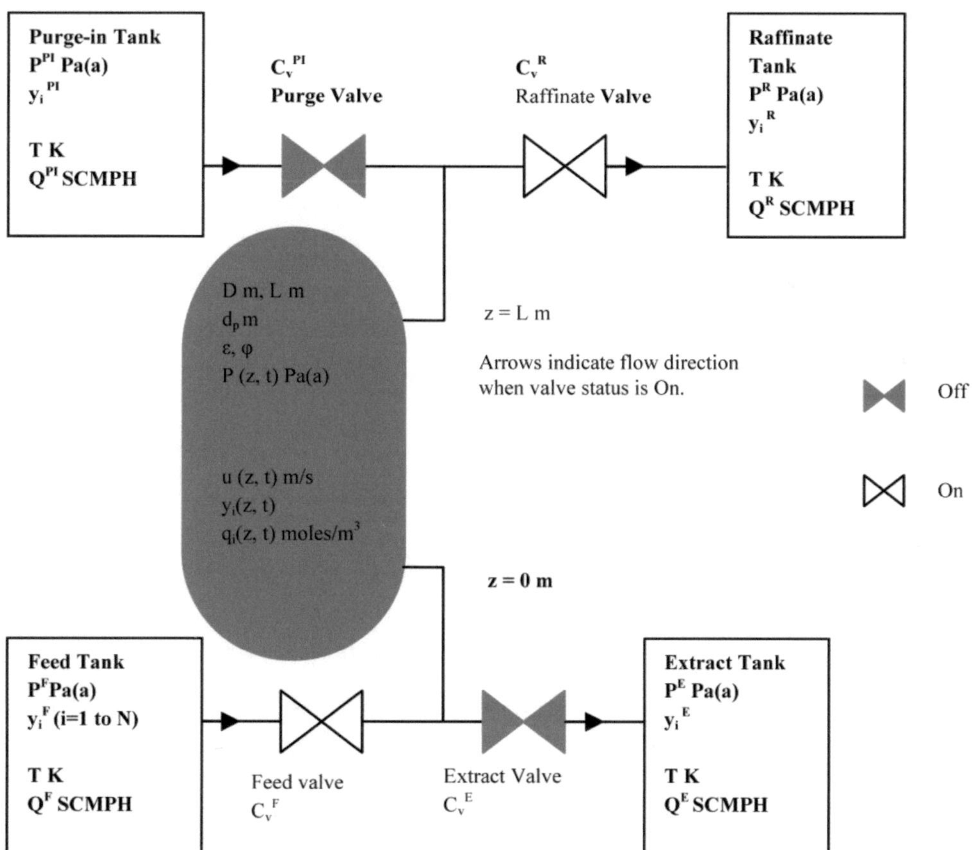

Figure 2. State of a bed in feed adsorption step

Step	Description	Feed valve	Raffinate valve	Extract valve	Purge valve
1	Feed pressurization	On	Off	Off	Off
2	Feed adsorption	On	On	Off	Off
3	Co-current depressurization to raffinate tank	Off	On	Off	Off
4	Counter-current depressurization to extract tank	Off	Off	On	Off
5	Counter-current purge by fraction of raffinate	Off	Off	On	On

Table 3. Steps, their sequence and the valve status for PSA cycle for Natural Gas treatment

In this work, an additional step of depressurization is added after adsorption step and before the counter-current depressurization step, thus creating a 5-step cycle. Stepwise description of the PSA cycle used in the current work is as follows:

Step 1: Feed pressurization (duration = t_1 seconds)

The bed pressure is much lower than the feed tank pressure. The valve connecting the feed tank to the bed is opened. The other valves are closed. The bed starts getting pressurized with the feed Natural Gas mixture and at the end of this step, the bed is pressurized to a value nearly equal to the feed tank pressure. Simultaneous adsorption/desorption that takes place, is considered in the PSA process model.

Step 2: Feed adsorption (duration = t_2 seconds)

The valve connecting the raffinate tank to the bed is opened at the start of this step. Since the raffinate tank pressure is less than that of the pressurized bed, material starts flowing out from the bed to the raffinate tank and fresh feed flows into the bed. The raffinate tank collects a gas that is relatively free of Propane and higher alkanes and Carbon Dioxide as these components get preferentially adsorbed in the adsorbent. Pressure in the bed remains reasonably constant in this step except developing small pressure drop across the bed due to flow.

Step 3: Co-current depressurization to Raffinate tank (duration = t_3 seconds)

At the start of this step, the feed tank connection is closed and the weaker adsorbates that may be trapped in the void space of the bed are released into the raffinate tank. Desorption of the species will also occur simultaneously with depressurization and the model considers it. This step is included to improve the recovery of the weaker adsorbates and reduce the velocity surges that would have occurred in Step 4 if the bed is directly depressurized counter-currently to the extract tank after finishing the adsorption step. This step also increases the chance of getting an extract more pure in the stronger adsorbate[3].

Step 4: Counter-current depressurization to Extract tank (duration = t_4 seconds)

At the start of this step, the raffinate tank connection is closed and the extract tank connection is opened. The extract tank is at a much lower pressure than the bed pressure. The bed starts depressurizing counter-currently. Desorption of heavier species takes place as the pressure decreases. The desorbed gas is collected in the extract tank.

Step 5: Counter-current purge with fraction of the raffinate (duration = t_5 seconds)

Some fraction from the raffinate tank is used for purging counter-currently the already depressurized bed. In the model, a separate purge-in tank is shown as supplying this purge gas. Its composition is set internally as the same as raffinate composition. The purge-in tank is at a higher pressure than the depressurized bed. The valve connecting the purge-in tank to the bed is opened at the start of this step. The purge gas enters the bed and starts flowing counter-currently through the bed. This step is used to flush the desorbed heavy components in the void spaces towards the extract end. The outgoing stream from this step is also collected in the extract tank. One cycle is completed at the end of this step (at $t_1+t_2+t_3+t_4+t_5 = t_{cycle}$) and a fresh cycle begins starting with Step 1.

2. Simulation model

PSA is a very versatile process and the process embodiments most suited for a given separation task depend entirely on the adsorbent-adsorbate system and desired separation performance. Optimal design and operation alone can help achieve the best out of this technology. To study various process options, an elaborate and costly experimental program needs to be launched and executed. Another equally effective, much less costlier and more exhaustive option that is possible is to systematically carry out simulation based design. A rigorous and predictive model of the process physics is at the core of this approach. The authors have developed a generic simulator for PSA process genre covering all its possible variations such as PSA, Vacuum Swing Adsorption (VSA), Pressure-Vacuum Swing Adsorption (PVSA), etc. The model allows putting together any PSA cycle and predict its effect on performance measured in terms of product purity, recovery and adsorbent throughput. The model has been presented in detail elsewhere [3]. A relevant summary along with the definitions of performance indices for the present case is presented here.

The model mainly assumes-(i) perfectly plug flow of the fluid, (ii) fluid behaves like an ideal gas, (iii) multi-component adsorption equilibrium behavior as per Extended Langmuir isotherm, (iv) isothermal operation, (v) mass transfer rate described by the linear driving force approximation model, (vi) flow at the ends of the bed described by a valve equation with the ability to vary the valve coefficient with respect to time in case of controlled opening or closing of the valve and (vii) pressure drop across the packed bed is described by the Ergun equation. The model can consider- (i) sorption occurs during the pressurization and depressurization steps (no frozen solid assumption), (ii) fluid velocity variations, (iii) fluid density variations and (iv) presence of inert or empty layer below or above the adsorbent layer. Equations (1)-(7) are the main equations of the model and are given in Table 4.

Overall mole balance for i[th] component

$$\frac{\partial(uc_i)}{\partial z}+\varepsilon\frac{\partial c_i}{\partial t}+(1-\varepsilon)\frac{\partial q_i}{\partial t}=0 \qquad i=1,2,..,N \tag{1}$$

Extended Langmuir isotherm

$$\frac{q_i^*}{q^{max}}=\frac{b_i c_i}{1+\sum_{j=1}^{N}(b_j c_j)} \qquad i=1,2,..,N \tag{2}$$

Linear Driving Force Approximation model for mass transfer

$$\frac{\partial q_i}{\partial t}=k_i^{LDF}(q_i^*-q_i) \qquad i=1,2,..,N \tag{3}$$

Ideal gas law

$$c_i=\frac{Py_i}{RT} \qquad i=1,2,..,N \tag{4}$$

Ergun equation for pressure drop

$$\frac{-\partial P}{\partial z}=150\frac{\mu(1-\varepsilon)^2 u}{g_c\phi^2 d_p^2\varepsilon^3}+1.75\frac{(1-\varepsilon)u^2\varrho}{g_c d_p\phi\varepsilon^3} \tag{5}$$

Valve equation to get velocity at the bed end where valve is open

$$u=C_v\frac{\sqrt{P^{US}-P^{DS}}}{\sqrt{\varrho}} \tag{6}$$

Stoichiometric condition in fluid phase

$$\sum_{i=1}^{N}y_i=1 \tag{7}$$

Model performance parameters

Recovery of Propane and heavier alkanes

$$\%Recovery=100\frac{\int_{t=t_1+t_2+t_3}^{t=t_{cycle}}(uPy_{UD})_{z=0}^{t}dt}{\int_{t=0}^{t=t_1+t_2}(uPy_{UD})_{z=0}^{t}dt} \tag{8}$$

Purity of Methane and Ethane in raffinate

$$\%Purity=100\frac{\int_{t=t_1}^{t=t_1+t_2+t_3}(uP(1\text{-}y_{UD}))_{z=L}^{t}dt}{\int_{t=t_1}^{t=t_1+t_2+t_3}(uP)_{z=L}^{t}dt} \tag{9}$$

Table 4. Important model equations and performance indices

The model performance parameters for given design and operating conditions used in this work are given by Equations (8) and (9) in Table 4. (See the nomenclature section for a description of the symbols in Table 4.)

1) Although getting a raffinate free of heavier hydrocarbons (C_3 and above) was an objective of this PSA, the same heavier components were also desired to be recovered to as high an extent as possible as they serve as LPG components. Recovery in this case is therefore defined as the percentage of these components from the feed gas which end up in the extract stream over a complete cycle. The definition is as in Equation (8). y_{UD} is the summation of mole fractions of all undesirable components in the raffinate, for example, the heavier hydrocarbons and CO_2.

2) The raffinate is considered as 100% pure if does not contain these heavy components (C_3 and above) and CO_2. Purity is therefore defined as the difference between 100 and mole percent of these undesirable components in the raffinate over a complete cycle. The definition is given as Equation (9).

3) Additionally, the feed rate was the average hourly volumetric flow rate of Natural Gas to a single bed in a complete cycle at Standard Temperature and Pressure conditions. This was also a performance parameter. It was estimated from the total moles of Natural Gas fed in a cycle to a bed and the cycles completed in an hour.

The mass balance equations in the form of differential algebraic equation (DAE) are discretized using finite difference technique. This converts the model equations into a set of simultaneous algebraic equations, which relate all the variables at an advanced time step to their values at the immediately previous time interval (initial conditions). A marching solution can thus be launched to model the bed performance during each step. Pressure and flow velocity, bulk and adsorbed phase concentrations for all the components at all grid points along the bed become the algebraic variables. For each step of the cycle, the boundary conditions as above are then specifications of relevant variables at the bed ends. After discretization and applying suitable boundary conditions, the number of variables becomes equal to the number of equations. Thus, degrees of freedom become zero. The number of such variables in each step of the cycle depend upon the number of divisions the bed is discretized into, and the number of components as given in Table 5.

Then the variables were normalized and non-dimensionalized. An upwind implicit scheme is used for the spatial discretization and a backward difference scheme is used for the temporal discretization. Both the schemes were of first order accuracy but simple and unconditionally stable. Discretization results in a system of simultaneous non-linear algebraic equations linking the variable values at an advanced time increment to the values at immediately previous time increment. The generation of the set of equations and its solution is repeated to march in time. Convergent temporal and spatial grid sizes were found by successively reducing them and simulating the cyclic steady state till differences in the concentration profiles vanished signifying convergence. Grid size reduction also helped in improvement in mass balance closures.

Step of the cycle	Number of variables (equations)	Remarks
Adsorption, purge and rinse	2+N+M(2N+2)	
Pressurization and depressurization	1+N+M(2N+2)	
Idle	N+M(2N+2)	Lowest no. of variables
Pressure equalization	N+(2M+1)(2N+2)	Highest no. of variables

Table 5. Number of variables for each step of the cycle as a function of number of components (N) and number of divisions (M) used for spatial discretization

The system of simultaneous non-linear algebraic equations is solved by using Powell's hybrid method. It is implemented in Fortran programming language. Initial guess for each unknown variable, maximum number of iterations, step size for Jacobian evaluation and tolerance have to be provided by the user. Powell's method is a combination of Newton's method and the steepest descent method and offers improved convergence as compared to Newton's method. As we march from one time step to another, the known conditions at the current time are provided as the initial guess for the respective unknowns of the advanced time step. The solver returns the converged solution with the stipulated accuracy. These values are used as the initial guess for the unknowns of the next time step and so on. The system of non-linear algebraic equations is converted to linear ones by estimating the Jacobian based on a finite difference approximation. The inverse of the Jacobian is calculated and used for giving the improved solution.

At start-up, the bed is assumed to be at the pressure of extract tank and contains pure CH_4 (most weakly adsorbed component) in both the phases. For simulation with component lumping, the bed is assumed to be filled with the pseudo-component having the lowest adsorption equilibrium constant. The adsorbed phase and the bulk phase are assumed to be in adsorption equilibrium at this pressure. After the first cycle, the initial conditions of the bed are the conditions of the bed at the end of the last step of the first cycle. Similarly, the initial conditions for any step are the conditions at the end of the previous step in the cycle. Each step is uniquely described by a boundary condition on velocity and the inlet fluid stream composition.

The PSA steps are simulated in a given sequence to simulate a PSA cycle. As the PSA cycles are simulated iteratively, a cyclic steady state (CSS) is achieved where the spatial-temporal concentration profiles of each component in each phase as well as the pressure and the velocity profiles in the bed during and at the end of each step in a cycle match the corresponding profiles in the corresponding step in the immediately previous cycle within the tolerance limit. This is the simulated cyclic steady state performance, which could then be evaluated analytically in terms of derived performance parameters such as product purity, recovery, and adsorbent throughput etc.

When the cyclic steady state is achieved, the time series of bulk phase concentrations and velocities entering and leaving the bed provide the transient detail of inputs to and outputs from the PSA system. Suitable numerical integration of these instantaneous profiles with respect to time at each of the bed ends summarize the input and output molar flows of all feed and product streams respectively over each step of a PSA cycle (integration limits are over the step duration). The PSA cycle configuration in terms of step durations used with this gives the net hourly input-output flow rate. The quality of the product streams can be defined in terms of the mole fraction of undesired and/or desired components in these streams. Process specific parameters such as recovery, purity can be suitably defined.

For determining the achievement of cyclic steady state by the system, the time averaged concentration in product tanks (such as raffinate and extract tanks) of each component over a cycle is compared with that over the previous cycle. In the present work, CSS is considered as achieved if the absolute difference between the two values for each component becomes less than or equal to a tolerance value of 10^{-4}. For rigorous quantification of numerical stability and convergence of the solution algorithm, apart from these gross convergence checks, mole balance check for each component is made for every cycle. At CSS, moles of ith component flowing into the bed over the entire cycle must be equal to the moles of that component flowing out from the bed. This should be true for all the components.

3. Simulation-based design

To begin with, simple adsorption breakthrough curves were simulated. In this, the bed was considered initially as devoid of any other adsorbate than the weakest adsorbate which is Methane. In other words, the adsorbent particles were considered as free of any adsorbed phase concentration. Feed was introduced in the bed at its pressure-composition-temperature conditions. Effluent specifications (13000-14000 SMCPH and content of less than or equal to 0.6% C_{2+} in raffinate) were observed. This was essential to get some feel of upper limit of adsorption step time in the eventual PSA cycle and also to gauge the amount of adsorbent requirement for specified capacity.

Desorption breakthrough of a fully saturated bed was also simulated by considering that the bed is purged with pure Methane (the weakest adsorbate) at the adsorption pressure and temperature to get an idea of the relative time scale of the regeneration time step duration. In actual cycle, purging step takes place with fraction of raffinate after the bed has been depressurized to the desorption pressure for systems where raffinate is the desired product such as this case.

These time steps would of course depend upon the adsorbent loading in the beds. Smaller the adsorbent loading, smaller would be the durations of both the steps.

These results with salient findings are presented below.

Typical adsorption breakthrough curves for a bed of 1.5 m diameter and 5 m height packed with 0.002 m spherical particles of Silica Gel adsorbent, 303.15 K temperature, 34 bara adsorption pressure and feed rate of 13800 SCMPH are shown in Figure 3 for Propane and

CO_2. The model predicts that Propane starts to come out of the bed from 515 seconds. Butanes will breakthrough later than Propane as they are adsorbed more strongly than Propane. So only the adsorption breakthrough behavior of Propane was followed out of the undesired alkanes (Propane and Butanes). The feed composition used and the isotherm parameters were as given earlier in Table 2. Mass transfer limitations were neglected due to the lack of data on the mass transfer coefficients. Initially the bed was assumed to be at the adsorption pressure of 34 bara and filled with pure CH_4 in both the phases. Adsorption breakthrough curve is a cycle having the step of only adsorption and desorption breakthrough curve is a cycle having the step of only desorption. Duration of these steps were entered for the respective cases. 15 divisions were used for the spatial discretization of the adsorbent layer and temporal step size was 1 second.

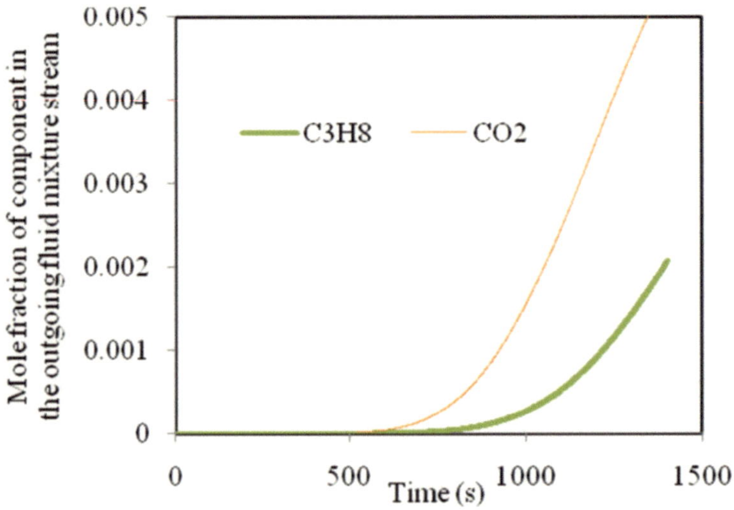

Figure 3. Model predicted adsorption breakthrough curve for Propane and CO_2

It can be seen that even if the bed is fully regenerated each time before the adsorption step begins, the adsorption step time cannot be more than 1800 seconds if product stream has to have less than 0.1% of C_{2+}. More time can be kept if product stream can have more than this content of C_{2+}. These limits on concentration were for the mixed cup composition of the bed effluent stream, or raffinate. From this, the breakthrough concentrations had to be integrated along with the flow rates. The integrated concentration against time chart is also shown for this case in Table 6.

Desorption breakthrough curves are similarly presented in Figure 4. Initially the bed was assumed to be at the adsorption pressure of 34 bara and filled with the Natural Gas composition in the fluid phase. The solid phase composition of each component was in equilibrium with the fluid phase composition of that component. It was assumed that pure CH_4 was fed to this bed for 1000 seconds. The feed rate of this stream was kept nearly equal to the feed rate of Natural Gas mixture in the adsorption breakthrough curve simulation. The bed size, adsorbent, adsorbent dimensions, temperature used in the desorption breakthrough and the feed rate of pure Methane were respectively same as that used in the adsorption breakthrough simulation. The stronger adsorbate (higher equilibrium constant) will require more time to be completely removed from the bed. This is observed in Figure 4 from the concentration profiles of Propane and Butanes.

Time (s)	% C_3H_8	% n-C_4H_{10}	%i-C_4H_{10}	% Total C_{2+}
2000	1.41E-01	1.54E-03	1.67E-04	1.43E-01
1800	9.96E-02	7.03E-04	6.36E-05	1.00E-01
1400	3.05E-02	7.78E-05	4.76E-06	3.06E-02
1200	1.15E-02	1.66E-05	8.29E-07	1.15E-02
1000	2.91E-03	2.32E-06	9.45E-08	2.91E-03

Table 6. Mixed-cup composition of raffinate against adsorption step time chart

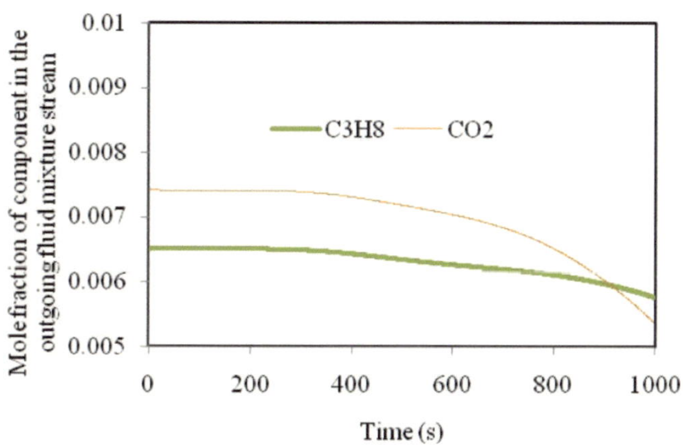

(a)

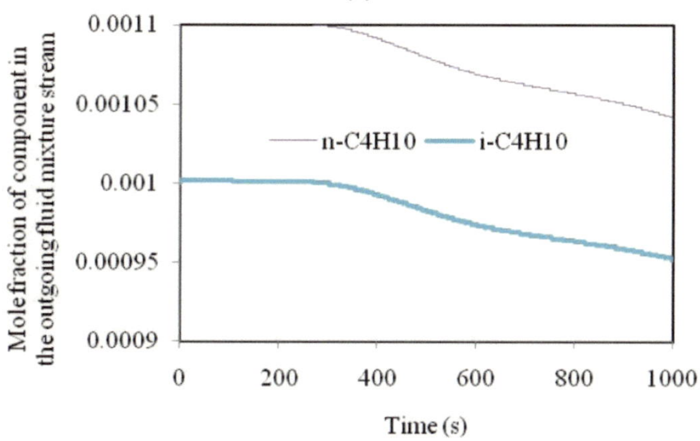

(b)

Figure 4. Model predicted desorption breakthrough curve for (a) Propane and CO_2 and (b) Butane isomers

In the actual PSA cycle, the regenerated bed is not entirely free of adsorbate and hence breakthrough would occur earlier. At the same time, during desorption, the bed is not initially fully saturated and desorption times could be smaller than observed above. The above adsorption-desorption comparison, nonetheless, gives an idea of two important steps of any PSA cycle, namely adsorption and regeneration. The total cycle time can be decided as follows. The feed rate and the purge rate are achieved by adjusting the raffinate valve and extract valve coefficient respectively. For a fixed feed rate, bed size and the valve coefficients, the time required to pressurize the bed from the desorption pressure to the adsorption pressure can be found by observing the model predicted pressure against time profile. Similarly, the time required to depressurize the bed back to the desorption pressure after the bed has completed the duration of the adsorption step can be found. The purge step duration can be tuned to get the desired purity. Thus the total cycle time can be obtained.

Most known PSA processes have matching adsorption and desorption step durations. Regeneration steps together must consume a total duration which is an integer multiple of total duration of adsorption steps [4]. If both the steps match in their durations, a 2 bed PSA system can be designed, with one bed in adsorption and one in desorption at any time. If desorption step is of a longer duration than adsorption duration, a 3 or 4 bed system can be conceived depending on whether the desorption time is twice or thrice of the adsorption time. 2 or 3 bed PSA systems are very common in this technology.

In the present case, the adsorption and desorption step times were of entirely different nature. The adsorption step was of much longer duration than the desorption. In fact, with the above initial simulations and more PSA simulations, it was seen that the range of adsorption-desorption steps is something like 600-100 seconds respectively. This would mean that the PSA system will have multiple beds, say 8, with 6 in adsorption step and 2 in desorption step at any time. These would further get fine tuned through simulations. However, it was apparent that an unconventional PSA cycle is in the offing.

The fact that we will need to have about 8 beds with 6 or 7 in adsorption step producing raffinate and 1 or 2 in desorption step producing extract, another important possibility emerged leading to the concept of surge-less PSA.

All known PSA plants use large drums as surge vessels. This is a necessity of PSA plants due to large surges in pressures and flows of raffinate and extract streams produced and feed used. The flow surges occur as a bed at low pressure is opened up to feed at high pressure and bed at high pressure is opened up to extract at lower pressure etc. At individual bed level, these pressure-flow surges an integral part of conventional PSA processes. Due to smaller number of beds employed, (2 or 3 bed cycles) these pressure surges also get transmitted to raffinate-extract-feed tanks. In the present case, as we anticipate to have 6 beds undergoing adsorption at various stages at any time, these beds produce raffinate streams at different flow rates and pressures simultaneously. The combined raffinate stream emerging from the beds undergoing adsorption is thus at a fairly uniform pressure and flow. There are pressure-flow surges during desorption as in the case of any 2 or 3 bed PSA plants with 1 or 2 beds undergoing desorption simultaneously. If these could also be controlled in some way, the PSA process would operate as a surge-less process. This concept was pursued further giving rise to the surge-less PSA design.

Before we go to the actual design emerging out of the simulation studies, the operational advantage of a 12 bed PSA system with highly asymmetric durations of adsorption and desorption steps is presented in Figure 5 graphically. Each bed passes through the same step

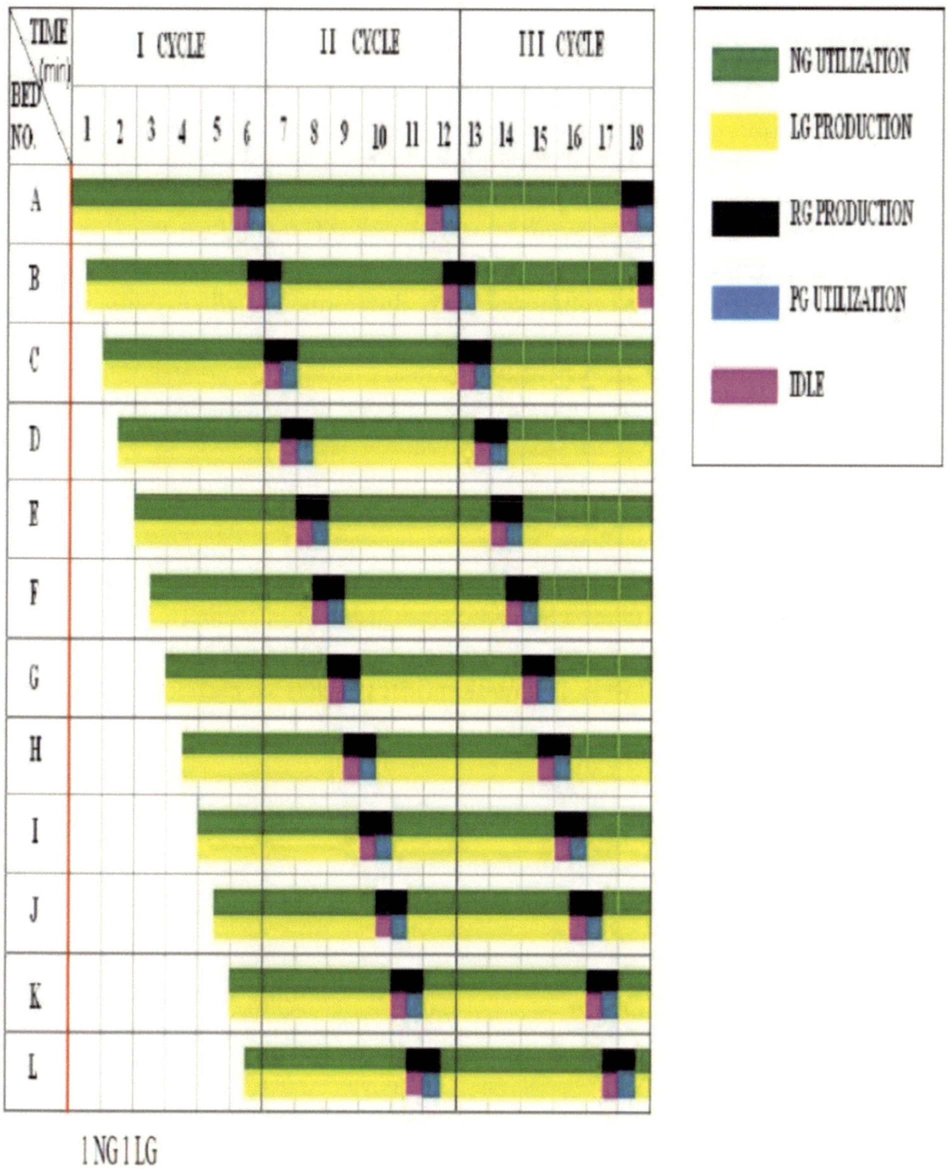

Figure 5. 12 bed system with t_{cycle} of 6 minutes and staggering time of ½ minute,(10 beds produce lean gas and 2 produce rich gas at any instant)

sequence. To get overall cyclic operation of the multi-bed system, the beds are staggered in their instants at which they start the first step of the PSA cycle. Staggering time is the total cycle time divided by the number of beds. Using the color code for the PSA steps as shown in Figure 5, the instantaneous position of the beds in this sequential execution of a PSA cycle employing 12 beds and a total cycle time of 360 seconds is summarized. The cycle described in this figure produces lean gas for 300 seconds and rich gas for 60 seconds. Natural Gas is fed for 300 seconds and the bed is purged for 30 seconds. The staggering time is 30 seconds. It can be seen that 10 beds are at various stages of saturation part of the PSA cycle while 2 beds are undergoing desorption/regeneration.

For the multiple-step cycle simulations, the feed composition, molecular weights or pure component and adsorption thermodynamics data was available as presented earlier in Table 2. The other required input data can be categorized as given by the captions of Tables 7-10. Feed gas had 98.4% of the desired components which are Methane and Ethane. If there are only ON/OFF valves, the flow rates to or from beds would depend on the instantaneous pressure differential available at the two ends of the valve. For example, at the start of a feed pressurization step, there is maximum pressure differential available and flow rates calculated using the valve equation could be very large. In actual practice also, the flow rates could be large causing the adsorbent bed to shake and cause undesirable attrition of particles. A control valve with appropriate time-adjusted valve opening could minimize these velocity surges. A 5-step PSA cycle having the steps of feed pressurization (10 seconds) + feed adsorption (560 seconds) + co-current depressurization(14 seconds) + counter-current depressurization (38 seconds) + purge (61 seconds) by fraction of raffinate was simulated on a column having 5 m adsorbent layer height. Two cases were studied. The first one had no controlled opening or closing of the valves unlike the valves for the second case. For the controlled case, instantaneous valve coefficient was given by Equation (10) as per linear valve characteristics. The value of valve coefficient at full opening (fo) or 100% opening (1st term in numerator of RHS of Equation (10)) is given for each valve in Table 10.

Sphericity	1
Particle diameter (mm)	2
Bed voidage	0.35
Inner Diameter of Bed (m)	1.5
Adsorbent Height (m)	5 (7 for lumping studies)
C_v of Feed Valve	0.01
C_v of Raffinate Valve	0.00029
C_v of Extract Valve	0.0038 (0.004 for lumped case)
C_v of Purge-in Valve	0.006

Table 7. Adsorbent, bed and valve properties used in the simulation

Adsorption temperature (K)	303.15
Feed Tank pressure (bara)	34
Raffinate Tank pressure (bara)	21
Extract Tank pressure (bara)	5.5
Purge-in Tank pressure (bara)	7.5
t_1 (s)	10
t_2 (s)	560
t_3 (s)	16
t_4 (s)	38
t_5 (s)	61
t_{cycle} (s)	685

Table 8. Operating conditions of the simulated cycle

Δz (m)	0.47
Δt for Step 1 (s)	0.10
Δt for Step 2 (s)	1.00
Δt for Step 3 (s)	0.10
Δt for Step 4 (s)	0.50
Δt for Step 5 (s)	1.00
Step size for Jacobian	0.001
Estimate of Euclidean distance of the solution from the initial estimate	100
Accuracy requirement	0.01
Max number of iterations	600000

Table 9. Numerical solver related data

Valve	C_v^{fo}
Feed	0.01
Raffinate	0.00029 (0.0003 for controlled opening and same feed rate as for no control case)
Purge-in	0.08
Extract	0.01 (0.00123 for controlled opening and same purge rate as for no control case)

Table 10. Coefficient at full opening of each valve

Time (s)	Raffinate valve	Purge-in valve	Feed valve	Extract valve
0	0	0	100	0
9.5	0	0	100	0
10	100	0	100	0
60	100	0	100	0
100	100	0	100	0
550	100	0	100	0
560	100	0	100	0
569.5	100	0	100	0
570	100	0	0	0
577	100	0	0	0
583.5	100	0	0	0
584	0	0	0	100
598	0	0	0	100
621.5	0	0	0	100
622	0	100	0	100
635	0	100	0	100
669	0	100	0	100
682.5	0	100	0	100
683	0	0	100	0

Table 11. % opening against time for each valve (No control)

Table 11 gives the percentage opening of each valve against time for the uncontrolled case. Table 12 gives a similar profile for the case of controlled opening and closing. In Table 11, it is seen that all the valves open at 100% opening immediately and completely close from full opening immediately. In Table 12, the feed valve opens with 10% opening and ramps up linearly to only 50% at the end of PF step. It ramps up linearly to 100% at the end of 60 seconds after the start of PF step. It starts to ramp-down from 550 seconds position to 60% at the end of 560 seconds and closes completely at the end of the adsorption step. Thus the ramp-up time of this valve is 60 seconds and ramp-down time is 20 seconds. The valve openings of the remaining three valves are as shown in Table 12. Again when a new cycle starts, it opens with 10% opening. Between 60 seconds to 550 seconds, it remains fully open. The percentage opening at intermediate time is found by interpolation. From the percent opening, the instantaneous coefficient of the valve is back-calculated as per valve characteristics say that given by the equation below. Equation (10) represents the characteristics of a linear opening valve and appropriate equation will have to be used if one has to use a control valve with other valve characteristics. The calculated valve coefficient is used in Equation (6) in the model equations.

$$C_V(t) = \frac{C_V^{fo}}{100} \%Opening(t) \tag{10}$$

The valve opening vs time profile is provided by the user as input to the simulation model. However, it can be deduced to begin with as follows. Consider the pressure of the bed after the end of PU step as P_5 and that of the feed gas as P_1. Let t_{PF} seconds be the duration of PF step. At the end of t_{PF} seconds, the pressure drop across the feed valve should ideally drop to zero from the starting value of (P_1-P_5).

Time (s)	Raffinate valve	Purge-in valve	Feed valve	Extract valve
0	0	0	10	0
9.5	0	0	45	0
10	10	0	50	0
60	100	0	100	0
100	100	0	100	0
550	100	0	100	0
560	100	0	60	0
569.5	100	0	10	0
570	100	0	0	0
577	100	0	0	0
583.5	10	0	0	0
584	0	0	0	1
598	0	0	0	100
621.5	0	0	0	100
622	0	10	0	100
635	0	100	0	100
669	0	100	0	100
682.5	0	10	0	10
683	0	0	10	0

Table 12. % opening against time for each valve (With control)

Even as the pressure drop across the feed valve drops with time, to get uniform draw of feed gas into the bed during this step, we should push same amount of gas every second into the bed. That is the velocity should be the same. Using Equation (6) and assuming constant flow velocity, we get Equation (11) which relates valve coefficients and pressures at two instances t_{1PF} and t_{2PF}. Valve Opening at (t_{2PF}) can be calculated using Equation (10).

$$\frac{C_V(t_{2PF})}{C_V(t_{1PF})} = \sqrt{\frac{(P_1\text{-}P(t_{1PF}))P(t_{2PF})}{(P_1\text{-}P(t_{2PF}))P(t_{1PF})}} \quad 0 \le t_{1PF} < t_{2PF} \le t_{PF,} P(t_{1PF}) < P(t_{2PF}) \le P_1 \qquad (11)$$

Thus, if feed pressure (P_1) is 34 bara, bed pressure is 18 bara at time t_{1PF} [i.e. P_1 - $P(t_{1PF})$ = 16] and 30 bara at time t_{2PF} [i.e. P_1-$P(t_{2PF})$ = 4], the square root of pressure differential has only halved between the two time instants. If flow should be the same at both the instants, the C_V of feed valve at time t_{2PF} should be approximately thrice of its value at time t_{1PF}. If the feed valve characteristics are as per Equation (10), it would mean that its percentage opening at time t_{2PF} should be also thrice of the percentage opening at time t_{1PF}. Same procedure can be applied for the remaining three valves to arrive at appropriate valve opening time series However, true surge-less operation will be achieved only if feed rate is ensured to be constant across pressurization and adsorption steps. Extract production rate is ensured to be constant across counter-current depressurization and purge steps. Raffinate production rate is ensured to be constant across adsorption and co-current depressurization steps and purge utilization rate is constant across purge step. This can be achieved only through percentage valve opening against time profile of control valves across each of the four valves.

Typical pressure and flow profiles at the two ends of any bed over a PSA cycle are shown in Figure 6 and Figure 7 respectively as given by the undashed lines for the case of controlled opening and closing and by the dashed lines for the case of uncontrolled opening and closing of the valves. There are significant surges in the velocity and the pressure for the case of uncontrolled opening and closing of the valves. However, the combined effect of all the beds together is much smoothened indicating the possibility of a surge-less PSA system or a PSA system with much smaller surge vessels. For operation with control, as seen from the undashed line in Figure 6, the bed has not reached the feed tank pressure, which it does after some duration of the adsorption step has elapsed. After that, the profiles are identical. In the depressurization step, controlled opening uses more time than the uncontrolled case to depressurize the bed to the desorption pressure value.

It is seen in Figure 7 that the superficial velocities for the case with no control have reached as high as 2.8 m/s in the pressurization step and then reach zero over half of its duration. It remains zero for the remaining half of the step. Comparatively the velocity variation for controlled opening case is less with the velocity magnitude not exceeding more than 1 m/s. Both the cases differ from each other in a similar manner in the depressurization step.

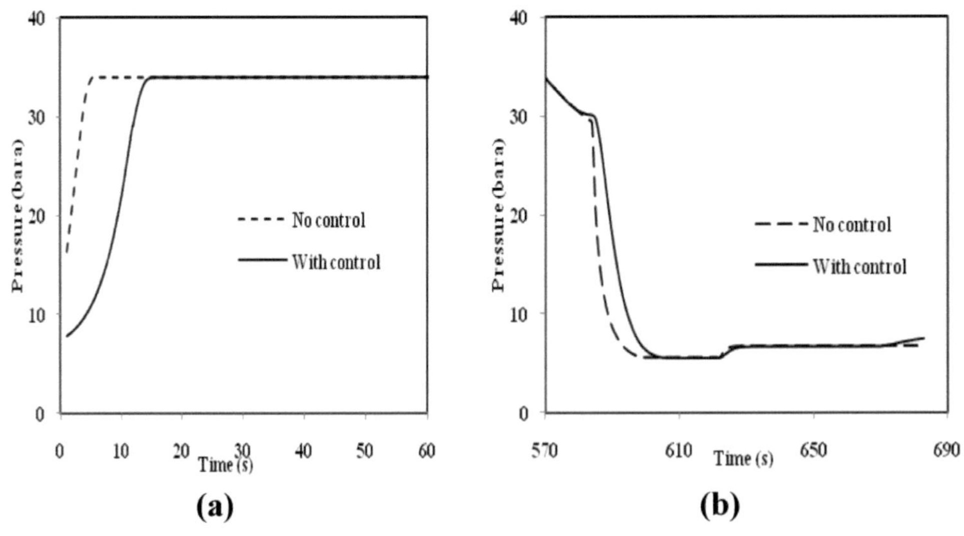

Figure 6. Pressure against time in (a) the feed pressurization step and the initial part of the feed adsorption step and (b) the depressurization steps and the purge step

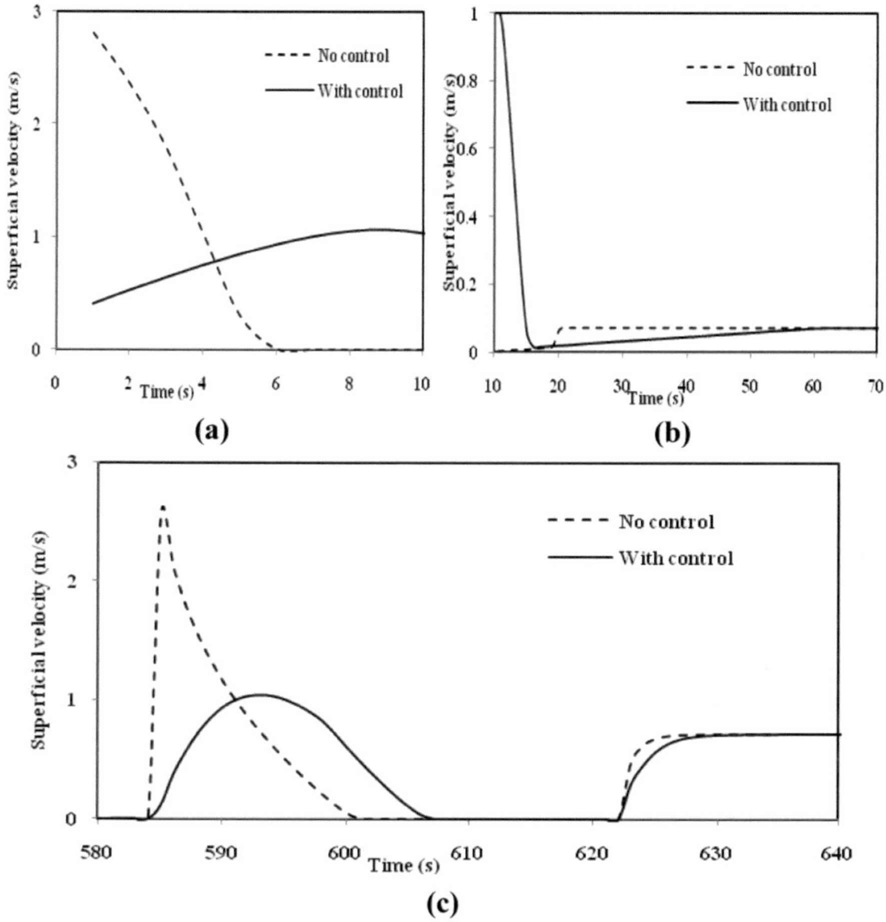

Figure 7. Superficial velocity against time for (a) the feed pressurization step, (b) first minute of the feed adsorption step and (c) the counter-current depressurization and the purge steps

Gradual pressurization is also preferred in kinetic separations like N_2-PSA [3]. Controlled opening is also better to prevent possible fluidization of the particles due to high velocity. Irrespective of whether the bed is getting pressurized or depressurized, controlled opening increases the time required to achieve the required pressure level, reduces the velocity variations and their magnitudes. It also helps to ensure that flow occurs during entire duration of the step smoothening the feed utilization or raffinate/extract production profiles.

Table 13 gives the comparison between the two cases based on important performance parameters. Separation performance is highly dependent on the purge rate or purge step duration. Using the same valve coefficients as that for the uncontrolled case, flow rates turned out to be lower for the first controlled case (middle column of Table 13). Since the purge rate is slightly lower for this controlled case, the purity is lower than that in the case where valve opening was not controlled (first column of Table 13). The last column in Table 13 gives the results for controlled case with feed rate and purge rates almost equal to that for no control case. The flow rates were matched by increasing the coefficient for raffinate and extract valves at 100% opening respectively. This case gives higher purity than that for the case in second column. The purity increased because the purge rate was increased. Thus controlled opening and closing of the valve can also change the flow rates and the separation performance of the process. The purity for the second controlled case was lower than that for the uncontrolled case. Correspondingly, the recovery for the former case was higher.

	No control	With control-1	With control-2
Feed rate (SCMPH)	12544	12156	12532
Purge rate (SCMPH)	2385	2188	2391
CO_2+C_{2+} recovery %	71.4		72.9
C_1+C_2 % in raffinate (purity)	99.83	99.53	99.57

Table 13. Comparison of both the cases based on performance parameters

Although the multi-bed PSA plant was expected to give a reasonably surge-less operation, further reduction in surges was achieved by actually providing a control valve at product and feed ends of each bed and arriving at the time series of % opening of these valves over a PSA cycle.

It is thus possible to get a surge-less PSA with programmed control valves regulating feed and product flows during individual steps of a PSA cycle. The surge-less concept eliminated the need of surge tanks for the PSA plant bringing in major cost savings. The controlled opening and closing helps to reduce the magnitude of the velocity, reduce the velocity fluctuations and improves the utilization of the duration of depressurization and pressurization steps.

The multi-bed PSA plant also offered another advantage. The plant has been designed and commissioned in a plant environment where processing capacity varied as per downstream requirement. PSA cycles were designed to operate a total of 7, 6, 5, 4 and 3 beds only with reduced plant capacity so as to give same product purity and recovery. This allowed the plant to operate at a large range of turndown ratios unlike conventional 2-3 bed processes.

The simulation model is very rigorous and involves extremely heavy computational effort. Design is an iterative activity and would involve hundreds of such simulations. The scale of computations and array sizes depend upon the number of components. It was therefore considered desirable if some sort of component lumping could be considered in simulations such that the simulation results capture the salient performance appropriately. Once such lumped models indicate optimal design, a rigorous simulation could be carried out only for those cases or operation window around it. Without this, the simulation based design would have been impossible. This is briefly discussed in the next section.

4. Utility of component lumping in the cycle simulation

There is hardly any work on suitable lumping strategy for PSA cycle simulations. The authors carried out extensive work on developing such strategies. The lumping strategies are covered in detail separately [5]. Mass transfer limitations were neglected due to lack of data of the mass transfer coefficients. Isotherm parameters of a pseudo-component are taken to be the weighted average of the isotherm parameters of all its member components. These and other properties such as molecular weights are given in Table 14. Lumping CH_4 with CO_2 or Ethane and heavier alkanes into a pseudo-component required modifying the definitions of the separation performance parameters. Salient features and relevant lumping for the present case are summarized below. A procedure to select the best strategy of component lumping is given in Figure 8. However before using the lumping strategy, care was taken to ensure that the step durations and bed dimensions were neither over-designed nor under-designed by observing the variable profiles for a case considering no lumping of components. Another point to be remembered is that in the simulations the number of components may get reduced by using various lumping rules, but to get the properties of the pseudo-components, the properties of each member component are required.

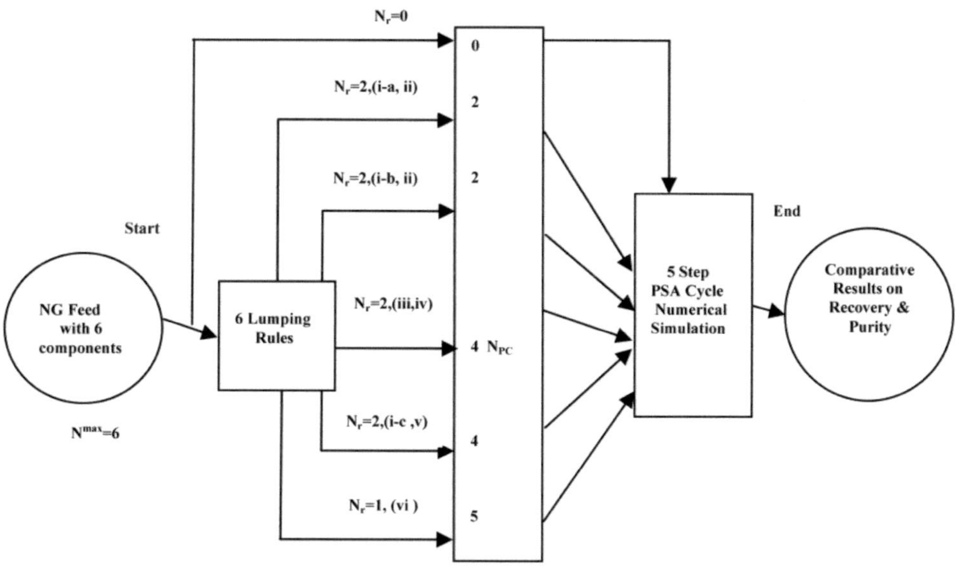

Lumping rules

(i)	Desired product is (i-a) C_1+C_2 or (i-b) C_1 or (i-c) C_1+CO_2	(iv)	Components have almost equal values of adsorption isotherm equilibrium constants
(ii)	Ratio of mole fraction of the desired product to the mole fraction of undesired or other components in the feed is greater than 15	(v)	Components have equal number of C atoms
(iii)	Components have equal values of molecular weights	(vi)	Keep distinct identity for inorganics, organic non-isomers having equal no. of C atoms and isomers

Figure 8. Flow-chart for the Lumping Strategy

Taking care of these two considerations, an extensive simulation program was launched. The bed height was increased to 7 m in these studies to keep a safety factor and the valve coefficients are as given in Table 7. Lumping based simulations were done only for the uncontrolled case. The PSA cycle configuration is as per Table 8. The cases of lumping are given in Table 14 which gives the member components of a pseudo-component of a case and the simulation required properties of that pseudo-component. The computational and separation performances are summarized in Table 15 and Table 16. The impact of changing-the feed composition, the operating temperature and the cycle configuration on the lumping strategy was also assessed. More details of these are given in our earlier published work [5]. Still the use of 2 pseudo-components with first one having Methane and Ethane as its member components and the second one having the remaining four components, predicted the separation performance closely with that predicted by using no pseudo-components (Case 1) with significant reduction in computation time and array sizes. Components of Natural Gas form a homologous series. However lumping strategy has to be used on a case to case basis.

Case	Pseudo-component	Member components	y_i	MW_i (kg/kmole)	b_i (m^3/mole)
2	1	C_1 and C_2	0.984	16.27	0.0006
	2	C_3, i-C_4, n-C_4 and CO_2	0.016	45.79	0.0089
3	1	C_1	0.9653	16	0.00055
	2	C_2, C_3, i-C_4, n-C_4 and CO_2	0.0347	37.22	0.006121
4	1	C_1	0.9654	16	0.00055
	2	C_2	0.0188	30	0.0025
	3	C_3 and CO_2	0.0138	44	0.007334
	4	n-C_4 and i-C_4	0.002	58	0.017233
5	1	C_1 and CO_2	0.9727	16.21	0.00067
	2	C_2	0.0188	30	0.0025
	3	C_3	0.0065	44	0.0084
	4	n-C_4 and i-C_4	0.002	58	0.017233
6	1	C_1	0.9653	16	0.00055
	2	C_2	0.0188	30	0.0025
	3	C_3	0.0065	44	0.0084
	4	n-C_4 and i-C_4	0.002	58	0.017233
	5	CO_2	0.0074	44	0.0064

Table 14. Member components and their properties for the lumped cases at 303.15 K

Compare results in Table 13 with results in Table 16. For the same feed valve coefficient, longer bed will require more time to get pressurized by the same pressure ratio. This implies that more flow will occur as the pressure drop across the valve remains higher than zero for a longer time in the longer bed. Apart from the adsorbent layer height, the purge-in and extract valve coefficients are lower in the cases considered for lumping studies compared to those used for getting the results in Table 13. So the average feed rate used in lumping exercise (around 14000 SCMPH) was higher and the average purge rate (around 1600 SCMPH) was lower than the respective values used in comparing control valve approach with no control valve approach. As a result the separation performances differ. The purity has dropped to 99.4%. Velocity at the bed ends will depend on the pressure differential across the valve which will be higher for the longer bed. A longer bed will also require higher coefficient of the extract valve to depressurize it in the same duration and by the same pressure ratio. Else it will decrease the purity. The recovery has also dropped to 69%.

Case	Equations solved in adsorption step per call to solver	Range of CPU time (s) per cycle	Q^F (SCMPH)
1	218	313-332	13981
2	94	36-41	14034
3	94	44-79	13993
4	156	88-164	13984
5	156	88-164	14129
6	187	154-220	13983

Table 15. Equations, CPU time and feed rate for each case

Case	1	2	4	6
% moles of C_1+C_2 in Raffinate (Purity)	99.40	99.42	99.41	99.40
% Recovery of CO_2+C_{2+}	69.27	69.79	69.07	69.27

Table 16. Comparison of Case 2, Case 4 and Case 6 with Case 1

5. Conclusions

1. Pressure Swing Adsorption based process has been designed for the removal of Propane and higher alkanes from Natural Gas. The feed gas had six components and was available at a pressure as high as 34 bara. The raffinate requires to have as low content of Propane and higher alkanes as far as possible and come at a pressure higher than 20 bara for downstream use. The operating temperature was ambient value and the operating adsorption pressure was also equal to 34 bara. The adsorbent physically adsorbs the components in the following order: Butanes>Propane>Ethane>Methane. Natural Gas pre-treatment thus belongs to the type of raffinate PSA systems wherein the desired product is the weakly adsorbed component. Extended Langmuir adsorption isotherm parameters for each of the six components of the feed mixture were estimated from the pure component Langmuir isotherm data available in the literature.
2. A generic and rigorous mathematical model was developed to describe the discrete-continuous nature of the PSA cycle. This required solving large number of simultaneous non-linear algebraic equations numerically. The number of equations to be solved at any instant and the computational array sizes depend upon the number of components in the system. This is what makes the study of component-lumping

strategy in the cycle simulation worth considering for the purpose of reduction in computation time and array sizes.

3. The adsorption and desorption breakthrough curves were simulated using this model, actual feed mixture composition, industrial bed sizes and industrial flow rates. The starting duration of adsorption step in the cycle can be decided based on the adsorption breakthrough curve, predicted mixed cup composition of the impurity in the raffinate up to the chosen value of adsorption step duration and the allowable content of Propane and higher alkanes in the raffinate.
4. Like other raffinate PSA systems, this system also used the conventional four step Skarstrom cycle. Adding the step of co-current depressurization before the step of counter-current depressurization also helped to reduce the velocity variation during depressurization step by reducing the pressure difference. Thus a five step PSA cycle was able to give the desired separation performance with lesser velocity surges. As the amount of strongly adsorbed components were present in fraction less than 2% in the feed, the cycle did with 7 to 8 times higher adsorption step time than the desorption step time. The asymmetric distribution of the adsorption step duration and the desorption step duration is another distinguishing feature of this system. The separation performance and variable profiles for the case of uncontrolled opening were observed. Significant variations in velocity exist due to uncontrolled opening. The magnitude of velocity was also high for this case particularly during the initial instants of the pressure changing steps. The variable profiles were studied to ensure that the bed dimensions and the step durations are properly designed. These are to be ensured before the component-lumping strategy is used in the cycle simulation.
5. Then the cycle simulations were done using controlled opening and closing of the valves. Equations were derived to find the appropriate instantaneous opening of the valve required to get the desired pressure difference between the tank and the bed end at constant velocity in a given duration. The velocity and pressure profiles for the case of controlled opening and closing of valves were observed. This case showed significant reduction in the magnitude and variations in the velocity at the expense of more time being required to pressurize or depressurize the bed to the target pressure. The step duration was completely utilized for flow. Thus selecting appropriate step sequence in the cycle and controlling the valve opening/closing will help in minimizing the surge-tank costs and also improve the life of the adsorbent particles. The cycle used in this work uses more beds than in conventional 2-bed or 3-bed cycles. These beds operate in a staggered sequence and are able to ensure continuous feeding/production at any instant without needing surge tanks. A bed can be taken off-line for maintenance or due to reduction in product demand/feed gas availability. This allows flexibility in the capacity which is impossible in conventional 2-bed or 3-bed cycles.
6. To reduce the computation time and array sizes, up to six rules for lumping were applied. These enabled the use of as low as 2 pseudo-components to describe the original mixture of 6 components. The procedure to find the isotherm parameters and molecular weights of the pseudo-components was also developed. To find these, the respective properties of each member component are required. Separation performance

predicted by using the lumped case was compared with that predicted by using no lumping. The number of equations, computational times and array sizes were also compared and the lumped cases had significant savings in each of these against those of the non-lumped case. The feed mixture of 6 components was accurately described by a mixture of 2 pseudo-components where in Methane and Ethane are the members of the 1st pseudo-component and the remaining four components are the members of the 2nd pseudo-component. However, this strategy has to be applied on a case to case basis but is worth considering for multi-component mixture.

Nomenclature

b	Adsorption equilibrium constant of a component on the solid surface at the operating temperature, m^3/mole
c	Concentration of a component in the fluid phase, moles/m^3
C_v	Valve coefficient, unit-less
C_v^E	Coefficient of the extract valve, unit-less
C_v^F	Coefficient of the feed valve, unit-less
C_v^{fo}	Coefficient of the valve at 100 % (full) opening, unit-less
C_v^{max}	Maximum value from the valve coefficients, unit-less
C_v^{PI}	Coefficient of the purge-in valve, unit-less
C_v^R	Coefficient of the raffinate valve, unit-less
D	Inner diameter of the bed, m
d_p	Diameter of a spherical adsorbent particle or equivalent diameter for a non-spherical particle, m
g_c	Gravitational conversion factor, unit-less
i	Component index or Cycle step index, unit-less
j	Component index, unit-less
k^{LDF}	Coefficient of linear driving force approximation model for a component on the adsorbent, 1/s
L	Height of the adsorbent layer packed inside the column or spatial coordinate at the outlet of the bed in adsorption step, m
M	Number of divisions used for the spatial discretization of the adsorbent layer, unit-less
MW	Molecular weight of a component, kg/kmole
N	Number of components, unit-less
N^{max}	Maximum number of components considered in the feed mixture, unit-less
N_{PC}	Number of pseudo-components, unit-less
N_r	Number of lumping rules applied, unit-less

P	Absolute Pressure, Pa(a) or bara
P_1	Absolute pressure of Feed gas, Pa(a) or bara
P_5	Absolute Pressure at the end of purge step, Pa(a) or bara
P^{DS}	Pressure downstream of the valve, bara
P^E	Extract tank pressure, bara
P^F	Feed tank pressure, bara
P^{PI}	Purge-in tank pressure, bara
P^{RA}	Raffinate tank pressure, bara
P^{RI}	Pressure of the rinse-in tank, bara
P^{US}	Pressure upstream of the valve, bara
q^{max}	Maximum monolayer saturation capacity in the adsorption isotherm, moles /m^3 particles
Q^E	Volumetric flow rate of extract from the bed averaged over the duration of the extract production steps, SCMPH
Q^F	Volumetric flow rate of feed to the bed averaged over the adsorption step duration, NLPM or SLPM or NCMPH or SCMPH
q	Average concentration of a component adsorbed in the volume of adsorbent particles, moles/m^3 particles
q^{max}	Maximum monolayer adsorption capacity of a component on the adsorbent surface, moles/m^3 particles
q^*	Adsorption equilibrium concentration of a component in the solid phase, moles/m^3 particles
Q^{PI}	Volumetric flow rate of purge-in to the bed averaged over the duration of the purge step, SLPM or NLPM or SCMPH
Q^R	Volumetric flow rate of raffinate from the bed averaged over the duration of the raffinate production steps, SCMPH
R	Ideal Gas Law constant, Pa-m^3/mole/K
t	Time or Temporal coordinate, s
t_i	Duration of ith step of the cycle, s
t_{1PF}	An instant of the feed pressurization step, s
t_{2PF}	An instant of the feed pressurization step, s
T	Operating Temperature, K
t_{cycle}	Cycle time, s or minutes (1 minute = 60 seconds)
t_{PF}	Duration of feed pressurization step, s
u	Linear superficial velocity of the fluid, m/s
w	Mass fraction of a component, unit-less
y^{PI}	Fluid phase mole fraction of a component in the purge-in tank, unit-less
y	Fluid phase mole fraction of a component, unit-less
y^E	Fluid phase mole fraction of a component in the extract tank, unit-less

y^R	Fluid phase mole fraction of a component in the raffinate tank, unit-less
y^F	Fluid phase mole fraction of a component in the feed tank, unit-less
y_{UD}	Fluid phase mole fraction of the undesired product, dimensionless
z	Spatial coordinate, m
Δt	Size used for the temporal discretization of the duration of a step of the cycle, s
Δz	Size used for the spatial discretization of the adsorbent layer, m
ε	External voidage in the bed packed with the adsorbent particles, unit-less
μ	Dynamic Viscosity of the fluid mixture at the operating temperature, Pa-s
ϕ	Sphericity of the adsorbent or inert particle, unit-less
ϱ	Density of the fluid mixture at the operating conditions, kg/m^3
C_1	Methane
C_{1+}	Ethane and heavier hydrocarbons
C_2	Ethane
C_{2+}	Propane and heavier hydrocarbons
C_3	Propane
C_4	Normal Butane and Iso-Butane
CPU	Central Processing Unit
CSS	Cyclic steady state
DAE	Differential Algebraic Equation
i-C_4	Iso-Butane
LG	Lean Gas (Raffinate)
LPG	Liquified Petroleum Gas
n-C_4	Normal Butane
NG	Natural Gas (Feed)
PG	Purge-in Gas
RG	Rich Gas (Extract)
SCMPH	Cubic meters per hour at 273.15 K and 1 bara or10^5 Pa absolute

Author details

Pramathesh R. Mhaskar and Arun S. Moharir*
Department of Chemical Engineering, IIT Bombay, Mumbai, India

6. References

[1] Daiminger U, Lind W, Mitariten M. Adsorption added value. Hydrocarbon Engineering 2006; 11(2) 83-86.

* Corresponding Author

[2] Olivier M, Jadot R. Adsorption of Light Hydrocarbons and Carbon Dioxide on Silica Gel. Journal of Chemical and Engineering Data 1997; 42 (2) 230-233.

[3] Mhaskar PR. Synthesis Strategies for Discrete-Continuous Adsorptive Separation Systems. PhD Thesis. IIT Bombay India; 2012.

[4] Yang RT. Adsorbents: Fundamentals and Applications. John Wiley & Sons Inc. USA; 2003.

[5] Mhaskar P, Moharir A. Multi-component Adsorptive Separation: Use of Lumping in PSA process simulation. Adsorption 2011; 17(4) 701-721.

Chapter 3

Review on Natural Gas Thermopysical Property Measurement Techniques

Mert Atilhan and Santiago Aparicio

Additional information is available at the end of the chapter

http://dx.doi.org/10.5772/48647

1. Introduction

Since it is impossible to measure the thermodynamics properties of all systems in nature, we must rely upon mathematical models to extrapolate the available experimental data. In order to develop such models, very accurate experimental data are necessary for selected complex mixtures, such as those that exist in natural gas. For this reason researchers must collect the most important and fundamental thermodynamics properties for such systems. Two of the most important thermodynamics properties are the pressure volume (density) and temperature ($P\rho T$) surface and the phase equilibrium properties of mixtures. Accurate volumetric property data are used in custody transfer operations for natural gas. Also accurate $P\rho T$ data are necessary for calculating energy functions. On the other hand, phase equilibria data are needed mostly for design calculations involving separation processes. Additionally, very accurate phase equilibrium knowledge is necessary for natural gas transfer through pipelines to avoid condensation in the pipelines. Atilhan et. al. [1] have shown that even widely used equations of state (EOS) such as Peng-Robinson or Redlich-Kwong (RK) cannot predict the retrograde condensation region for simple natural gas-like mixtures that do not contain heavy fractions.

When natural gas rises from the reservoir to the ocean floor at offshore platforms, the stream temperature can drop quickly (perhaps 5 to 10 °C) until it reaches the surrounding ocean temperature. This rapid temperature drop at high pressure along with moisture in the natural gas stream make conditions favorable for gas hydrate formation in the pipeline. Hydrates can cause several serious problems such as: plugging the pipeline and blowouts [2]. Similar problems are as well common in natural gas compression facilities at offshore and onshore processing plants. Such problems can be avoided by increasing the temperature and insulating the stream that comes from the ocean bed, or by lowering the pressure of the pipeline. Another possible solution is lowering the dew point of water in the stream by adding polar solvents to the line such as methanol or glycols. In order to apply all these methods, accurate knowledge on PρT behavior of the natural gas stream is necessary.

Because of their importance, density measurements essential for both industrial applications and scientific research. Very accurate PρT data is required not only to calculate custody transfer of natural gas in pipelines but also to develop new EOS for industrial and scientific use. Experimental PρT data is employed to calculate thermal properties of fluids required for industrial process design calculations. Loss of accuracy from density predictions directly impacts processes; therefore only exceptionally good density values ensure good thermal properties [3].

Knowledge of temperature, pressure and composition enables determination of the density from an EOS. The equation most widely used in custody transfer of natural gas is the Detailed Characterization Method or AGA8-DC92 EOS developed by American Gas Association (AGA) in 1992 [4]. This EOS was derived using an extensive and reliable experimental PρT database that included real natural gas mixtures as well as high order hydrocarbon mixtures (mostly binary mixtures of natural gas components). AGA8-DC92 EOS has different accuracy regions as shown in Figure 1.

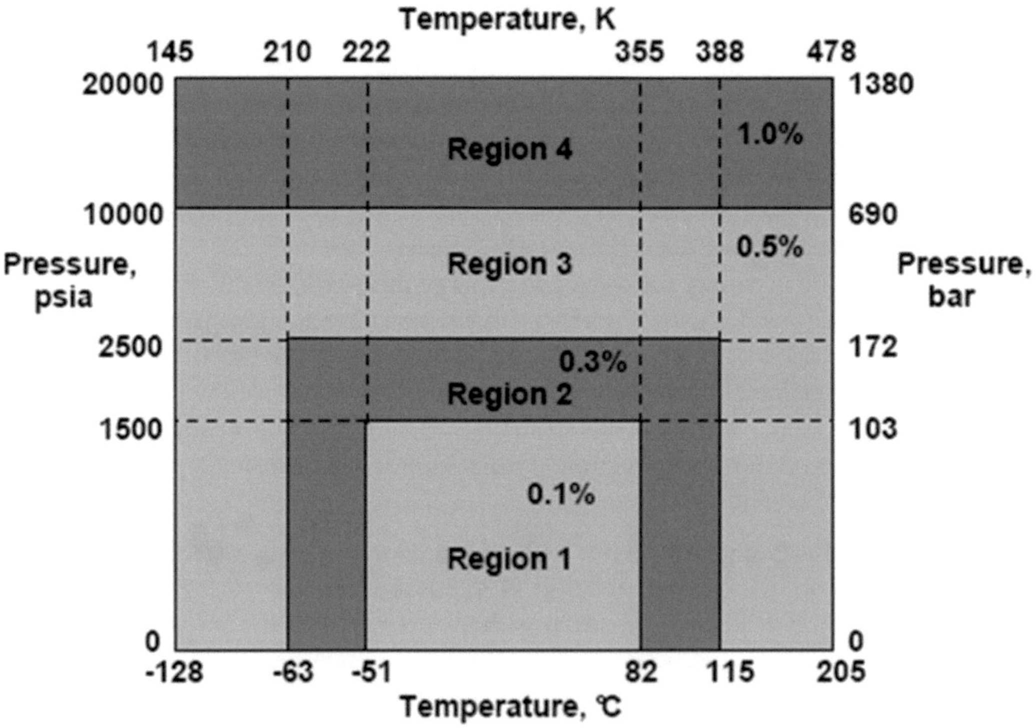

Figure 1. AGA8-DC92 EOS uncertainty regions [5].

As seen in Figure 1, the maximum uncertainty claimed for the EOS is 0.1% in region 1, 0.3% in region 2, 0.5% in region 3 and 1.0% in region 4. However, the equation is valid only for lean natural gas mixtures over this wide range of conditions, and its ability to describe rich natural gases is untested. The equation cannot perform equilibrium calculations, as it is only valid for gas phase calculations. Also, use of the equation is not recommended near the critical point. The deviations from density measurements are 2 to 2.5% for North Sea natural gas samples at 8 to 17 MPa and 40 to 80 °C [6]. Following this, the Gas Research Institute (GRI) and National Engineering Laboratory (NEL) started a collaborative, joint industry project to extend the range of applicability of AGA8-DC92 EOS for natural gas mixtures to include the gas compositions observed in the North Sea. GERG-2008 EOS is the latest benchmark EOS that addresses the drawbacks of AGA8-DC92 EOS and it has proven to have better uncertainty in PρT predictions [7].

2. Density measurement techniques

2.1. Introduction

This section contains reviews of several experimental methods for density measurements, and discusses the relative strengths and weakness of each method. An EOS can describe the thermodynamic state or vapor liquid equilibrium (VLE) of pure fluids and mixtures with accuracy that depends upon the application. The accuracy of an EOS depends upon the experimental data used during development of the equation. Historically, the quality of predictions obtained from EOS has improved dramatically as advanced technologies and new instrumentation have become more common for experimental methods. Among the thermodynamic properties, density is the most directly predicted property using EOS. The measured densities should be approximated by suitable EOS and the measurements should be traceable to the International System of Units [8]. According to Kleinrahm et. al. [9], the following considerations are important when deciding upon a density measurement technique:

i. Large pressure and temperature range for wide operations.
ii. Low total uncertainty and high accuracy of the method for the overall range.
iii. Simplicity in design and ease of maintenance and operation.
iv. Little time required for each data point measurement.

2.2. Density measuring devices

Several different density-measurement techniques are described in this section, including: speed of sound methods, vibrating body techniques, continuous weighing method and buoyancy-based densimeters.

2.2.1. Speed of sound methods

Speed of sound measurements can be used to determine the performance of an equation of state for thermodynamics property predictions. By correlating the speed of sound to thermodynamics properties, one can build experimental devices and investigate solid, liquid and gas thermodynamics properties for pure components and mixtures. Densities and isothermal and isentropic compressibility factors result from speed of sound measurements experiments [10]. Based upon a pulse technique described by Daridon et. al. [11], a cylindrical-shaped cell is used to measure ultrasonic waves. In the pulse technique, effects of pressure upon piezo-electric materials are isolated by separating piezo-electric elements from the fluid studied. The speed results from the measurements of the transit time through the sample and the length of passage, which is a function of temperature and pressure. Density comes from:

$$\rho(P,T) = \rho_0(P_0,T_0) + \int_{P_0}^{P} u^{-2} dP + \int_{P_0}^{P} \left(\alpha^2 / C_p\right) dP \tag{1}$$

In above equation u is the sound speed, α is the isobaric coefficient of thermal expansion, C_p is the isobaric heat capacity and P_0 is the atmospheric pressure. The sum of these terms gives the density with as a function of pressure at different temperatures. The first integral, where can be expressed as a polynomial in pressure with coefficients expressed as polynomials in temperature, can be evaluated along the isotherms considered. By using thermodynamics relations for α and C_p, the second integral can be calculated iteratively. The second integral is a few percent of the first integral. More detailed discussion on numerical evaluations for such measurements appears in [10] and [11].

2.2.2. *Vibrating devices*

Vibrating tubes and vibrating forks are common density measuring techniques. These devices measure the fluid density of interest by determining the oscillation frequency of the vibrating element in the fluid. These instruments provide accurate results quickly. However, frequent calibration is necessary for this apparatus to maintain its accuracy [12]. Moreover, when the density of the fluid is vastly different from air or pure water (frequently used as reference fluids because of their well-known thermophysical properties) the uncertainty of the measurements increases as reported by Kuramoto et al. [8].

2.2.2.1. *Vibrating wires*

In vibrating wire densimeters, a wire carrying a diamagnetic weight is suspended in the fluid to be monitored. The wire is placed in a robust position in a uniform magnetic field provided by permanent magnet in both vertical and horizontal directions. When an alternating current passes through the power source to the wire, interaction starts with the current and the magnetic field. This leads to induced harmonic motion that is orthogonal to the magnetic field and the wire. If the mass, density and the dimension of all the solid components of the system are known, the resonant frequency of the wire can be determined experimentally under vacuum conditions. If the viscosity of the fluid of interest is known, experimental measurement of resonant frequency of wire velocity provides the fluid density [13; 14]. Although the vibrating wire technique is suitable for a wide range of pure fluid and mixture gas density applications, it suffers from problems such as surface tension on the wire, adsorption on the weight, detailed knowledge need of exact dimensions of the wire and the assembly. However, the device is used widely as a primary densimeter device because it has a simple operating principle and allows development of an exact physical model. Density, in principle, can be calculated directly from the theory.

2.2.2.2. Vibrating tubes

Vibrating tube densimeters consist of an assembly that includes two thin walled metallic or glass tubes bent in Y or V shapes as shown in figure 2. A permanent magnet and drive coil reside between these two tubes. Generally, a drive coil and a permanent magnet are placed in the middle of the two tubes. The drive coils and magnet are mounted on the opposite legs of the tubes. Each coil and magnet on the side leg forms a pick-off circuit. Alternatively, attractive and repulsive magnetic fields between the coils and magnets are provided by sending alternating current to the drive coil.

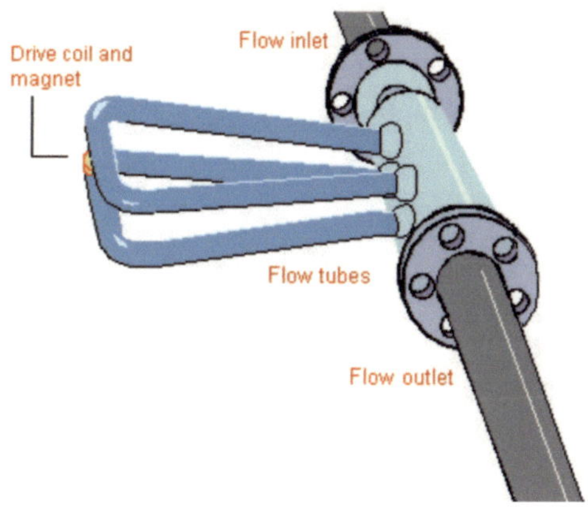

Figure 2. Vibrating tube densimeter scheme.

Because the drive coils and the magnets are installed on the opposing side of the tubes, a sine wave generated by the two pick-up circuits represents the motion of one tube relative to the other. The sine waves are in phase if there is no flow in the tube. The density of the fluid is:

$$\rho = K\tau^2 + L \quad (2)$$

where, tube parameters K and L are both pressure and temperature dependent and τ is the period of vibration. Because it is not possible to determine the temperature and pressure dependence of K and L, measurements are performed at the same temperature and pressure conditions with the sample. A reference fluid of well-known properties is used for this reason. Finally, the corresponding density difference equation is:

$$\rho - \rho_r = K\left(\tau^2 - \tau_r^2\right) \quad (3)$$

where subscript r stands for the reference fluid. ρ and τ are not exactly linear, and this must be taken into consideration. The vibrating tube densimeters are designed for rapid operation and they perform very precise measurements of density differences. If one assumes linearity between ρ and τ, the highest levels of accuracy expectations are not achieved.

2.2.3. *Expansion devices: Burnett method*

Another well-established and widely-used density measurement device is the Burnett apparatus. Burnett [15] suggested a technique to measure the densities of sample fluids without measuring the mass or volume directly. An expansion device contains two cells. During operation, the sample is charged initially into the first cell and, after pressure and temperature measurement, expanded into the second cell. The ratio of the final volume to the original volume equals the ratio of densities before and after the expansion. Only pressure and temperature are measured before and after expansion of the sample from a single volume (V_A) into the combination of the original volume and a second volume (V_A+V_B). Some of the gas goes through a sequence of isothermal expansions into a chamber, which is evacuated every time the expansion takes place. Both virial coefficient and gas density can be calculated with this method. The ratio of the densities before and after the expansion is calculated for each expansion:

$$\rho_i = \rho_m \pi_\infty^{m-i} \prod_{j=i+1}^{m} \left[\frac{1+\gamma_{ab} P_j}{1+\gamma_a P_{j-1}} \right] \tag{4}$$

where

$$\pi_\infty = \lim_{P \to 0} \frac{V_a + V_b}{V_a} \tag{5}$$

In equation 4, γ_a and γ_{ab} are the pressure distortions of the volumes V_a and $(V_a + V_b)$ respectively, π_∞ is the zero pressure cell constant, ρ_m is the density at the lowest pressure and subscripts i and superscript m indicate the value after the i-th and m-th (last) expansions, respectively. A serious problem that can affect the Burnett apparatus is adsorption of the sample gas on the inner surfaces of the measuring cell [16]. Also a Burnett apparatus is difficult to automate fully because of frequent valve operations. Because of error accumulation, very high precision is necessary in the pressure measurements, which necessitates use of high-quality dead-weight gauges. The adsorption affect can be minimized by the using two cells with a surface area ratio almost equal to the volume ratio [17]. Eubank et al. have formulated new adsorption correction schemes, based upon the BET adsorption isotherm [18].

2.2.4. *Continuously weighed pycnometer method*

In the continuously weighed pycnometer method, the mass of the sample is determined by direct weighing of the cell. A typical pycnometer consists of a weight measurement system, constant temperature bath, temperature control system and data acquisition system, a volume bellows cell for changing pressure and density without transferring mass, and a high vacuum system [19]. The major component of this method is a constant volume pycnometer suspended from a digital balance. The pycnometer can be filled and evacuated with an extension tube that enables faster measurements and reduces operator errors. The mass of the pycnometer when empty and when filled with fluid is measured by a digital balance. The density of the fluid being measured at constant temperature and pressure is calculated from the measured mass value of the fluid and the known volume of the pycnometer. One disadvantage of this method is that the long feed tube exposes part of the sample to ambient temperature making it impossible to measure mixture densities when the sample exists as one phase at the cell set point temperature and at room temperature.

2.2.5. *Hydrostatic buoyancy methods*

The hydrostatic buoyancy force technique is based upon Archimedes' Principle. Basically, Archimedes' Principle states "when a solid body is immersed in a fluid, it displaces a volume of fluid the weight of which is equal to the buoyancy force exerted by the fluid on the sinker." This means that the buoyancy force is proportional to the density of the fluid in the measuring cell under pressure. This principle can be applied to determine the gas density of any pure fluid or mixture. Historically, improvements have appeared in the application of buoyancy method based densimeters.

2.2.5.1. *Classical methods*

In classical hydrostatic buoyancy densimeters, an object (sinker hereafter), usually a sphere or cylinder, is suspended from a commercial digital balance by a thin wire. The fluid is kept in a pressure cell at constant temperature using a temperature control mechanism. The sinker is submerged in the fluid and weight of the sinker is constantly monitored. According to Archimedes' principal, the apparent loss in the true weight of the sinker is equal to the weight of the displaced fluid. Density of the fluid results from:

$$\rho = \frac{m_V - m_a}{V_S(T,P)} \tag{6}$$

In above equation m_v is the 'true' mass of the sinker in vacuum, m_a is 'apparent' mass of the sinker in the fluid and V_s is the calibrated volume of the sinker, which is a function of temperature and pressure. In such densimeters, several corrections are necessary to reduce the effect of surface tension between the sample liquid and the immersed part of the wire, and the effect of the buoyant force of air on the masses of the analytical balance. Zero shift of balance readings, buoyancy forces on auxiliary devices, adsorption effects and surface tension may reduce the accuracy of such measurements [20].

2.2.5.2. *Magnetic suspension devices*

To overcome limitations in achievable accuracy, the need for frequent calibration of the apparatus with reference fluids, complexity of operation, limitations on temperature and pressure, Kleinrahm and Wagner [9] introduced an MSD based upon magnetic levitation of the sinker in the measuring cell. The novelty of the magnetic suspension coupling was that it used non-physical-contact force transmission between the sinker in the pressurized cell and the weighing balance at atmospheric pressure, thus allowing a cell design that covered a very wide temperature and pressure range [21]. Then, Kleinrahm and Wagner [20] modified the hydrostatic buoyancy force method by introducing an alternative force transmission method in which they levitated two sinkers through a magnetic suspension coupling. By compensation for surface tension, buoyancy, adsorption effects and shifts in zero-point of the balance, a two-sinker MSD improved the accuracy of the density measurements.

Operation of a two-sinker MSD is rather complex and its advantage is not required for medium or high-density measurements encountered in many practical applications. To extend the instrument range towards higher temperatures and pressures, Wagner et al. [16] have developed a single-sinker densimeter. Although the single sinker design is much simpler than that of the two-sinker densimeter, it is still possible to perform high-accuracy density measurements at relatively low gas densities by applying some of the advantageous features of the two sinker device [22]. The single-sinker densimeter also operates based upon Archimedes' principle and the force transmission comes from levitation of the sinker in the measuring housing of the high-pressure cell. Klimeck *et al.* [23] have concluded that the accuracy of density measurement from a single-sinker densimeter is lower than that from a two-sinker densimeter especially at low densities because it lacks compensation for the adsorption effect. Moreover, the force transmission error has more effect on total density measurement uncertainty than observed in a two-sinker densimeter. Also for small densities, having the load compensation system outside of the measuring cell is less effective than having it inside as with the two-sinker densimeter.

3. Viscosity measurement techniques

3.1. Introduction

Viscosity is a remarkable property for natural gas because of its influence on flow behavior, which is especially important for reservoir conditions. Natural gas viscosity is several orders of magnitude lower than for oil or water, and thus, natural gas mobility in reservoirs is larger than for water or oil. Moreover, gas flow is predominantly laminar in reservoirs, and thus, the influence of viscosity is especially important. The upstream gas industry faces new challenges for precision monitoring of gas supplies, for which accurate and reliable knowledge of the natural gas viscosity is a prerequisite. Davani et al. [24] and Denney [25] analyzed the gas viscosity estimation errors on the gas recovery from a high pressure-high temperature (HPHT) reservoir, showing that a -10 % error in gas viscosity estimation can produce a relative 8.22 % error in estimated cumulative gas production, and a + 10 % error can lead to a relative 5.5 % error in cumulative production. Moreover, uncertainty in gas viscosity data has a direct effect for inflow performance relationship curves [26], Davani et al. [27] showed that a 1 % uncertainty in gas viscosity data leads to a 1 % uncertainty in gas flow rate.

Sanjari et al. [28] reported in recent study that common prediction methods for natural gas viscosity lead to large deviations when applied for high pressure – high temperature ranges, and thus, these methods should be applied with caution for reservoir estimations. This lack of accuracy for current predictive viscosity models, and the large economical impact of viscosity uncertainties, show the need of experimental accurate natural gas viscosity to analyze *i*) the effect of natural gas composition on viscosity for wide pressure – temperature ranges, and *ii*) the predictive performance of current viscosity analyze the predictive performance of current viscosity predictive models and developing more accurate predictive approaches.

Experimental accurate measurement of gas viscosity is very difficult, requires properly designed equipment, especially for high pressure–high temperature conditions, and is very costly both in time and resources. Moreover, possible natural gas mixtures under experimental study is very large considering that natural gas composition depends strongly on the origin, age, and depth of the reservoir [29]. Likewise, conditions of interest, especially the high pressure – high temperature conditions found in many new reservoirs that can be explored with current technologies [30], has a remarkable effect on natural gas viscosity, and discard the use of traditional measurement techniques. The analysis of published viscosity data in the open literature shows its scarcity both in the number of studied systems and the experimental conditions (high pressure data are almost absent) [28,31]. Therefore, systematic studies on natural gas viscosity has to be carried out in wide pressure-temperature ranges and as a function of mixture compositions, for selected mixtures representative of key reservoirs. Our group is involved in a multilaboratory international research project in which natural gas viscosity is measured, and other relevant thermophysical properties, using state-of-the-art equipments [31,32]. As a result of our research, we report in this section a detailed analysis of the available experimental methods for measurement of natural gas viscosity.

3.2. Viscosity measuring devices

3.2.1. Rolling ball viscometers

Rolling ball viscometers measuring principle is based on the travelling time of a metal or glass ball through a known distance to measure the viscosity of the fluid, Figure 3. The ball of known diameter rolls down a tube of known length, at a known inclination, filled with the fluid under study, under isothermal-isobaric conditions.

The viscosity of the fluid is proportional to the ball travelling time, considering that the fluid flow around the ball is laminar. Sage and Lacey [33] proposed corrections to be applied in case of viscosity measurements under turbulent flow.

The working principle of rolling ball viscometers may be summarized in equation (7):

$$\eta(P,T) = k_1(P,T,\theta) \times t(\theta) \times (\rho_b - \rho) + k_2(P,T,\theta) \tag{7}$$

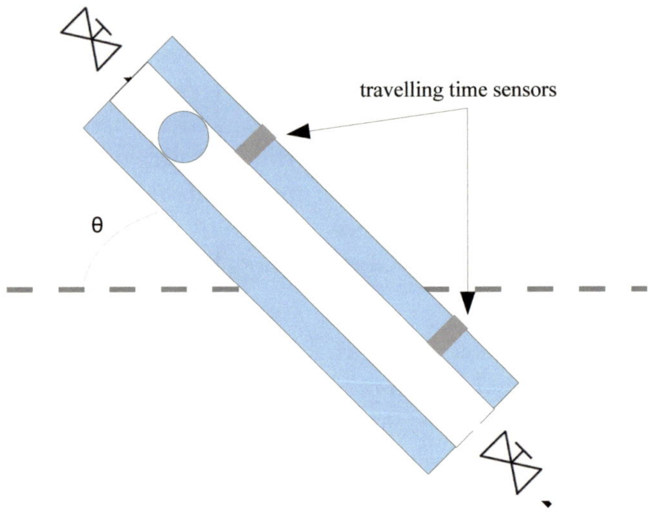

Figure 3. Scheme of a rolling ball viscometer.

Where ρ_b and ρ stand for the ball and fluid density, respectively, t for the ball travelling time, and k_1 and k_2 for the calibration constants, which are function of the measurement angle (θ). Calibration constants are obtained with fluids of known viscosity for the required pressure-temperature ranges. It should be remarked that rolling ball equipments are kinematic viscometers, and thus, and additional instrument is required for density measuring in the studied pressure-temperature ranges.

The use of rolling ball viscometers for gas measurements is very scarce, and high pressure data are almost absent. Sage and Lacey [33] measured viscosity data for methane and two hydrocarbon gases up to 2900 psi, Bicher and Katz [34] measured the viscosities of methane-propane mixtures up to 5000 psi. The uncertainty of viscosity measurements using rolling ball viscometers it is claimed to be ± 2 % when applied for measurements in the liquid state [35], although no detailed uncertainty analysis is available for gas measurements. Dolan et al [36] carried out gas phase measurements for n-butane using a capillary viscometer and showed large difference with rolling ball viscometer data by Sage et al. [37], who claimed a ± 5 % for their gas phase viscosity measurements.

3.2.2. Capillary tube viscometers

The working principle of classical capillary viscometers (so-called Rankine viscometers) is based on the introductions of a pellet of mercury into a tube filled with the gas under study, with the mercury pellet completely filling the cross section of the tube. The mercury pellet will reach a steady descending velocity will for any inclination of the tube. The descending pellet will act as a moving piston, which will force the gas through an adjacent capillary, leading to a constant pressure difference across the fine capillary. Gas viscosity is obtained from the measurement of the mercury pellet falling time between two fixed points. Capillary viscometers have been widely studied for gas measurement because of their operational simplicity and the small amounts of required gas. Heath [38] analyzed the performance of capillary viscometers for gas phase measurements, and Kobayashi [39] carried out a detailed analysis of the application of Hagen-Poiseuille's law for gases to the Rankine viscometer. Giddings et al. [40] developed a high-pressure capillary viscometer, which was applied for measurements of methane, propane and their mixtures up to 544.4 atm with a 0.25 % claimed reproducibility. Lee et al. [41] carried out systematic measurements of the viscosity of natural gas mixtures using a capillary viscometer with an estimated accuracy of ±2.00. In a recent work, May et al. [42] obtained reference viscosity data for several gases, including methane, using a modification of capillary viscometer.

3.2.3. Vibrating wire viscometers

The falling body and the capillary tube methods for viscosity measurements require hydrodynamic corrections and approximations for ends, edges, and walls. These applied corrections are not always known, and in most cases are sources of important errors. These corrections are avoided by the use of vibrating wire based viscometers. These apparatus are based on the damping of the vibrations of a wire in the fluid under study. An applied external field disrupts the wire immersed in the fluid, leading to periodic oscillations. In the free-decay mode the damping of the oscillations depends on the viscosity and density of the fluid [43]. This technique can be applied in a straightforward manner to low viscosity fluids, in fact most of the available contemporary natural gas viscosity data have been measured using vibrating-wire based approaches. Tough et al. [44] and Trappeniers [45] were the first to apply vibrating wire devices for the measurement of low viscous fluids and gases at high pressures. Assael et al. [46] carried out viscosity measurements of a natural gas mixture with a claimed uncertainty of ± 1 % up to 15 MPa. Langelandsvik et al. [47] measured viscosity of three natural gas mixtures with a ± 1% estimated uncertainty up to 5.0 MPa. Schley al. [48] measured viscosity of methane and two natural gases up to 29 MPa with a ±0.3 % and ±0.5 % estimated uncertainties for methane and gas mixtures, respectively. Likewise, vibrating-wire viscometers are considered as quasi-primary measurement methods [43].

3.2.4. Falling body viscometers

The experimental setup for these instruments is very similar to that for rolling body viscometer with the exception that the ball is replaced with a piston. These equipments have been widely for measuring liquid viscosity but measurements for low viscous gases are scarce. In most cases the viscometer arrangement is vertical, which is a serious limitation for gas measurements considering the very short travelling times. Heidaryan et al. [49] used a falling body viscometer for measuring methane viscosity up to 140 MPa.

A modification of traditional falling body viscometers has recently proposed to withdraw the aforementioned disadvantages and to apply this equipment to large pressure temperature ranges. In this equipment a piston is placed inside a cylindrical measuring chamber filled with the fluid (gas or liquid) under study, the piston is driven electromagnetically by two coils located at opposite ends and the time taken by the piston to complete one motion is correlated to the viscosity of the fluid in the measuring chamber by a working equation. A scheme of the equipment installed in our laboratory is showed in Figure 4. The measurement chamber is placed at 45º inclination, and the setup allows measurements from 0.02 to 10000 cp, simply changing the used piston.

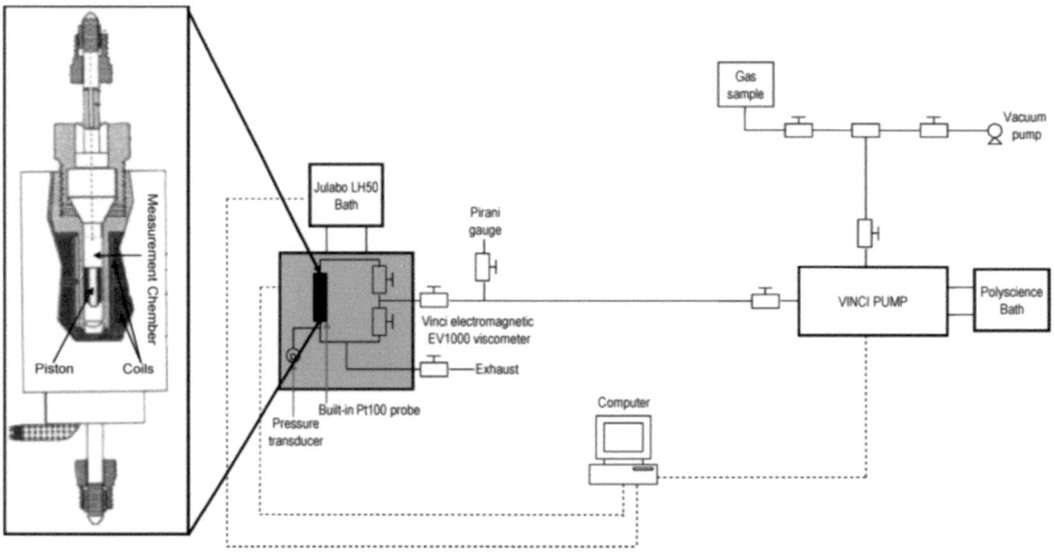

Figure 4. Electromagnetic viscosity measuring device [31].

The main advantage of this equipment is the ability to carry out fast measurements in wide pressure-temperature ranges. Nevertheless, Thomas et al. [50] and Viswanathan [51] reported problems for gas viscosity measurements using electromagnetic viscometers, mainly raising from the poorly defined pressure-temperature dependence of the measurement chamber properties. Therefore, we proposed a new calibration method using this equipment to avoid these problems, which was validated against reference viscosity data, leading ± 0.1 % reproducibility and to ± 2.5 % and ± 4.0 % uncertainties, for pressures lower than 30 MPa and higher than 30 MPa, respectively. Nevertheless, we should remark that the uncertainty values are derived from the uncertainties obtained from the calibration fluids. This is a remarkable problem for viscosity measurements, the use of secondary methods requires the knowledge of highly accurate viscosity data for reference fluids, in wide pressure – temperature ranges, which is not currently available [43]. Nevertheless, the use of electromagnetic based viscometer allow to obtain wide collection of viscosity data for gas mixtures, in wide pressure – temperature ranges, with acceptable accuracy and at moderate costs.

4. Phase equilibria measurement techniques

4.1. Introduction

The prediction of natural gas vapor–liquid equilibria is of primary importance for industrial purposes [52]. This property, together with the knowledge of PVT behavior, is required in all the stages of natural gas production / transportation chain. The vapor–liquid equilibria of natural gas mixtures is commonly analyzed using the pressure–temperature projection, so–called phase envelope, Figure 5, in which a curve separating the two–phase (vapor–liquid equilibria) and single–phase (vapor or liquid) regions is plotted.

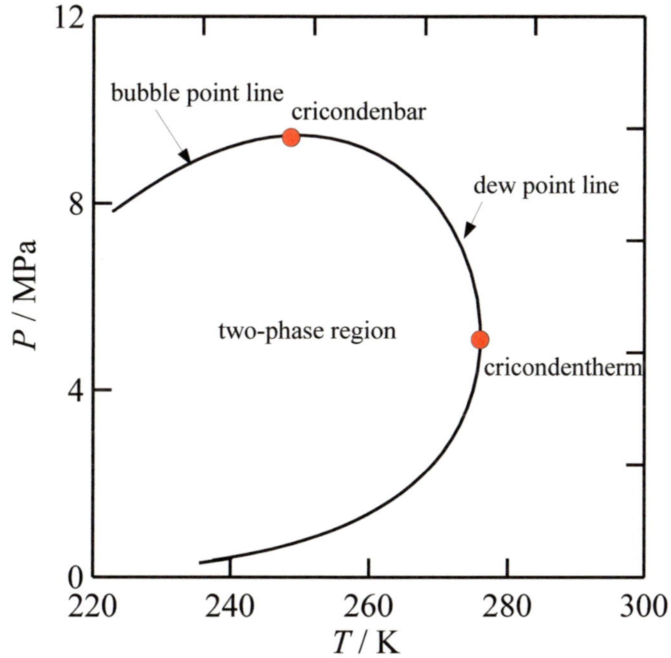

Figure 5. Phase envelope of a typical natural gas.

In this curve, three points are of remarkable importance: *i*) the critical point, where bubble and dew points curves meet, *ii*) the cricondenbar, the maximum pressure at which vapor and liquid phases may coexist, and *iii*) the cricondentherm, the highest dew point temperature. Therefore the importance of an accurate knowledge of natural gas phase envelopes could be summarized in three main areas: *i*) reservoir engineering, *ii*) custody transfer and *iii*) gas transportation through pipelines [53].

For reservoir engineering purposes, care must be taken to maintain the gas in single phase conditions because during exploration a pressure depletion may lead to an erroneous determination of reservoir composition (and thus to incorrect field–development designs), because dew point curve is crossed and liquid is formed, and during production liquid drop-out may lead to valuable hydrocarbons left behind in the reservoir [54]. If liquid dropout appears in the wellbore, it may lead to liquid loading concerns. For custody transfer purposes, the inaccuracy in the knowledge of phase envelopes may lead to contractual disputes rising from the contract sales gas dew point specifications between the company that sales the gas and the purchaser. For transportation through pipelines [29], it is essential to maintain the fluid pressure above the cricondenbar to protect the compressors which may be damaged by the presence of liquid drops [55]. Moreover, the sizing of the compressors depends on the value of the predicted cricondenbar, and thus, inaccuracy in the knowledge of this value may lead to an oversizing of the expensive compressors (if the real value is lower than the predicted ones) or to a two–phase flow (if the real value is above the predicted one) along the pipelines. Hence, accurate knowledge of phase envelopes allows reducing the design margins, to optimize the pipelines capacity, and thus, to improve the economic viability of new pipelines projects [56].

Therefore phase envelopes must be known before production / transportation operations are designed for any new gas–field. The most accurate and reliable way to obtain the phase envelope for any natural gas mixtures is its determination through experimental measurements [12] Nevertheless, these measurements require state–of–the–art apparatus, which are very costly, both in time and resources, and, considering that the composition, and thus properties, of natural gases can vary widely depending upon the reservoir from which the fluid comes, it is almost impossible carry out measurements for all the possible mixtures over the wide temperature and pressure ranges required. Thus, the common way to obtain phase envelopes for design purposes in the gas industry is through available theoretical models, using gas chromatography determined compositions, and thus, the reliability of the designs stands on the accuracy of the used models. In the natural gas industry, equations of state (EOS) are the common choice for phase equilibria predictions, and cubic EOS are mainly used because of their simplicity. Multiparametric EOS developed in the last years, such as AGA8–DC92 [4], which are used with high accuracy for density predictions are not applicable for phase equilibrium calculations or for liquid properties, and thus other EOS such as cubic ones or new multiparametric GERG2008 [7] are used to characterize natural gas. Nonetheless, the predictive ability of available models for phase equilibria have to be studied against accurate experimental data to test their reliability, and thus, experimental measurements for selected natural gas mixtures using accurate methods are required [57].

Experimental methods for measurement of vapor-liquid equilibria in high pressure conditions may be classified as *i*) analytical and *ii*) synthetic [58].

4.2. Phase equilibria measuring methods

4.2.1. Analytic methods

For the analytic method, pressure and temperature are fixed and then phase separation is produced, sampling for each phase under equilibrium is carried out and then composition is determined by chromatographic methods. This technique may carried out using static or dynamic circulation approaches. The experimental results from this type of experimental equipment are usually isothermal or isobaric phase diagrams. Natural gas mixtures usually contain small fractions of heavy components (C_{6+} fractions), which are difficult to measure accurately through chromatographic methods, and thus, this approach is not very useful for the study of natural gas mixtures.

4.2.2. *Synthetic methods*

In the synthetic method a mixture of known composition is prepared and its behavior observed as a function of pressure or temperature. The experimental results from this type of equipment are isopleth phase diagrams. The synthetic method may be subdivided into visual and non-visual.

4.2.2.1. Visual synthetic methods

Chilled mirror dew point meters is the simplest and most widely applied method of hydrocarbon dew point measurement in the gas industry. These instruments have a metallic mirror surface inside a high-pressure sample cell. The instruments are also equipped with a glass viewing port through which operator can observe the mirror surface. Mirror dew point meters normally are used for periodic spot check measurements [59].

The principle of the mirror dew point meter is to observe the very first signs of condensate, Figure 6. Therefore, cooling samples for several pressures allows to build the phase envelope, Figure 7.

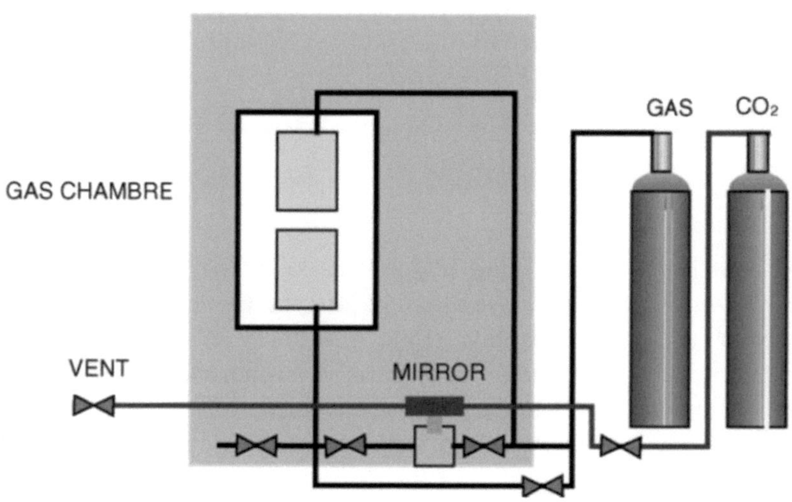

Figure 6. Scheme of a chilled mirror equipment for phase envelope measurements.

The main difficulties in making such a measurement lie in the characteristics of hydrocarbon condensates. The natural gas condensates are colorless and have low surface tension. This means that the liquid film that forms as the sample cools through the dew point temperature is almost invisible to operators. At the same time, the decreasing temperature measurement reading must be observed as the mirror temperature drops. To achieve the best sensitivity and repeatability of measurement, the rate of mirror cooling is critical and should be as slow as possible through the region in which the dew point is likely to be found [60]. The subjective nature of such a measurement technique can result in large uncertainty in natural gas dew point measurement [59].

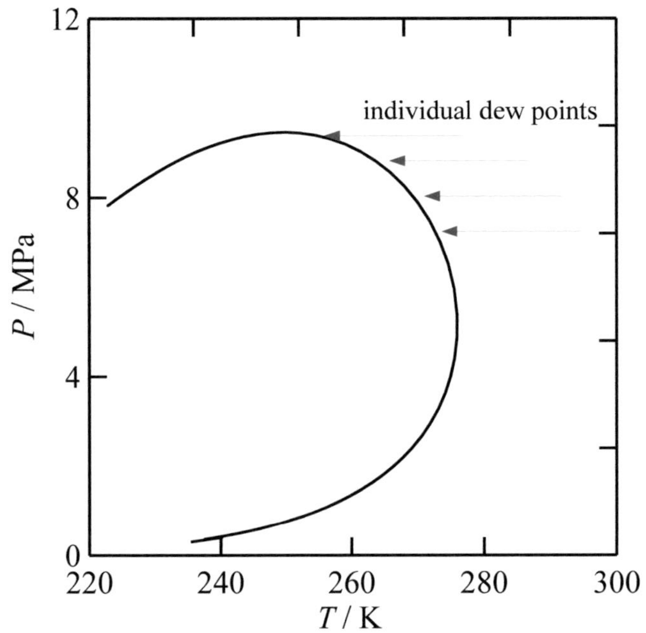

Figure 7. Scheme of experimental procedure for measuring phase envelopes using visual chilled mirror method.

4.2.2.2. Non-visual synthetic methods

The method usually involves a blind *pVT* cell filled in the single-phase region with a mixture of known composition. Variation of one of the measured quantities allows measurement of a second quantity while the third is held constant. Therefore, measurements are made along either an isobaric or isochoric or isothermal path with the homogenous / heterogeneous boundary (phase envelope) determined from a discontinuity in the slope of the other two variables [61]. Among the non-visual methods, the isochoric technique is probably the most useful for the determination of phase envelopes in natural gas mixtures, its amenability to automation allow the collection of large quantities of accurate data at moderate costs.

The technique for determining phase loops using isochoric method utilizes the change of the slope of an isochore as it crosses the phase boundary, Figure 8.

It should be remarked that the change of slope does not occur at the cricondentherm, which has a collinear isochore [62]. The apparatus cell volume changes slightly with pressure and temperature, and thus, experimental data require an application of the cell distortion equation to correct the results to truly isochores. In a recent work Acosta et al. [63] developed a new computational method for the analysis of isochoric data to obtain phase envelope experimental data points, reporting that the phase boundary temperatures and pressures could be determined within 0.45% and 0.04%, respectively. Our research group has developed a systematic measurement program on selected natural gas-like mixtures using the isochoric method with the cell design reported in Figure 9 [64-66]. The reported results allowed to analyze the predictive ability of the equations of state commonly used in the gas industry for phase envelope predictions showing large deviations, which justify the need of very accurate experimental data for the analysis and prediction of natural gas phase equilibria.

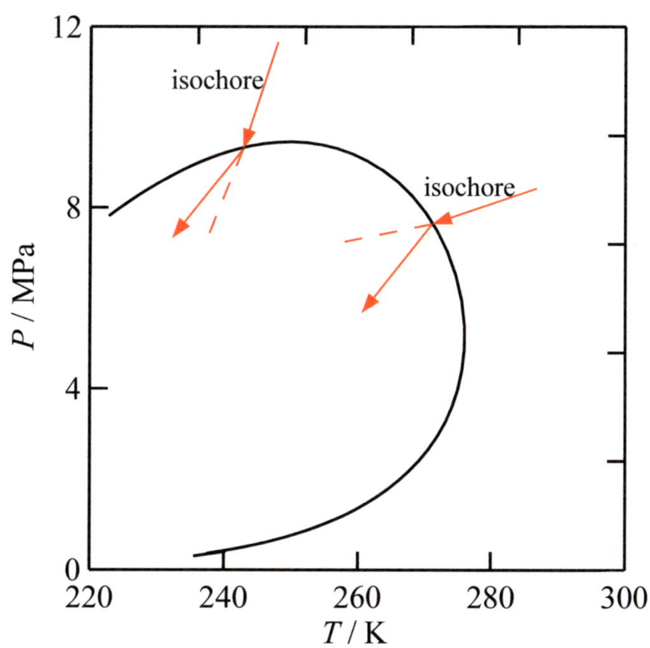

Figure 8. Method for determine phase envelope from isochoric measurements.

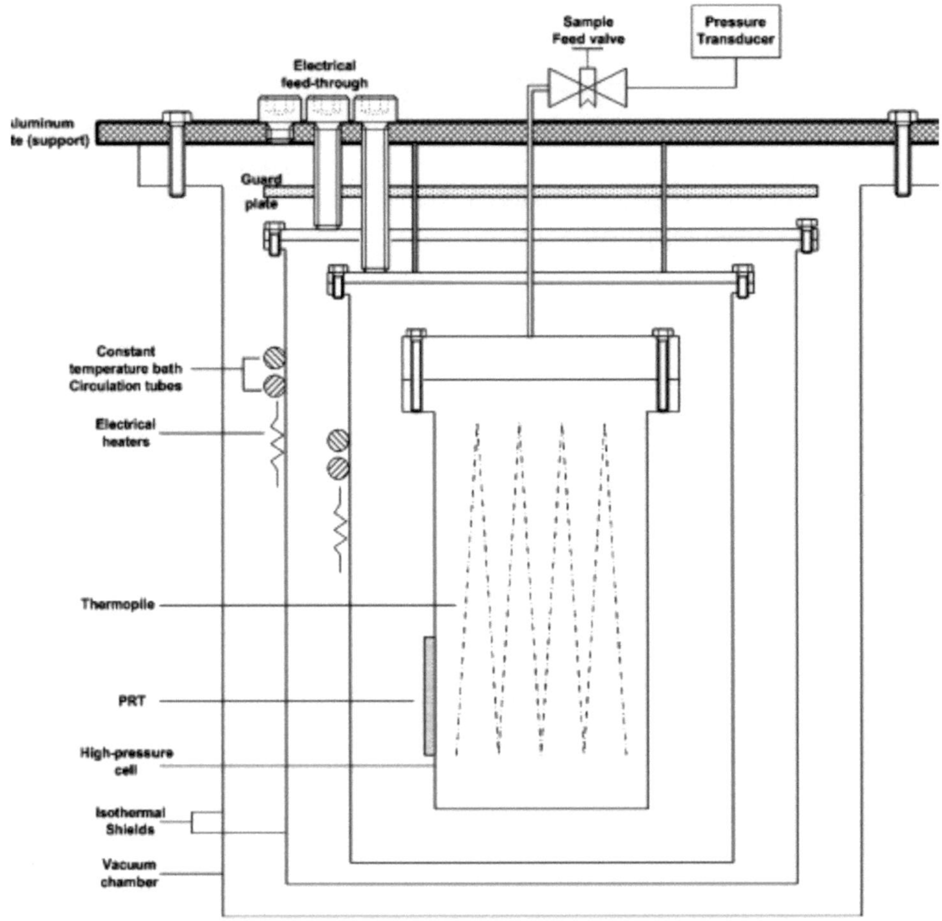

Figure 9. Scheme of isochoric cell design [65].

5. Conclusions

Techniques for the measurement of the most relevant thermophysical properties for the characterization of natural gas mixtures are reviewed showing the state-of-the-art, and the weaknesses and strengths of the current methods. Accurate determination of density, viscosity and phase equilibria are extremely important for the natural gas industry, both for technical and economical reasons, for production, processing and transportation purposes. Likewise, exploration and production using non-conventional reservoirs, including high pressure conditions, requires the accurate measurements in wide pressure-temperature ranges. Moreover, current predictive theoretical models using in the gas industry can not be used out of their tested pressure – temperature ranges, and thus, new accurate data are required to check the models performance and / or develop new theoretical predictive approaches. Therefore, this review could help to guide in the selection of the most suitable experimental approach for the determination of the required properties in wide pressure – temperature ranges.

Author details

Mert Atilhan
Qatar University, Department of Chemical Engineering, Doha, Qatar

Santiago Aparicio
University of Burgos, Department of Chemistry, Burgos, Spain

6. References

[1] M. Atilhan, S. Ejaz, S. Aparicio-Martinez, K.R. Hall, Prediction of Natural Gas Mixtures Properties from Cubic and Molecular-Based Equations of State, Sixteenth Symposium on Thermophysical Properties, Boulder, CO, 2006.
[2] X. Wang, J.G. Bomba, Hydrate Problems in Deepwater Flowlines, Proceedings of the 13th International Conference on Offshore Mechanics and Arctic Engineering, Houston, 1994.
[3] W. Wagner, R. Kleinrahm, Metrologia 41 (2004) S24-S39.
[4] American, Gas, Association, Compressibility Factors of Natural Gas and Other Related Hydrocarbon Gases, Compressibility Factors of Natural Gas and Other Related Hydrocarbon Gases, Washington DC, 1992.
[5] http://www.4uengineer.com/topic/FlowMeter/AGA8.gif.
[6] J.T.R. Watson, B. Millington, Project No DRG001, Report No 110/97, National Engineering Lab (1998).
[7] O. Kunz and W. Wagner, submitted for publication to J. Chem. Eng. Data, 2012.
[8] N. Kuramoto, K. Fujii, A. Waseda, Metrologia 41 (2004) S84-S94.
[9] R. Kleinrahm, W. Wagner, Progress Reports of the VDI Journals 3 (1984) 92.
[10] A.R.H. Goodwin, K.N.Marsh, W.A. Wakeham, (Eds.), Measurement of the Thermodynamics Properties of Single Phases, 2003.
[11] J.L. Daridon, A. Lagrabette, B. Lagourette, J. Chem. Thermodyn. 30 (1998) 607-623.
[12] J.J. Zhou, P. Patil, S. Ejaz, M. Atilhan, J.C. Holste, K.R. Hall, J. Chem. Thermodyn. 38 (2006) 1489-1494.
[13] V. Majer, A.A.H. Padua, Measurement of Density with Vibrating Bodies. in: A.R.H. Goodwin, K.N.Marsh, W.A.Wakeham, (Eds.), Measurement of the Thermodynamic Properties of Single Phases, Experimental Thermodynamics, Amsterdam, 2003.
[14] A.A.H. Padua, J.M.N.A. Fareleira, J.C.G. Calado, W.A. Wakeham, Review of Scientific Instruments 69 (1998) 2392-2399.
[15] E.S. Burnett, J. Appl. Mech. 3 (1936) A-136.
[16] W. Wagner, K. Brachthäuser, R. Kleinrahm, H.W. Losch, Int. J. Thermophys. 16 (1995) 399-411.
[17] R. Tillner-Roth, H.D. Baehr, J. Chem. Thermodyn. 24 (1992) 413-424.
[18] P.T. Eubank, L.L. Joffrion, M.R. Patel, W. Warowny, J. Chem. Thermodyn. 20 (1988) 1009-1034.

[19] W.R. Lau, C.A. Hwang, H.B. Brugge, G.A. Iglesias-Silva, H.A. Duarte-Garza, W.J. Rogers, K.R. Hall, J.C. Holste, B.E. Gammon, K.N. Marsh, J. Chem. Eng. Data 42 (1997) 738-744.

[20] R. Kleinrahm, W. Wagner, J. Chem. Thermodyn. 18 (1986) 739-760.

[21] H.W. Lösch, Development and Design of New Magnetic Suspension Balances for Non-Contact Measurements of Vertical Forces, VDI Verlog, Dusseldolf, 1987.

[22] W. Wagner, R. Kleinrahm, H.W. Lösch, J.T.R. Watson, Hydrostatic Balance Densimeters with Magnetic Suspension Coupling. in: K.N. Marsh, W.A. Wakeham, (Eds.), Measurements of the Thermodynamic Properties of Single Phases, Elsevier, Amsterdam, 2003.

[23] J. Klimeck, R. Kleinrahm, W. Wagner, J. Chem. Thermodyn. 30 (1998) 1571-1588.

[24] E. Davani, K. Ling, C. Teodoriu, W. D. McCain, G. Falcone, More accurate gas viscosity correlation for use at hpht conditions ensures better reserves estimation. Paper SPE 124734, SPE Annual Technical Conference and Exhibition, New Orleans, LA, 2009.

[25] D. Denney, J. Pet. Technol. 62 (2010) 76–78.

[26] T. Billiter, J. Lee, Proceedings of the Presentation at the SPE Eastern Regional Meeting, Paper SPE 59759, Canton, OH, 2001.

[27] E. Davani, K. Ling, C. Teodoriu, W.D. McCain, G. Falcone, Proceedings of the Latin American and Caribbean Petroleum Engineering Conference, Paper SPE 122827, Cartagena de Indias, Colombia, 2009.

[28] E. Sanjari, E.N. Lay, M. Peymani, J. Nat. Gas. Chem. 20 (2011) 654-658.

[29] S. Mokhatab, W.A. Poe, J. G. Speight, Handbook of Natural Gas Transmission and Processing; Gulf Professional Publishing, Burlington, MA, 2006.

[30] T.S. Dyman, R.E. Wyman, V.A. Kuuskraa, M.D. Lewan, T.A. Cook, Nat. Resour. Res. 12 (2003) 41–56.

[31] M. Atilhan, S. Aparicio, R. Alcalde, G.A. Iglesias-Silva, M. El-Halwagi, K.R. Hall, J. Chem. Eng. Data 55 (2010) 2498-2504.

[32] M. Atilhan, S. Aparicio, G.A. Iglesias-Silva, M. El-Halwagi, K. R. Hall, J. Chem. Eng. Data 55 (2010) 5117-5123.

[33] B.H. Sage, W.N. Lacey, *Trans. Am. Inst. Mining Met. Engrs.* 127 (1938) 118-134.

[34] L.B. Bicher, D.L. Katz, *Ind. Eng. Chem.* 35 (1943), 754-61

[35] A.S. Pensado, M. J.P. Comuñas, L. Lugo, J. Fernández, J. Chem. Eng. Data 50 (2005) 849-855.

[36] J.P. Dolan, K.E. Starling, A.L. Lee, B.E. Eakin, R.T. Ellington, J. Chem. Eng. Data 8 (1963) 396-399.

[37] B.H. Sage, W.D. Yale, W.N. Lacey, Ind. Eng. Chem. 31 (1939) 223-226.

[38] H.R. Heath, Proc. Phys. Soc. B 66 (1953) 362-367.

[39] R. Kobayashi, A Study on the Rankine Gas Viscometer, Research Reports of Faculty of Engineering, Tokyo University, 2001.

[40] J.G. Giddings, J.T.F. Kao, R. Kobayashi, J. Chem. Phys. 45 (1966) 578-586.

[41] A.L. Lee, M.H. Gonzalez, B.E. Eakin, J. Petrol. Technol. 18 (1966) 997-1000.

[42] E.F. May, R.F. Berg, M.R. Moldover, Int. J. Thermophys., 28 (2007) 1085-1110.

[43] C.A.Nieto de Castro, F.J.V. Santos, J. M.N. A. Fareleira, W.A.Wakeham, J. Chem. Eng. Data 54 (2009) 171-178.

[44] J.T. Tough, W.D. McCormick, J.G. Dash, *Rev. Sci. Instrum.* 35 (1964) 1345-1348.

[45] N.J. Trappeniers, P.S. van der Gulik, H. van den Hooff, *Chem. Phys. Lett.* 70 (1980) 438-443.

[46] M.J. Assael, N.K. Dalaouti, V. Vesovic Int. J. Thermophys, 22 (2001) 61-71.

[47] L.I. Langelandsvik, S. Solvang, M. Rousselet, I.N. Metaxa, M.J. Assael, Int. J. Thermophys., 28 (2007) 1120-1130.

[48] P. Schley, M. Jaeschke, C. Kuchenmeister, E. Vogel, Int. J. Thermophys., 25 (2004) 1623-1652.

[49] E. Heidaryan, J. Moghadasi, A. Salarabadi, J. Nat. Gas. Chem. 19 (2010) 552-556.

[50] F.B. Thomas, E. Stephani, D.B. Bennion, J. Rushing, J. Can. Petrol. Technol. 42 (2003) 35–39.

[51] A. Viswanathan, Viscosities of Natural Gases at High Pressures and High Temperatures. M.Sc. Thesis, Texas A&M University, College Station, TX, 2007.

[52] K.S. Pedersen, P.L. Christensen, Phase Behavior of Petroleum Reservoir Fluids. CRC Press, Boca Raton, FL, 2007.

[53] White Paper on Liquid Hydrocarbon Drop Out in Natural Gas Infrastructure. Natural Gas Council for the Federal Energy Regulatory Commission, NGC+ Liquid Hydrocarbon Dropout Task Group, 2005.

[54] D. Denney, Identifying Condensate Banking With Multiphase Flowmeters—A Case Study. J. Petrol. Technol. 39 (2007) 73.

[55] M. Atilhan, J. Zhou, S. Ejaz, D. Cristancho, J. Holste, K.R. Hall, Phase Behavior Concerns for Multicomponent Natural Gas – Like Mixtures. Proceedings of the 1st Annual Gas Processing Symposium. H. Alfadala, G. V. Rex, M. M. El-Halwagi, (Eds.), Elsevier, 2009.

[56] E. Skouras, E. Thermodynamics in Gas Processing–Phase Envelope Predictions and Process Design. 1st Trondheim Gas Technology Conference, 2009.

[57] K. Nasrifar, O. Bolland, M. Moshfeghian, Predicting Natural Gas Dew Points from 15 Equations of State. Energy & Fuels 19 (2005) 561.

[58] U. Deiters, G. Schneider, Fluid Phase Equilibria, 29 (1986) 145-160.

[59] A.J. Benton, Hydrocarbon Engineering, 7 (2002) 59-64.

[60] H.R. Warner, E.E. Leamer, A.P. Spence, R.L. Bone, R.A. Hubbard, J. Bernos, and W. A. Kriel, "Hydrocarbon Dewpoint Determination of Lean Natural Gases," *80th Annual Convention Presentations*, Gas Processors Association, Tulsa, OK (2001).

[61] P.T. Eubank, K.R. Hall, J.C. Holste, A Review of Experimental Techniques for Vapor-Liquid Equilibria at High Pressures, *Phase Equilibria and Fluid Properties in the Chemical Industry: Proceedings, 2nd International Conference*, H. Knapp and S. I. Sandler, eds., Great Neck, NY 675 (1980).

[62] P.T. Eubank, M. A. Barrufet, General conditions of collinearity at the phase boundaries of fluid mixtures. AIChE J. 33 (1987) 1882–1887.

[63] P.L. Acosta, D.E. Cristancho, I. D.Mantilla, K. R. Hall, G. A. Iglesias-Silva, Fluid Phase Equilibr. 283 (2009) 17-21.

[64] M. Atilhan, S. Ejaz, J. Zhou, I. Mantilla, J. Holste, K. R. Hall, J. Chem. Eng. Data 55 (2010) 4907-4911.

[65] M. Atilhan, S. Aparicio, S. Ejaz, D. Cristancho, K. R.Hall, J. Chem. Eng. Data 56 (2011) 212-221.

[66] M. Atilhan, S. Aparicio, S. Ejaz, D. Cristancho, K. R.Hall, J. Chem. Eng. Data 56 (2011) 3766-3774.

Chapter 4

Environmental Radioactivity of TE-NORM Waste Produced from Petroleum Industry in Egypt: Review on Characterization and Treatment

M. F. Attallah, N. S. Awwad and H. F. Aly

Additional information is available at the end of the chapter

http://dx.doi.org/10.5772/CHAPTERDOI

1. Introduction

At present time the different environmental compartments suffer from excessive accumulation of various toxic pollutants, hazardous fallout contaminants and several naturally occurring radionuclides, including potassium-40, thorium and uranium with the natural decay series of Th and U as well as several other man-made radionuclides. It is now of common practice to regulate any uncontrolled release of hazardous wastes in different environmental compartment [1]. Toxic hazardous wastes are defined as containing chemicals posing substantial hazards to human health or to the environment when improperly treated, stored, transported, or disposed. Scientific studies show that these wastes have toxic, carcinogenic, mutagenic, or teratogenic effects on human or other life forms. The majority of hazardous waste is generated by the chemical manufacturing, petroleum, pesticides and coal processing industries. Hazardous wastes may enter the body through ingestion, inhalation, absorption, or puncture wounds [2].

Naturally Occurring Radioactive Materials (NORM's) are those materials that contain radioactive elements what are found naturally in the earth's environment. Examples of these radioactive elements are the ^{238}U, ^{235}U, ^{232}Th series and their respective decay daughter, as well as ^{40}K. NORM's exist in soil, water, plants, animals, human, coal, lignite, petroleum, phosphate ores, geothermal wastes, wastewater…etc., in small but varying amounts almost everywhere [3].

On the other hand, nearly all the naturally occurring radioactive materials are considered in balance state. However, in several industrial processes e.g., mining of minerals (U, Th, steel, rare earth's metals), phosphate, oil and gas production, concentration of the natural radionuclides may be altered than its physical state, and exists in concentrations over than that exists naturally.

Wastes associated with the various industrial activities, with enhanced levels of the natural radioactivity as a result of industrial process, causes what is called, "Technological Enhanced-Naturally Occurring Radioactive Materials", to be named as acronym word "**TE-NORM**". For instance, TE-NORM scales may build up inside of oil field production tubing and may concentrate considerable quantities of radioactive material that has the potential to expose humans to relatively high dose of radioactivity. TE-NORM is often precipitated as sludge and scales. The human body cannot sense or detect TE-NORM, so, it can be detected and measured indirectly through their ionizing radiation using specialized instrumentation [4]. Uncontrolled release of activities associated with enhanced levels of NORM can contaminate the environment and pose a risk to human health. These risks can be alleviated by the adoption of controls to identify where NORM is present; and by the control of NORM-contaminated equipment and waste while protecting workers.

1.1. Discovery TE-NORM in industry

The history of TE-NORM in oil and gas production follows closely in history of the discovery of radioactivity in the first part of 20th century. We must remember that, the discovery of radioactivity is more than one hundred years old. In 1918, a Canadian paper was published on radioactivity in natural gases. In the 1930's an elevated radium level were detected in the Russian oilfields. In 1953, the US geological society published a paper on uranium and helium in gas formations. In 1973, (EPA) performed a study on the presence Rn-222 natural gases [5-8].

A number of major oil companies helped sponsor a study on radon in natural gas products that was completed in 1975. The thrust of this study was the potential effect that radon would have on the consumer of natural gas products. Radon contamination of natural gas has been known for nearly 100 years [9]. These studies concluded that, radon in natural gas products does not present any hazard to the consumer. However, that radon could be a problem for different processing industries and some research efforts have focused on this concern [9-10].

In 1981, scale produced on offshore oil platforms in the North Sea was found to contain TE-NORM in significant quantities. These findings were presented in a 1985 offshore technology conference paper on radioactive scale formation in Houston, Texas, USA. Consequently, industry and government officials were aware of the possibility that TE-NORM scale could be present in US domestic operations. In 1986, significant TE-NORM scale was found in Laurel, Mississippi (USA). Some rather alarming press headlines and featured articles followed shortly in both Mississippi and Louisiana after the presence of TE-NORM in the oil path became better known. These articles stressed the fact that there are no current regulations, either by USA or other federal governments controlling this radioactive waste and called for their creation and enforcement. Since TE-NORM was first re-discovered domestically, the oil and gas industry has responded progressively to TE-NORM issues by notifying appropriate state agencies, initiating field surveys and studies to characterize and locate occurrence of TE-NORM in conjunction with the American Petroleum Institute, informing other oil and gas operations, employees and contractors, and reviewing operating practices [5].

TE-NORM contamination of oil and gas industry petroleum equipment has been identified world- wide, e.g. USA (Alaska, Gulf of Mexico region), the North Sea region, Canada, Australia, several Middle East countries (Egypt, Saudi Arabia...etc.). Since 1918 till 1980, most researches were focused on the TE-NORM contamination of natural gas facilities, and the contamination is attributed to ^{226}Ra as well as ^{222}Rn gas and its decay products, e.g. ^{218}Po, ^{214}Pb, ^{214}Bi, and ^{210}Pb. Within the text, the activity concentration ranges from background level to several hundreds Bq/g (^{226}Ra). Doses to workers involved in handling, contaminated equipment, or waste are usually very low, and the main problem related to radioactive deposits is waste disposal [11]. The presence of TE-NORM or naturally occurring radionuclides in the product materials from oil and gas facilities, give rise to deposits with enhanced levels of these radionuclides in the processing equipment [12].

1.2. Origin and formation of TE-NORM

In nature, there are three naturally occurring radioactive decay series. The first series, known as thorium series, consists of a group of radionuclides related through decay in which all the mass numbers are evenly divisible by the number four, (4n) series. This series has its origin radionuclides Th-232, its abundance is 100%, specific activity is 2.4×10^5 dpm/g, which undergoes α -decay with a half-life of 1.41×10^{10} y. The terminal nuclide in this decay series is the stable Pb-208. In this series, the transformation from the original parent Th-232 to the finial product Pb-208 requires 7α and 4β -decays. The long-lived intermediate is 6.7 y for Ra-228, Fig. (1).

The second series is the uranium series, which consists of group radionuclides that, the parent radionuclide in this series is U-238 (4n+2) abundance = 99.27 %, which undergoes α - decay with a half-life of 4.47×10^9 y. The stable product of the uranium series is Pb-206, which is reached after 8α and 6β -decay steps, Fig. (2). This is a particularly important series in nature since it provides the more important isotopes of elements Ra, Rn and Po, which can be isolated in large amounts in the processing of uranium minerals. Each ton of uranium is associated with 0.34 g of Ra-226.

The third radioactive decay series, (4n+3) known as the actinium series, the head of this series is U-235, which has an abundance of 0.72 % and a half-life of 7.1×10^8 y for α -decay. The stable end product of this series is Pb-207, which is formed after 7α and 4β -decay steps. The specific activity of U-235 is 4.8×10^6 dpm/g, Fig. (3) [13].

Other important radionuclide that exists in the nature is potasium-40 ($t_{1/2.}$ 1.28×10^9 y, isotopic abundance 0.0118 %). K-40 is found in plants, animals and in human bones. It is widely distributed in nature with volume concentrations ranging from 0.1 to 3.5 % in carbonates (limestones). The bones of an average human body contain concentrations of ~ 17 mg of K-40. The average radiation dose received from K-40 is 0.25 mSv/y to tissue and ~ 0.36 mSv/y to bone.

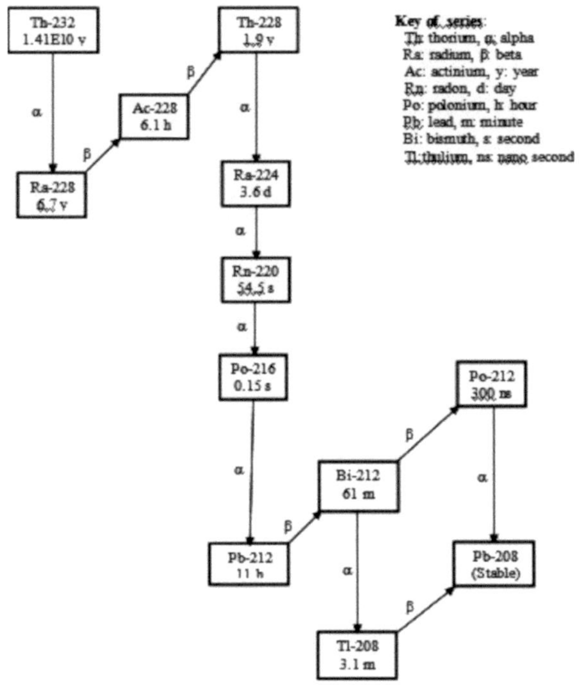

Figure 1. Scheme of the thorium decay (Th-232) series.

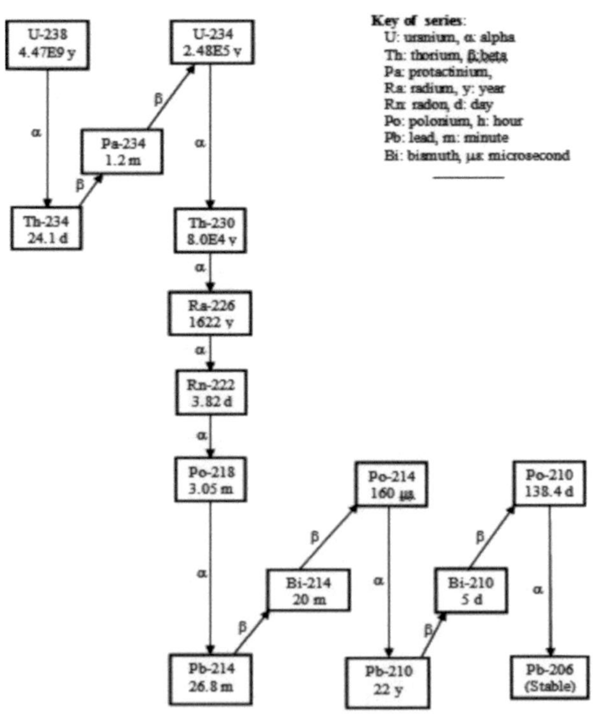

Figure 2. Scheme of uranium (U-238) decay series.

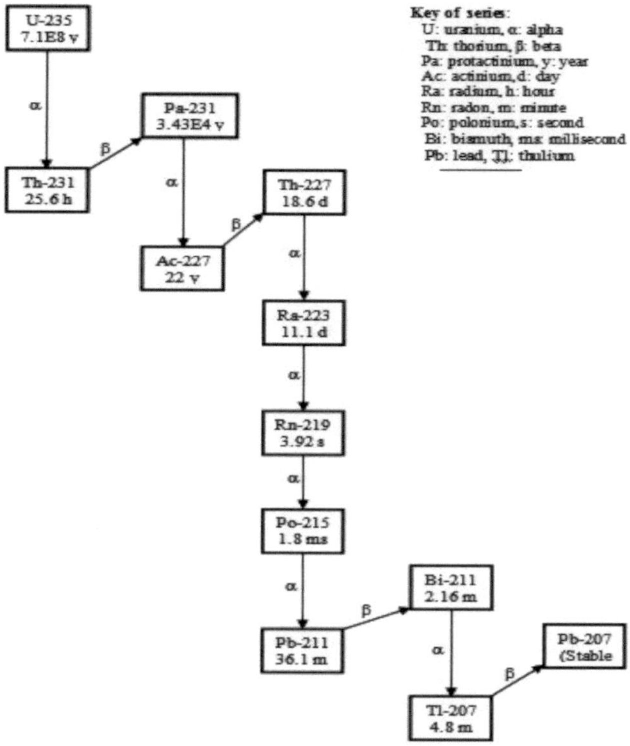

Figure 3. Scheme of actinium (U-235) decay series.

Radioactive materials such as uranium and thorium were incorporated in the Earth's crust when it was formed; these normally exist at low concentrations in rock formations. Decay of these unstable radioactive elements produces other radionuclides that, under certain conditions (dependent upon pressure, temperature, acidity *etc*) in the subsurface environment are mobile and can be transported from the reservoir to the surface with the oil and gas products being recovered. During the production process, NORM flows with the oil, gas and water mixture and accumulates in scale, sludge and scrap materials. It can also form a thin film on the interior surfaces of gas processing equipment and vessels. The level of NORM accumulation can vary substantially from one facility to another depending on geological formation, operational and other factors. To determine whether or not a facility has NORM contamination, NORM survey, sampling and analysis needs to be conducted. Table (1) gives the physical information for selected natural occurring radionuclides.

The TE-NORM waste occurs though the extraction and treatment of liquid and gases hydrocarbons and is generally accompanied by the formation and accumulation of radioactive scales, sludges and films. The petroleum waste (scale or sludge) was produced by two mechanisms either incorporate or precipitate into the production equipment such as pipelines, tank storage, pumps etc. The TE-NORM waste as scale and sludge generated in oil and gas equipments is due to the precipitation of alkaline earth metals as sulphates, carbonates, and/or silicates [9].

Nuclide	Half-life	Decay mode (energy)+	Γ * (R/hr)/Ci	Formation method	Uses
^{40}K	1.28×10^{9} y	β-(1.31 MeV) γ (1.46 MeV)	0.0817	Primordial NORM	Geologic dating
^{222}Rn	3.8 days	α (5.49 MeV) γ (512 KeV)	2.73×10^{-4}	^{238}U decay series	Patient medicines
^{226}Ra	1600 yrs	α (4.78 MeV) γ (186 KeV)	0.0121	^{238}U decay series	Luminous products
^{232}Th	1.4×10^{10} yrs	α (4.01 MeV) γ (12 KeV)	0.0684	Primordial NORM	gas lantern mantles
^{235}U	7.04×10^{8} yrs	α (4.40 MeV) γ (186 KeV)	0.339	Primordial NORM	Nuclear reactor fuel,
^{238}U	4.47×10^{9} yrs	α (4.20 MeV) γ (13 KeV)	0.0652	Primordial NORM	Military uses (armor, shells)
^{208}Tl	3.05 min	β (1.79 MeV) γ (2.6 MeV)	1.70	^{232}Th decay series	none

+ beta energies given are maximum decay energy, alpha and gamma energies are for the most probable decay energy

* given gamma constant reflects radiation dose in air as distance of one meter. The gamma constant (Γ) can be used to determine radiation dose from a gamma emitting radionuclide. The higher the value of gamma constant, the higher radiation dose. So, for example, the radiation dose from ^{235}U will be higher than the radiation dose from ^{238}U.

Table 1. Physical information for selected naturally occurring Radionuclides [14]

Some investigations were performed concerning the composition and characterization of the formed radioactive TE-NORM waste produced during oil and gas processing. Nuclear spectroscopic analysis showed that, the main radionuclides present in these wastes are traces of ^{238}U and ^{232}Th as well as their decay daughters. These studies indicated that the incorporation and co-precipitation of these natural radionuclides with alkaline earth metals (e.g., Mg, Ca, Sr, Ba) and some quantities of lead as sulphates, carbonates and/or silicates [15]. Some other studies were also performed under controlled production conditions, proved that, the factors that are greatly responsible for the formation of radioactive scale and/or sludge are water composition. The way in which the scales are deposited is connected to the pipes, superficial features, fluid-dynamic phenomena and crystallization kinetics. Testa et al. found that, variations in sulphates and carbonate solubility can give rise to scale formation, which are connected to some physical and chemical factors, e.g., temperature variation, pressure changes, pH-balance, evaporation in the gas extraction pipes and injection of incompatible sea waters. Also, the re-injected water into the reservoirs to maintain the production pressure during the field exploration seemed to be a principal cause for the scale formation [16]. As a consequence of the physical and chemical processes during the extraction of oil, besides the production water additional scale is obtained. Scale production in gas and oil field equipment is due to precipitation of alkaline earth metal sulphates or carbonates according to the following chemical reactions:

$$Ca^{+2} + CO_3^{-2} \rightarrow CaCO_3$$

$$Sr^{+2} + SO_4^{-2} \rightarrow SrSO_4$$

$$Ba^{+2} + SO_4^{-2} \rightarrow BaSO_4$$

Therefore, the observed levels of activity concentrations both in the separated sludge and solid scale are much higher than those observed in the produced water from the oil industry. Many of the physical characteristics of oil formations, high temperature pressure,..etc. tends to increase the radionuclide solubility in production fluids. This due to the complex process for sludge formation. Generally, uranium and thorium are relatively insoluble and remain stationary in the reservoir, while, radium is more soluble and may become mobilized in the produced water phase of the reservoir. It has been estimated that: 25000 tones of TE-NORM contaminated scale, and 225000 tones of TE-NORM contaminated sludge, are generated each year by the petroleum industry. The available data indicate that, the total radium levels to extreme measurements of 15.17 kBq/g scale and 25.9 kBq/g in sludge [17].

Many industries produce wastes that might contain natural radionuclides. Some of industries producing TE-NORM are coal, petroleum and natural gas, tungsten, phosphate fertilizer production, mineral processing, zirconium minerals or sand, welding and ceramics and building materials industry. In this review, we are interested in petroleum and phosphate industries that are the important industries in Egypt.

1.3. Characterization and radiological assessment

Some analytical and radiometric techniques are used to identify the chemical composition and radioactivity level of samples. The main phase composition of samples has been carried out using a powder X -ray diffractormeter equipped with a copper target and a nickel filter. The micro and trace-elemental analysis of samples have been performed by X -ray fluorescence (XRF) spectrometer. The IR spectra of the samples under study were done using FT-IR spectrometer. The different naturally occurring radionuclides present in the tested samples, were identified and the concentration of their radioactivity levels were detected and determined using a typical non-destructive nuclear spectroscopic technique. This has been performed using γ -ray spectrometer equipped with a High Purity Germanium (HPGe) detector.

The test sample containing different radionuclides was measured by γ -ray spectrometry, the specific activity (Bq/kg) of this unknown sample will be calculated using the following equation [18,19]:

$$A = \frac{C_{i,e}}{\varepsilon_{i,e} \times P_{\gamma i,e} \times G \times m \times t}$$

Where:

A : is the specific activity of the parent radionuclide or its daughters assuming the secular equilibrium in the test sample (Bq/kg or Bq/L),

$C_{i,e}$: is the net count of radionuclide (i) at γ -energy line (e) in keV,
$\varepsilon_{i,e}$: is the absolute efficiency of the used γ -ray spectrometer,
$P_{\gamma\, i,e}$: is the photopeak intensity (%),
G : is the geometry factor which is equal to unity when all test samples were counted under the same conditions,
m : is the weight of constant sample (kg) and
t : is the counting time in seconds.

Evaluation of the radiological characteristics, the parameters that are determined include radon emanation fraction released (EF), radium equivalent index (Ra-eq) and total absorbed dose rate ($D_{\gamma r}$) for the test samples.

To determine the radon (^{222}Rn) emanation fraction released (EF), the waste samples were initially counted (C_1) for 2 h after sealing the sample container, and counted again (C_2) after reaching the radioactive equilibrium between ^{222}Rn decayed from ^{226}Ra and its respective short-life daughters. The ^{222}Rn EC was determined using the formula [20,21]:

$$EF = \frac{N}{A_0 + N}$$

Where ^{222}Rn EF is the radon emanation fraction; A_0 is the amount (net count rate, cps) of ^{222}Rn existing at the sealing time of the sample container; N is the amount of (net count rate, cps) of ^{222}Rn emanated at the radioactive equilibrium.

To represent the activity levels of the materials containing ^{226}Ra, ^{232}Th and ^{40}K, by a single quantity, which takes into account the radiation hazards associated with them, a common radiological index has been introduced. This index is called radium equivalent (Ra-eq) activity and is mathematically calculated according to the following formula [22,23]:

$$\text{Ra-eq (Bq/kg)} = A_{Ra} + 1.43\ A_{Th} + 0.077\ A_K$$

Where: A_{Ra}, A_{Th} and A_K are the activities (Bq/kg) of ^{226}Ra (^{238}U-series), ^{232}Th and ^{40}K, respectively.

The absorbed dose rates ($D_{\gamma r}$) due to γ -radiations in air at 1m above the ground surface for the uniform distribution of the naturally occurring radionuclides (^{226}Ra, ^{232}Th and ^{40}K) were calculated based on guidelines provided by UNSCEAR [24,25]. Therefore, $D_{\gamma r}$ in outdoor air at 1 m above the ground was calculated as follows:

$$D_{\gamma r}\ \text{(nGy/h)} = 0.462\ A_{Ra} + 0.621\ A_{Th} + 0.0417\ A_K$$

2. TE-NORM in petroleum industry

In this part, the waste generated from oil and gas production that were investigated in Egypt and other countries are reviewed. The waste samples namely, Scales, sludge and produced water have been characterized.

2.1. TE-NORM in scales and sludge

The activity concentration of ^{226}Ra in the TE-NORM waste at three different sites for petroleum and gas production in Egypt has been determined using gamma spectroscopy. **El Afifi et al.** [26] chose three sites of oil and gas production for radiological assessment; (1) in the South Sinai Governorate, (2) in the Suez Gulf area, (3) in the Matrouh Governorate. Seven waste samples with an intermediate structure (mixed between granular and massive) were taken from Abu Rudeis (AR) onshore oil and gas field in the Suez Gulf. Nine samples representing two waste types were from Gabal El Zeit (GEZ) offshore oil and gas field at the Suez Gulf. Four samples had a granular structure and 5 samples had a massive one. Eighteen samples with granular structure were taken from Badr El Din (BED) offshore oil and gas field, respectively (Fig. 4). The TE-NORM waste samples taken for analysis, are a mix of scale and sludge formed in the production equipment and removed during the periodical maintenance. The samples were taken from the TE-NORM waste stacks accumulated around the petroleum and gas plants using a stainless steel template of 25×25 cm^2 with a thickness of 5 cm.

The results showed that the average activity concentrations of ^{226}Ra changed between 5.9 and 68.9 kBq/kg (dry weight) in the waste samples from GEZ and AR fields, respectively. In Gabal El Zeit field (GEZ), granular and massive wastes were investigated. The lower activity concentrations (28.6 kBq/kg) of ^{226}R were found in granular samples (GS), while higher values (56.6 kBq/kg) were found in the massive samples (MS). The activity concentrations of ^{226}Ra in all investigated waste samples from different region in Egypt can be ordered as follows: AR > BED > GEZ.

The mean activity concentrations of ^{232}Th were 25.4, 2.6, 7.2 and 61.3 kBq/kg and those of ^{40}K were 1.3, 0.96, 2.3 and 5.9 kBq/kg, for AR, GEZ (GS), GEZ(MS), and BED TE-NORM waste samples, respectively [26]. The lower ^{226}Ra activity concentration in the waste samples may be due to the lower Ra content in the subsurface formation. Therefore, the quantity of ^{226}Ra leached by oil and gas during exploration was lower in this case. Generally, the variation in the ^{226}Ra activity concentrations in TENORM wastes of different origins, can attributed to the differences in the Ra/Ba ratio in the formation waters scale and/or sludge formation processes on the exterior surfaces of the casing material and the amount of ^{226}Ra in the subsurface [19]. The Ra content depends on the amount of Ra present in subsurface formation, formation water chemistry, extraction and treatment processes and the age of the waste after production.

The average ^{222}Rn emanation fraction released from the TE-NORM wastes investigated ranged from 0.053 to 0.081 in the massive samples (MS) from GEZ field and the granular samples (GS) from BED field, respectively. The ^{222}Rn EF released from MS is lower than that of GS although the activity concentration of ^{226}Ra in the MS is higher. The variation of EF is independent of the ^{226}Ra content and is strongly correlated to the grain surface density [7,8,19]. The smaller the grain size the higher the EF [26].

The main types of scale encountered in oil & gas facilities are sulphate scale which results from drilling clay, it called Barite slurry. It is usually colorless or milky white, but can be almost any color, depending on the impurities trapped in the crystals during their formation. Barite is relatively soft, measuring 3-3.5 on Mohs' scale of hardness. It is unusually heavy for a non-metallic mineral. The high density is responsible for its value in many applications. Barite slurry is generally used as mud in drilling oil. Barite is chemically inert and insoluble. Radium is chemically similar to barium (Ba), strontium (Sr) and calcium (Ca), hence radium co-precipitates with Sr, Ba or Ca scale forming radium sulphate, radium carbonate and – in some cases – radium silicate. The mixing of seawater, which is rich in sulphate, with the formation water, which is rich in brine, increases the scaling formation tendency. Also the sudden change in pressure and temperature or even acidity of the formation water, as it is brought to the surface, contributes to scale build-up. This phenomenon has significant implications for the production of oil; in this case the capacity of the pipe to transfer oil would be reduced significantly. The activity concentration of ^{226}Ra and ^{228}Ra in hard scales in Egypt and some other countries are mentioned in Table 2.

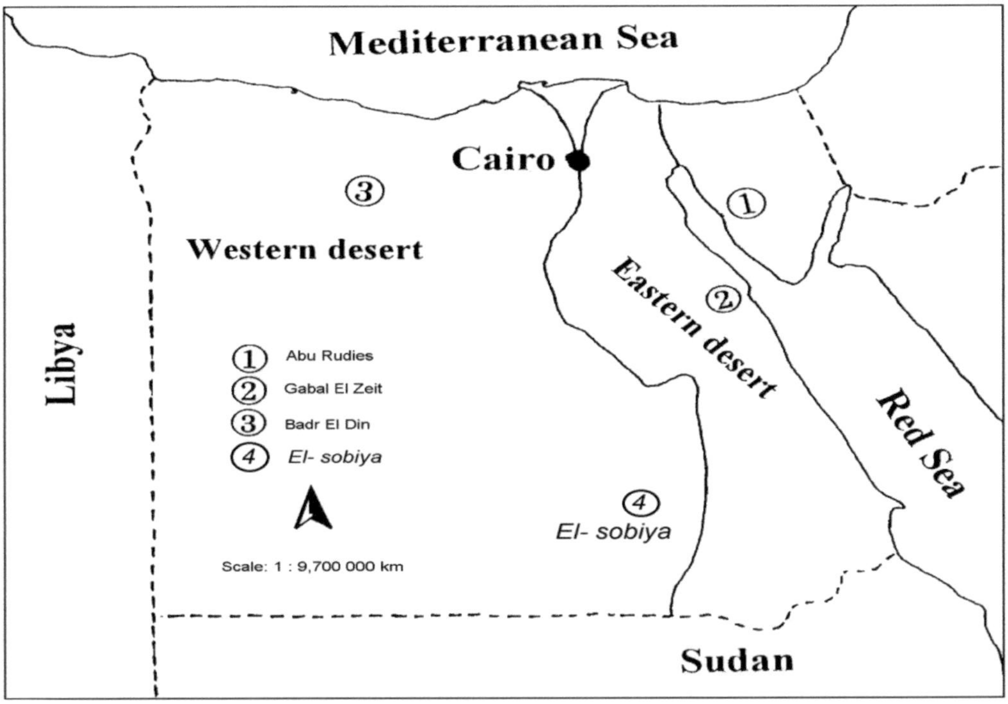

Figure 4. Map for some sites of the TE-NORM wastes associated with phosphate and petroleum and gas production in Egypt.

El Afifi and Awwad [27] characterized TE-NORM from Abu Rudeis region (onshore oil field, Suez gulf area) in the North Sinai Governorate, Egypt. The mineralogical analysis by X-ray techniques (XRF and XRD) has been carried out. Table 3 represents the chemical analysis of the TE-NORM waste samples using the XRF technique. The data showed major elements (Si, Fe, Al and Na) and alkaline earth elements (Mg, Ca, Sr and Ba) [27].

Country/region	Activity (Bq/g)		Ref.
	^{226}Ra	^{228}Ra	
Egypt/oil field			
Abu Rudeis	68.9	24	[27]
Gabal El Zeit	14.8	4.3	[28]
Badr El Din	31.4	43.3	[28]
Red Sea	195	897.8	[29]
Western desert	59.2	244.5	[29]
South Sinia	6.99	1-1.9	[30]
South Sinia	506	32-50	[30]
Other countries			
Australia	20-70		[31]
USA	70.8		[21]
Algeria	1-950		[32]
Tunisia	4.3-658		[33]
UK	1-1000		[6]

Table 2. Activity of ^{226}Ra (U-series), and ^{228}Ra (U-series) in the TE-NORM in Egypt and some other countries [6, 21, 27-33].

Sample code	Concentration (%)							
	Mg	Ca	Sr	Ba	Na	Al	Fe	Si
B.F.[a]	2.44	13.53	0.91	3.25	4.13	3.63	27.12	44.05
A.F.[b]								
F2 (0.1–0.2) mm	2.51	17.94	0.69	2.14	3.82	4.66	29.15	48.75
F4 (0.3–0.5) mm	1.93	11.23	1.27	4.63	4.22	3.67	29.69	44.33
F6 (1.0–1.6) mm	1.53	7.78	1.53	5.47	4.00	3.76	29.96	47.21
F8 (2.0–2.6) mm	1.31	8.65	1.66	6.37	4.58	3.91	27.61	47.39

[a]: TE-NORM waste in its original state (before fractionation).

[b]: After fractionation.

Table 3. Results of XRF analysis of the TE-NORM waste before fractionation and after sieve fractionation for some selected fractions [27].

The radioactivity of naturally occurring radionuclides of ^{238}U, ^{235}U, ^{232}Th-series and ^{40}K in the sediment samples of the TE-NORM waste from Abu Rudeis region before fractionation (B.F.) and after fractionation (A.F.) are given in Table 4. The bulk waste was fractionated into nine homogeneous fractions with different particle sizes (< 3.0– < 0.1 mm) to show the effect of particle size on the activity distribution. Moreover, radiation hazardous indices including the radium equivalent activity (Ra-eq), radon (^{222}Rn) emanation coefficient (EC) and absorbed dose rate ($D\gamma r$) were also estimated of TE-NORM waste. The radon emanation coefficient (EC) is a very important radiological index used to evaluate the amount of the 222Rn emanated fraction released from the waste materials containing naturally occurring radionuclides (e.g. ^{238}U, ^{235}U and ^{226}Ra). In this study, the assessment of Rn EC is related only to ^{222}Rn decayed from its parent ^{226}Ra content in the waste. Since ^{222}Rn and its respective decay progenies (e.g. ^{210}Pb and ^{210}Po) have longer half lives than that of other radon isotopes, ^{222}Rn is considered to be more radiological hazardous to human health than radionuclides coming from other radon isotopes. Fig. 5 represents the ^{222}Rn EC released from the bulk TE-NORM waste and the different fractions. The amount of the ^{222}Rn fraction emanated from the bulk waste was 0.066. It is found that the grain size has an effect on the amount of ^{222}Rn EC. There is a gradual increase in the ^{222}Rn EC with the waste particle size. This was observed in fine particle sizes from less than 0.1 mm up to 2 mm. In this range, ^{222}Rn EC increased from 0.041 to 0.086. There is no effect of the waste particle sizes on the ^{222}Rn EC released from large particle sizes as shown in the grain sizes between 2 and 3 mm. The radon EC found in this range of particle sizes was ranging between 0.093 and 0.095 [27].

Reported levels of the ^{226}Ra and ^{228}Ra activity concentrations observed in the solid scale and sludge in different countries are listed in Table 5.

Sample	**^{238}U-series**			**^{235}U-series**	**^{232}Th- series**			**^{40}K (Bq/g)**
	^{238}U (Bq/g)	^{226}Ra (Bq/g)	^{210}Pb (Bq/g)	^{223}Ra (Bq/g)	^{228}Ac (Bq/g)	^{212}Pb (Bq/g)	^{208}Tl (Bq/g)	
B.F.	7.1	86.9	4.4	2.7	24.0	22.4	25.2	1.3
A.F.*								
F1	7.5	60.4	4.3	3.5	25.8	24.7	22.9	1.9
F2	4.5	43.0	4.2	1.7	19.1	17.0	20.3	1.1
F3	6.2	55.4	4.3	2.6	21.5	21.0	20.5	1.3
F4	9.2	81.1	5.3	3.7	34.6	33.3	36.5	2.5
F5	11.2	96.5	4.3	5.2	41.4	37.2	39.0	3.4
F6	9.6	85.4	4.4	3.5	39.7	36.2	40.1	3.6
F7	7.1	78.3	4.5	4.0	35.4	36.4	32.9	4.3
F8	11.8	102	6.5	5.7	43.7	38.5	39.0	4.8
F9	7.1	54.5	2.7	3.4	22.8	20.1	22.1	2.5

*F1 (< 0.10) mm, F2 (0.1–0.2) mm, F3 (0.2–0.3) mm
F4 (0.3–0.5) mm, F5 (0.5–1.0) mm, F6 (1.0–1.6) mm
F7 (1.6–2.0) mm, F8 (2.0–2.6) mm, F9 (2.6–3.0) mm

Table 4. Results of gamma-spectrometric analysis of the TE-NORM waste before (B.F.) and after (A. F.) dry fractionation [27].

The radium equivalent activity (Ra-eq) is a radiation index, used to evaluate the actual radioactivity in the materials containing naturally occurring radionuclides, e.g. ^{238}U and ^{232}Th series, and/or ^{40}K. Values of Ra-eq activity for the bulk TENORM waste and the waste fractions were calculated. It is clear that Ra-eq exceeds the maximum permissible radium activity (Ref. value is 370 Bq/kg) as reported by Zaidi et al. [23]. The higher Ra-eq activity reached about 164.9 kBq/kg in fraction F8, whilst the lower value amounted 70.4 kBq/kg in F2 (Fig. 6). The higher the radioactivity level in the waste, the higher is the radiological impacts, especially when considering the potential of operators to be exposed via internal contamination by ingesting the dust during waste processing. The total absorbed dose rate due to g-emissions was estimated and the obtained values are presented in Fig. 6 [27]. It is recommended that the acceptable total absorbed dose rate by the workers in areas containing g-radiations from ^{238}U and ^{232}Th series and their respective decay progenies, as well as ^{40}K, must not exceed 0.055 mGy/h [24]. It is obvious that the calculated total absorbed dose rates for all waste samples are higher than the recommended dose level that are acceptable (Fig. 8). The low total absorbed g-dose rate is 31 mGy/h in fraction F2, while the high value is 72.7 mGy/h in fraction F8. It is clear that the absorbed dose rate depends on the activities of g-emitters (e.g. ^{226}Ra, ^{232}Th ^{40}K), while it is independent of the waste particle size. Therefore, the total absorbed dose rates increases with the activity concentration, and consequently enhances the radiological impact on the workers surrounded by the wastes.

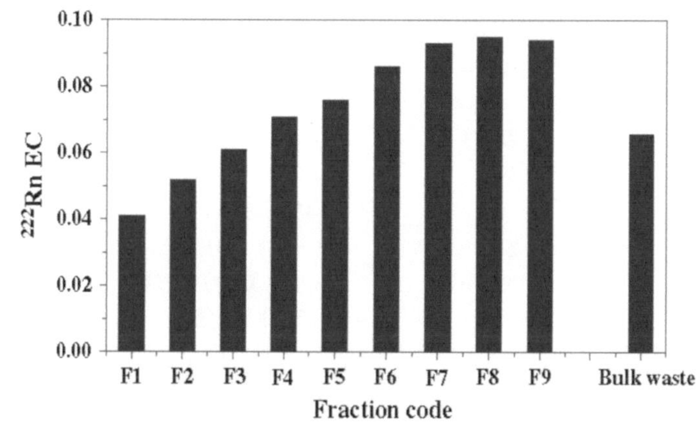

Figure 5. Effect of the waste particle size on the radon emanation fraction release on TENORM of petroleum in Egypt.

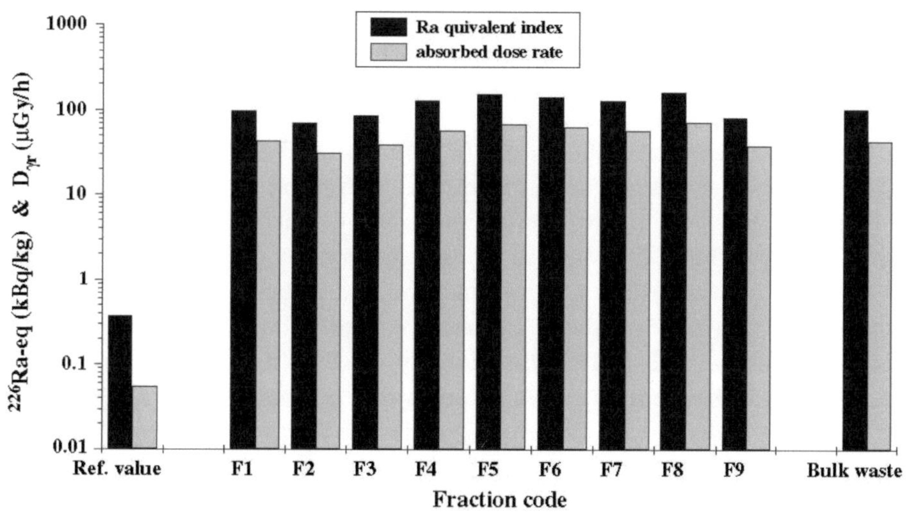

Figure 6. Variation of radium equivalent index and total absorbed dose rate with waste fractions and bulk Waste of TENORM of petroleum in Egypt.

Field	Sample	^{226}Ra (Bq/kg)	^{228}Ra (Bq/kg)
Algeria [32]	Scale	1000-950.000	
Australia [34]	Scale	21.000-250.000	48.000-300.000
Brazil [35]	Scale	19.100-323.000	4210-235.000
Brazil [36]	Scale	121.000-3.500.000	148.000-2.195.000
Brazil [37]	Scale	77.900-2.110.000	101.500-1.550.000
Congo [38]	Scale	97-151	
Egypt [27]	Scale	68.900	24.000
Egypt [29]	Scale	7541-143.262	35.460-368.654
Egypt [30]	Scale	493-519	32-50
Italy [38]	Scale	< 2.7-2890	
Kazakhstan [39]	Scale	510-51.000	200-10.000
Malaysia [40]	Scale	114.300-187.750	130.120-206.630
Norway [41]	Scale	300-32.300	300-33.500
Saudi Arabia [42]	Scale	.08-1.5	
Tunisia [38]	Scale	31-1189	
Tunisia [33]	Scale	4300-658.000	
UK [6]	Scale	1000-1.000.000	
USA [43]	Scale	up to 3.700.000	
USA [21]	Scale	15.400-76.100	
Australia [34]	Sludge	25.000	30.000
Brazil [35]	Sludge	50.000-168.000	49.000-52.000
Brazil [36]	Sludge	< LLD-413.000	< LLD-117.900
Egypt [29]	Sludge	18.000	13.250
Egypt [30]	Sludge	5.27-8.68	1-1.9
Malaysia [40]	Sludge	6-560	4.520
Norway [41]	Sludge	100-4700	100-4600
Tunisia [38]	Sludge	66-453	

Table 5. Ranges of activity levels of ^{226}Ra in different scale and sludge samples

As shown in Table 5, the concentration levels of radium nuclides in scale vary within a wide range being much higher than those of the sludge. According to the latest Environmental Protection Agency (EPA) estimation, the average radium nuclide concentration is around 18,000 Bq/kg and 2800 Bq/kg in scale and sludge, respectively [44]. Elevated concentration activities of both radionuclides, exceeding the exemption level of 10,000 Bq/kg recommended by IAEA safety standards, were frequently found in the scale samples. A large uncertainty is observed in the estimations of the total amount of radioactive waste generated by oil industry, and the EPA assumes that 100 tons of scale per oil well are generated annually in the United States [44], while for the North Sea wells a somewhat lower value of 20 t is suggested [45] and only 2.25 t per year by one oil-producing well for Latin American oil producing countries [46]. It was also estimated that approximately 2.5×10^4 and 2.25×10^5 tons of contaminated scale and sludge, respectively, were generated each year from the petroleum industry in the middle of the previous decade [47,48]. This means that TENORM waste from the oil industry may generate radiation exposure levels which require attention and continuous monitoring during some routine operation in this industry. This exposure is caused by external radiation coming from the ^{226}Ra radionuclide and its progenies: ^{214}Pb and ^{214}Bi as well as by inhalation of α -emitting radionuclides: ^{222}Rn as well as ^{218}Po and ^{214} Po formed from ^{222}Rn escaping into the air adjacent to scale deposits, as reported in Table 6.

Country	Reported range (μSv/h)
Algeria [32]	Back Ground-100
United Kingdom [32]	10-300
Egypt [27]	50-100
Congo, Italy, Tunisia [38]	0.1-6
USA [49]	up to300

Table 6. Exposure rate levels in the oil industry

2.2. TE-NORM in produced water

Oil and gas reservoirs contain water (formation water) that becomes produced water when brought to the surface during hydrocarbon production. Oil reservoirs can contain large volumes of this water whereas gas reservoirs typically produce smaller quantities. In many fields, water is injected into the reservoir to maintain pressure and / or maximize production. In many offshore oil fields sea water is additionally injected to maintain pressure, and it mixes with the formation water. In such cases, in the exploited oil/water mixture, the content of the production water can reach even 95%. For this reason, the produced waters are typically saline and rich in Cl^- anions forming aqueous complexes with Ra that enhance the mobility of Ra nuclides from adjacent geological rocks into these waters [50]. Comprehensive older literature reviews of radium nuclide concentrations in formation and produced water indicated an average radium nuclide concentration in waters in excess of 1.85 Bq/dm3 and exceptionally up to ~ 1000 Bq/dm^3 [49,51,52]. As ^{226}Ra originated from the radioactive decay of ^{238}U, while ^{228}Ra from ^{232}Th, the $^{226}Ra/^{228}Ra$ ratio in the oil-field brines depends on the U/Th ratio of the reservoir rock and ranges from 0.1 to 2.0, but for the most cases its activities are comparable.

Typical ranges or average values of the radium radionuclide concentrations in the formation or produced water from different oil fields, including the recent data, are listed in Table 7 [29,32,34,38,51,53-62].

A critical review of the intense studies of the activity concentrations of ^{226}Ra, ^{228}Ra as well as ^{210}Pb and ^{210}Po in produced water in 2003 from Norwegian oil and gas platforms located in the North Sea were also reported [55]. The concentrations of ^{226}Ra and ^{228}Ra in produced water discharged from these offshore platforms vary between 0.1 Bq/dm^3 and about 200 Bq/dm^3 with the average values estimated to be 3.3 Bq/dm^3 and 2.8 Bq/dm^3, respectively. Slightly higher radium activities ~ 10 Bq/dm3 have been found for produced water outfalls in the Gulf of Mexico [63]. The European Commission (EC) derived the specific clearance levels at low activity for metals and buildings in radiation protection. The world wide average concentration of these radionuclides in produced water discharged to the environment is estimated at 10 Bq/l. These concentrations are approximately three orders of magnitude higher than natural concentrations of radium in drinking or sea water. Because the radium radionuclide concentrations in that waste water are usually below the clearance levels (Table 8), it is recognized as a low specific activity waste and they may be injected into underground formations or disposed into the sea.

Field	Sample	^{226}Ra (Bq/dm^3)	^{228}Ra (Bq/dm^3)
Algeria [32]	Formation water	5.1-14.8	
Australia [34]	Produced water	17[a]	23[a]
Brazil [53]	Produced water	0.01-6	0.05-12
Congo [38]	Produced water	5.1[a]	
Egypt [29]	Formation water	5-40	1-59
Italy [38]	Produced water	0.2-2	
Norway [54]	Formation water	0.3-10.4	
Norway [55]	Produced water	3.3[a]	2.8[a]
Norway [56]	Produced water	0.5-16	0.5-21
Syria [57]	Produced water	9.9-111.2	8.8-60.4
UK [58]	Produced water	1.7[a]	
USA [51]	Produced water	0.1-60	
USA [59]	Produced water	0.15-21.6	0.7-1.7
USA [60]	Oilfield brine	12.6[a]	15.1[a]
USA [61]	Produced water	22-30	25-30
Egypt [62]	Produced water	26.5-217.5	
	Crude oil	31-2669	

[a] Mean activity concentration.

Table 7. Ranges of activity levels in produced water from the oil fields

Radionuclide	Concentration (Bq/g)
^{40}K	100
^{226}Ra	10
^{232}Th	1

Table 8. The European Commission Clearance levels.

A comprehensive evaluation of discharges from the oil industry to the sea was done for European waters during the European Commission Marina Project [64].

The annual release of ^{226}Ra and ^{228}Ra with produced water from off-shore fields in Europe in the 1990s stabilized at around 5 and 2.5 TBq per year, respectively. The commonly used two steps model of the radionuclide dispersing and diluting in the water in the vicinity of the oil platforms predicts a diluting factor up to 10^3 within minutes and within a few meters of the discharge source [65]. Therefore, additional radium nuclide concentrations in seawater of the local zone could be estimated as equal to around 5–10 Bq/m^3, in comparison with the natural concentration of around 1 Bq/m^3 for ^{226}Ra. The radium activities in the produced water for the North Sea, a yearly release of 0.65 TBq for ^{226}Ra and 0.33 TBq for ^{228}Ra was appraised for offshore oil production from Argentina and Brazil [66].

Produced water contains a complex mixture of inorganic (dissolved salts, trace metals, suspended particles) and organic (dispersed and dissolved hydrocarbons, organic acids) compounds, and in many cases, residual chemical additives (e.g. scale and corrosion inhibitors) that are added into the hydrocarbon production process. Feasible alternatives for the management and disposal of produced water should be evaluated and integrated into production design. These alternatives may include injection along with seawater for reservoir pressure maintenance, injection into a suitable offshore disposal well, or export to shore with produced hydrocarbons for treatment and disposal. If none of these alternatives are technically or financially feasible, produced water should be treated before disposal into the marine environment. Treatment technologies to consider include combinations of gravity and / or mechanical separation and chemical treatment, and may include a multistage system, typically including a skim tank or a parallel plate separator, followed by a gas flotation cell or hydrocyclone. There are also a number of treatment package technologies available that should be considered depending on the application and particular field conditions. Sufficient treatment system backup capability should be in place to ensure continual operation and for use in the event of failure of an alternative disposal method, for example, produced water injection system failure. Where disposal to sea is necessary, all means to reduce the volume of produced water should be considered, including:

- Adequate well management during well completion activities to minimize water production;
- Recompletion of high water producing wells to minimize water production;
- Use of down hole fluid separation techniques, where possible, and water shutoff techniques, when technically and economically feasible;
- Shutting in high water producing wells. To minimize environmental hazards related to residual chemical additives in the produced water stream, where surface disposal methods are used, production chemicals should be selected carefully by taking into account their volume, toxicity, bioavailability, and bioaccumulation potential [67].

The average worldwide activity levels of uranium (U) , thorium (Th) and potassium (K) [68] and the exemption activity levels of NORM as recommended in the IAEA basic safety standards [69], were given at Tables 9&10. The average worldwide levels of the most common radiological indices [68] was given at Table 11. These indices include radium equivalent (Ra-eq), total absorbed dose ($D_{\gamma r}$) and effective annual dose rate (EDAR).

Radionuclide	**U**	**Th**	**K**
Activity level (Bq/Kg)	50	50	500

Table 9. The average worldwide activity levels of U, Th and K [68].

Radionuclide	^{238}**U**	^{226}**Ra**	^{222}**Rn**	^{232}**Th**	^{228}**Ra**	^{224}**Ra**
Exemption level (Bq/g)	1	10	10	1	10	10

Table 10. The exemption activity levels of NORM as recommended in the IAEA basic safety standards [69].

Radiological indices (Unit)	**Ra-eq (Bq/Kg)**	**$D_{\gamma r}$ (nGy/h)**	**EADR (mSv/yr) for worker**	**EADR (mSv/yr) for Public**
Activity level (Bq/Kg)	370	55	20	1

Table 11. The average worldwide levels of the most common radiological indices [68].

2.3. Investigation of TE-NORM treatment in petroleum industry

In the last decade, attention was focused on the environmental and health impacts from the uncontrolled release of TENORM wastes [19,28,70,71]. Therefore, the treatment of these wastes is of increasing interest because accumulation of large amounts with a significant activity may cause a health risks to the workers through exposure, inhalation of radon (^{222}Rn) decayed from radium and/or ingestion of waste dust during the periodical maintenance of the equipment used. The trials towards the treatment of TENORM wastes from many industries are still limited. In this concern, removal of ^{226}Ra from TENORM wastes produced from oil and gas industry was carried out by a simple extraction process using saline solutions (i.e., seawater) and chemical solutions [70]. The chemical treatment process of TE-NORM sludge has been carried out by suspending the clay fraction content in the solid waste in suitable leaching solutions was reported. It was found that, the maximum removal % of ^{226}Ra is ~ 85% [70]

El-Afifi, studied the treatment of radioactive waste containing ^{226}Ra from oil and gas production, using different chemical solutions, in terms of a simple and sequential techniques based on suspending ^{226}Ra through the clay fraction in the waste. More than 50% of ^{226}Ra was removed through the treatment using moderate acids and salts solutions, while more than, 75% of ^{226}Ra was removed based on successive treatment or using some strong chelating reagent solutions [28].

The development of the treatment of a sludge TENORM waste produced from the petroleum industry in Egypt, using selective leaching solutions based on two approaches 'A' and 'B' has been investigated by El-Afifi et al. [71]. The results obtained showed that treatment of the waste through main four successive leaching steps removed ~ 78 and 91% of ^{226}Ra, 65 and 87% of ^{228}Ra as well as 76 and ~ 90% of ^{224}Ra using approaches A and B, respectively [71]. El-Afifi et al. [72] reported some data about the radiological characterization for phosphogypsum waste and phosphate rocke samples by γ -ray spectrometer. El-Didamony et al.[73] used the solvent extraction technique in treatment of phosphogypsum waste obtained as a byproduct of phosphoric acid production from phosphate ore.

3. Conclusion

The naturally occurring radioactive materials (NORM) are found everywhere. We are exposed to it every day. NORM represent an integral part of the planet, our bodies, the food we eat, air we breath, the places where we live and work, and within products we use. However , in the exploration and extraction processes of oil and gas, the natural radionuclides ^{238}U, ^{235}U and ^{232}Th, as well as the radium-radionuclides (^{223}Ra, ^{224}Ra, ^{226}Ra and ^{228}Ra) and ^{210}Pb,.etc., are brought to the slurry surfaces and may contain levels of radioactivity above the surface background. The petroleum waste (scale or sludge) have been produced by two mechanisms: either incorporation or precipitation onto the production equipment such as: pipelines, tank storage, pumps, ..etc. The waste generated in oil and gas equipment is due to the precipitation of alkaline earth metals as sulfate, carbonates and/or silicates. It is clear that PG and sludge and scale wastes represent one of the major sources of ^{226}Ra in the environment. Since the concentration of ^{226}Ra found in both of them waste exceeds that permitted by the international regulations, it was found necessary to reduce the risks due to indoor radon and direct γ -radiation in each wastes to be used in different life aspects.

Author details

M. F. Attallah and H. F. Aly
Hot Laboratories and Waste Management Center, Atomic Energy Authority, Cairo, Egypt

N. S. Awwad
Hot Laboratories and Waste Management Center, Atomic Energy Authority, Cairo, Egypt;
Chemistry Department, Faculty of science , King Khalid university, Abha, KSA

Appendix

Nomenclature of symbols associated with this article is listed as follows.

NORM	Naturally occurring radioactive material
TE-NORM	Technically enhanced naturally occurring radioactive material
$t_{1/2}$	Half-life
hr	Hour
min	Minute
y	Year
α	Alpha decay
β	Beta decay
γ	Gamma decay
dpm	Disintegration per minute
g	Gram
Ci	Curie
mSv	Milly sievert
Γ	Gamma constant can be used to determine radiation dose

USEPA	United state of Environmental protection agency
Bq	Becquerel
AR	Abu Rudeis region
GEZ	Gabal El Zeit region
BED	Badr El Din region
cm	Centimeter
GS	Granular samples
MS	Massive samples
EF	Emanation fraction released
XRF	X ray fluorescence
XRD	X ray diffraction
B.F.	Before fractionation
A.F.	After fractionation
F	Fraction
mm	Milie meter
Ra-eq	Radium equivalent activity
$D\gamma r$	Absorbed dose rate
mGy	Gray
IAEA	International Atomic Energy Agency
UNSCEAR	United Nations Scientific Committee on the Effects of Atomic Radiation
WHO,	World Health Organization
Kg	Kilo gram
u	Combined uncertainty
CV	Coefficient variance
SD	Standard deviation
TDS	Total dissolved solids
EC	Electric conductivity
A_0	Net count rate of ^{222}Rn existing at the sealing time of the sample
N	Net count rate of ^{222}Rn emanated at the radioactive equilibrium

4. References

[1] A.A. Abed-Rassoul, Lectures Proc. of 2nd Arab Conf. on the peaceful uses of atomic energy, Cairo, Egypt, 1 (1994).

[2] H.F. David Liu and G. Bela Liptak, "Hazardous waste and solid waste", Lewis Publishers, USA (2000).

[3] F.A. Shehata, E.M. El-Afifi and H.F. Aly, Proc. of Int. Conf. in Radioactive waste management and environment remediation, Nagoya, Japan, CD-ROM, Sept. 26-30 (1999).

[4] M.F. Maged and E.A. Saad, Environ. Management and Health, 9(2), 30 (1998).

[5] K.J. Grice, Proc. of the 1st Int. Conf. on Health, Safety and Environment, Hague, Netherlands, Nov. 10-14, 559 (1991).

[6] Exploration & Production Forum, 1987. Low specific activity scale origin treatment and disposal. Report no. 6.6/127, Old Burlington Street, London WlX 1LB, pp. 25–28.
[7] R. Colle, R. J. Rubin, L. I. Knab, J. M. R. Hutchinson, NBS Technical note No. 1139, Washington DC, US Department of Commerce, National Bureau of Standards, 1981, p. 1.
[8] A. B. TANNER, Radon migration in the ground. A supplementary review, in: Natural Radiation Environment III, Vol. 1, T. F. GESELL, W. M. LOWDER (Eds), US Department of Energy Report CONF-780422, 1980, p. 5.
[9] F.S. Grimaldi, M.H. Fletcher and L.B. Jenkins, J. Anal. Chem., 29, 848 (1957).
[10] K. Sudhalatha, Talanta, 10, 934 (1963).
[11] T. Strand and I. Lysebo, Proc. of the 2nd Int. Symp. on the Treatment of NORM, Krefeld, Germany, Nov. 10-13, 137 (1998).
[12] J.E. Oddo, X. Zhou, D.G. Linz, S. He and M.B. Tomoson, Proc. of Exploration and production environmental Conf., Houston, Texas, USA, March 27-29, 207 (1995).
[13] G.R. Choppin and J. Rydberg, "Nuclear chemistry theory and application", 1st Ed., Pergamon Press, New York, USA, 65 (1980).
[14] Hamlat MS, Kadi H, Djeffal S, Brahimi H (2003) Radon concentrations in Algerian oil and gas industry. Appl. Radiat. Isot. 58:125–130
[15] P.J. Shuller, D.A. Baudoin and D.J. Weintritt, Proc. of Exploration and production environmental Conf., Texas, USA, March 27-29, 219 (1995).
[16] C. Testa, D. Desideri, F. Guerra, M.A. Meli and C. Roselli, Proc. of the Int. Solv. Extraction Symp., Moscow, Russia, June 21-27, 416 (1998).
[17] K.P. Simth, D.L. Blunt, G.P. William and L.L. Tebes, Proc. of Explo-ration and production environmental Conf., Houston, Texas, USA, March 27-29, 231 (1995).
[18] M.A. Hilal, "Nuclear spectroscopic measurements for materials and environmental control", M. Sc. Thesis (Physics), Faculty of Science, Zagazig University (Banha branch), Egypt (1998).
[19] M. F. Attallah, Chemical studies on some radionuclides in industrial wastes. M. Sc. Thesis, Benha University, Egypt (2006).
[20] S. Stoulos, M. Manolopoulou and C. Papastefanou, "Measurement of radon emanation factor from granular samples: effects of additives in cement", J. App. Radiat. Isot., 60, 49 (2004).
[21] G. J. White, A. S. Rood, Radon emanation from NORM contaminated pipe scale and soil at petroleum industry sites. J. Environ. Radioact., 54 (2001) 401-413.
[22] J. Beretka and P. J. Mathew, J. Health Phys. Society, 48, 87 (1985).
[23] J.H. Zaidi, M. Arif, S. Ahmed, S. Fatima, and I.A. Quresh, 1991. Determination of natural radioactivity in building materials used in the Rawalpindi/Islamabad area by g-ray spectrometry and instrumental neutron activation analysis. Applied Radiation and Isotopes 51, 559–564.
[24] UNSCEAR, United Nations Scientific Committee on the Effect of Atomic Radiation: Sources and effects of ionizing radiation, United Nations, New York (1993).
[25] UNSCEAR, United Nations Scientific Committee on the Effect of Atomic Radiation: Sources and effects of ionizing radiation, United Nations, New York (2000).

[26] E. M. El Afifi, S. M. Khalifa, H. F. Aly, Assessment of the ^{226}Ra content and the ^{222}Rn emanation fraction of TE-NORM wastes at certain sites of petroleum and gas production in Egypt, Journal of Radioanalytical and Nuclear Chemistry, Vol. 260, No. 1 (2004) 221.224.

[27] E. M. El Afifi, N. S. Awwad, (2005) , Characterization of the TENORM waste associated with oil and natural gas production in Abu Rudies, Egypt, J. Environ. Radioact. 82 ,7-19.

[28] El Afifi, E.M., (2001) Radiochemical studies related to environmental radioactivities. Ph.D. thesis (Chemistry), Faculty of Science, Ain Shams University, Cairo, Egypt, 98.

[29] Shawky, S., Amer, H., Nada, A.A., Abdel Maksoud, T.M., Ibrahim, N.M., (2001) Characteristics of NORM in the oil industry from eastern and western deserts of Egypt. Applied Radiation and Isotopes 55, 135–139.

[30] M. Abo-Elmagd, H.A. Soliman, Kh.A. Salman, N.M. El-Masry, Radiological hazards of TE-NORM in the wasted petroleum pipes, J. Envron. Radio. 101, 51-54 (2010)

[31] Holland, B., 1998. Experience with Operations Involving NORM in the UK and some Other Regions.Australian Nuclear Science and Technology Organization, Lucas heights, Sydney, March 16–20.

[32] Hamlat, M.S., Gjeffal, S., Kadi, H., 2001. Assessment of radiation exposures from naturally occurring radioactive materials in the oil and gas industry. Applied Radiation and Isotopes 55, 141–146.

[33] Heaton, B., Lambley, J., 1995. TENORM in the oil, gas and mineral mining industry. Applied Radiation and Isotopes 46, 577–581.

[34] Guidelines for naturally occurring radioactive materials (2002) Australian Petroleum Production & Exploration Associated Ltd. Report ABN 44000292773, March 2002, Canberra

[35] Godoy MJ, da Cruz RP (2003) 226Ra and 228Ra in scale and sludge samples and their correlation with the chemical composition. J Environ Radioact 70:199–206.

[36] Gazineu MHP, de Araujo AA, Brandao YB, Hazin CA, Godoy JM (2005) Radioactivity concentration in liquid and solid phases of scale and sludge generated in the petroleum industry. J Environ Radioact 81:47–54.

[37] Gazineu MHP, Hazin CA (2008) Radium and potassium-40 in solid wastes from the oil industry. Appl. Radiat. Isot. 60:90–94.

[38] Testa C, Desideri C, Meli MA (1994) Radiation protection and radioactive scales in oil and gas production. Health Phys 71:34–38.

[39] Kadyrzhanov KK, Tuleushev AZ, Marabaev ZN (2005) Radioactive components of scales at the inner surface of pipes in oil fields of Kazakhstan. J Radioanal Nucl Chem 264:413–416

[40] Omar M, Ali HM, Abu MP (2004) Distribution of radium in oil and gas industry wastes from Malaysia. Appl.Radiat Isot 60:779–782.

[41] Lysebo J, Birovliev A, Strand T (1996) NORM in oil production – occupational doses and environmental aspects. In: Proc of the 11th Congress of the Nordic Radiation Protection Society, 26–30 August 1996, Reykjavik, p 137.

[42] Al-Saleh FS, Al-Harshan GA (2008) Measurements of radiation level in petroleum products and wastes in Ryad City refinery. J Environ Radioact 99:1026–1031.

[43] Scot ML (1998) Naturally occurring radioactive materials in non-nuclear industry. In: Proc of the 2nd Int. Symp. on the Treatment of Naturally Occurring Radioactive Materials NORM II, 10–13 November 1998, Klefeld, Germany, pp 163–167.

[44] Oil and gas production wastes. http://www.epa.gov/rpdweb00/ tenorm/oilandgas.html.

[45] European Commission Externe (1994) Externalities of energy. Vol. 4: Oil and gas. Report EUR 16524EN. ECE, Luxembourg.

[46] Steinhäusler F, Paschoa AS, Zaborowski W (2000) Radiological impact due to oil-and gas extraction and processing: a comparative assessment between Asia- Pacific, Europe and South America. In: Proc of the 10th IRPA Association Congress, 14–19 May 2000, Hiroshima, P-61-285, pp 1–7.

[47] Smith GE, Fitzgibbon T, Karp S (1995) Economic impact of potential NORM regulations In: Proc of SPA/EPA Exploration and Production. Environmental Conference,27–29 March 1995, Houston, USA, pp 181–231.

[48] Firyal Bou-Rabee, Abdallah Z. Al-Zamel, Rana A. Al-Fares, Henryk Bem, (2009) NUKLEONIKA ,Review paper;54(1):3–9

[49] Jonkers G, Hartog FA, Knappen AAI, Lance PFJ (1997) Characterization of NORM in the oil and gas production (E&P) industry. In: Proc of the NORM I, Amsterdam, pp 23–47

[50] Fischer SR (1998) Geologic and geochemical controls on naturally occurring radioactive materials (NORM) in produced water from oil, gas, and geothermal operations. Environ Geosciences 5:139–159.

[51] Snavely ES (1989) Radionuclides in produced water. Report to the American Petroleum Institute. Publication no 5404. API, Washington, DC, pp 1–86.

[52] White GJ (1992) Naturally Occurring Radioactive Materials (NORM) in oil and gas industry, equipment and wastes: a literature review. Report DOE/ID/01570-T158.Bartlesville.

[53] Vegueria JSF, Godoy JM, Miekeley N (2002) Environmental impact studies of barium and radium discharges by produced waters from the "Bacia deCampos" oil field offshore platforms, Brazil. J Environ Radioact 62:23–38.

[54] Strand T, Lysebo I (1998), NORM in oil production activity levels and occupational doses. In: Proc of the 2nd Int Symp on the Treatment of Naturally Occurring Radioactive Materials NORM II, 10–13 November 1998, Klefeld, Germany, pp 137–141

[55] Norwegian Radiation Protection Authority (2005) Natural radioactivity in produced water from the Norwegian oil and gas industry in 2003. Report no 2. NRPA, Ǿsteras.

[56] Eriksen DO, Sidhu R, Strälberg E (2006) Radionuclides in produced water from Norwegian oil and gas installations-concentrations and bioavailability. CzechoslovakJ Phys 56:D43–D48.

[57] Al-Masri MS (2006) Spatial and monthly variations of radium isotopes in produced water during oil production. Appl Radiat Isot 64:615–623.

[58] United Kingdom Off-Shore Operations Association (1992) UK North Sea oil and gas industry; environmental inputs, impacts and issues. A report prepared by Environmental and Resource Technology Ltd, London.

[59] Stephenson MT, Supernow IR (1990) Offshore Operators Committee 44 Platform study radionuclide analysis results. Offshore Operation Committee Report, New Orleans, Louisiana.

[60] Swan C, Matthews J, Ericksen R, Kuszmaul J (2004) Evaluation of radionuclides of uranium, thorium, and radium associated with produced water fluids, precipitates and sludge from oil, gas and oilfield brine injections wells in Mississippi. US DOE Report; DE-FG26-02NT 15227.

[61] Zieliński RA, Budahn JR (2007) Mode of occurrence and environmental mobility of oil-field radioactive material at US Geological Survey research site B. Appl Geochem 22:2125–2137.

[62] W.F. Bakr, Assessment of the radiológical impact of oil refining industry, J. Enviro. Radio. 101, 237-243 (2010).

[63] Stephenson MT (1992) Components of produced water. J Pet Technol 548–603.

[64] Betti M, Aldave de las Heras L, Janssens A (2004) Results of the European Commission Marina II Study Part – effects of discharges of naturally occurring radioactive material. J Environ Radioact 74:255–277.

[65] Brandsma MG, Smith JP, O'Reilly JE, Ayers RC, Holmquist AL (1992) Modelling offshore discharges of produced water. In: Ray JP, Englehart FR (eds) Produced water. Plenum Press, New York, pp 59–71.

[66] Hanfland C (2002) Radium-226 and Radium-228 in the Atlantic Sector of the Southern Ocean. Ber Polarforsch Meeresforssch, p 43125. Heaton B, Lambley JG (1995) TENORM in the oil and gas industry. Appl Radiat Isot 46:577–581.

[67] Environmental, Health, and Safety Guidelines Offshore Oil and Gas Development, APRIL 30, 2007.

[68] UNSCEAR, 1994, United Nations Committee on the Effect of Atomic Radiation: Sources and NCRP. Exposure of the population in the United Sates and Canada from natural background radiation. NCRP report no.94. national Council on Radiation Protection and Measurement , Bethesda , Maryland.

[69] IAEA, 2001. International Atomic Energy Agency. Report of analysis on Determination of thorium and uranium naturally occurring radioisotopes in IAEA reference materials. IAEA laboratories (Chemistry Unit-01-10), Seibersdorf Austria , p.1-5.

[70] E. M. El Afifi, S. A. El-Reefy, H. F. Aly, Treatment of solid waste containing Ra-226, Arab J. Nucl. Sci. Appl., 39 (2006) 35-47.

[71] E.M. El Afifi, N.S. Awwad, M.A. Hilal (2009) Sequential chemical treatment of radium species in TENORM waste sludge produced from oil and natural gas production, Journal of Hazardous Materials 161 907–912.

[72] M. El Afifi, M. A. Hilal, M. F. Attallah, and S. A. El Reefy, Characterization of Phosphogypsum Wastes Associated with Phosphoric Acid and Fertilizers Production, J. Environ. Rad., 100, (2009) 407.

[73] El-Didamony, M. M. Ali, N. S. Awwad, M. M. Fawzy, M. F. Attallah, Treatment of phosphogypsum waste using suitable organic extractants, J.Radioanal Nucl Chem, DOI 10.1007/s10967-011-1547-3,2011

Section 2

Storage and Distribution

Chapter 5

A Raster Based Geospatial Model for Natural Gas Transmission Line Routing

V. Yildirim, T. Yomralioglu and R. Nisanci

Additional information is available at the end of the chapter

http://dx.doi.org/10.5772/45814

1. Introduction

The most important step in planning activities of Natural Gas Transmission Line (NGTL) is an applicable route selection [1]. Because obtaining the optimum route on a surface is a very complex process. Because many factors must be considered at the same time, and it is an important step for the NGTL projects. Defined route has influence on the project in every stage in economical, environmental, sociological and temporal way [2]. To reduce unfavorable effects as much as possible in terms of flora and fauna, and the environment and to complete the project at least cost base on the efficiency of the route. Determining the best route depends on examination, query and analysis of many complex data together [3]. Route determination requires spatial data from different organizations and state institutions, and it also needs to be carefully chosen, saved, queried and analyzed. Today, this type of analysis and a quick result are possible with Geographical Information System (GIS). GIS is an effective engineering tool for systematically organizing factors affecting route determination. When these factors are identified based on the length of the project, a GIS should be used to evaluate these factors simultaneously. Additionally the GIS based visualization technologies and cartographic abilities are generally adequate to determine the effective routes [4,5,6,7,8].

Route problems, including route selection, route planning, finding the optimal route, corridor analysis and site selection, can be solved using network analysis based on GIS. GIS technologies have significantly improved recently. As one of the most important current information technologies, it is used as an effective tool in network analysis. Network analysis can be carried out on both vector-based data, such as roads, streams or pipelines, and raster-based data from non-defined space. Nevertheless, applications of network analysis for route determination of linear engineering structures must be carried out with raster-based data because they do not have a defined space [9,10,11]. Route determination with raster-based data is advantageous because it is simple to perform cost calculation, design, and modeling and to obtain Remote Sensing (RS) data directly in raster format [12,13,14,15].

Raster data model is the most useful data format to carry out arithmetic operations among pixels of same coverage or different coverage on the same geographical location. Today, lots of GIS software extensively uses specific functions such as surface analysis, determination of the least cost route, forming arithmetic operations among the coverage, thanks to advantages provided by raster data format. In raster operations, surface crossing among pixels is more effective than crossing among lines on the vector based network analysis. Because obtaining the optimum route or the least cost route can be possible with the movement among the pixels on raster based continuous data, not on vector based finite line. Many researchers, realizing the importance of the taster approach in route determination process, have carried out various studies and have provided solutions to eliminate the deficiencies of this method continuing from past to present [16,5].

Many factors in NGTL studies must be examined together. These factors exist in a complex and dispersed structure. Generally, decision makers randomly determine factors which they will use and effect rates of these factors to the route. However, required factors in determining route must be organized in a specific systematic manner in importance order. As a result of statistical examinations, required factors should be determined and the number of factors must be minimized as possible as. Besides these factors, the effect rates to the route need to be determined by using Multi Criteria Decision Making (MCDM) methods. In this sense, raster based routing model which will be generated should include the factor and determination process dynamically [17].

In route determination of NGTL, as not to be able to determine potential landslide areas, protected areas, flora/fauna areas, wetlands, rocky areas, soil types, other infrastructure lands, agricultural activity lands and may factors affecting the route in advance cause environmental and economic problems in these projects. Especially in Turkey, there are many present projects in which the pipelines, passing through the landslides, have been reconstructed; as coming across the breeding place of a specific type of animal, NGTL is waited for breeding time for days; pipelines as also coming across hard rocks or steep slope were reconstructed by taking back to avoid sharp turning and there is extra cost for crossing unnecessary streams and wetlands [18]. In other countries, besides these problems, there are projects which NGTL has been passing through fault lines [19], ore beds [20] and residential areas [21] causing the deaths of many people. Additionally, there are also projects that have been cancelled without completing the estimated operating time of the pipelines or have been repaired again with excessive maintenance costs because of not having determined the factors (groundwater causing corruption and abrasion) that affect the occupancy of NGTL [22].

2. NGTL routing model

Being carried out the process of NGTL route determination with GIS based models provides many advantages economically, environmentally and sociologically [23,24]. Like in all GIS applications, efficiency of results is proportional to the data quality. In determining route studies of NGTL projects, many data should be collected, stored and analyzed depending on the size of the field. Requiring location determination, the route determination problem might be basically perceived as a location problem. Because in these problems, every factor, affecting the route, corresponds to a location data set.

The deficiency of location data, used in studies including large areas that different corporations in Turkey require, has negatively influence on GIS studies which will be held on a regional scale. Maps used by many public corporations are 1:25.000 scaled topographic maps produced by General Command of Mapping (GCM). Corporations provide needed data (road, river, residential areas, natural sources, etc.) with these maps. However updating of these maps throughout the country requires a long time process. These map produced for defense purposes are not enough for the usage of different disciplines. The first stage of GIS based route determination method is to obtain the necessary location data/information by taking into account the factors affecting to the route [25]. In this process, the majority of the data that will be used as base is the location data (Figure 1).

Land Cover	**Other Linear Engineering Structures**
Rock Areas	Highway
Forest Area	Railway
Agriculture Area	**Settlement**
Flora / Fauna	Residential Areas
Recreation Area	Ownership
National Park	Industrial Areas
Protected Area	Oil pipeline
Wetlands	Natural Gas Pipeline
Geology	Power Transmission line
Fault Lines	***Spatial***
Lithology	***Data***
Elevation	Irrigation/Drying Canals
Ore Beds	Boundaries
Soil	**Hydrology**
Landslide	Streams
	Lakes
	Irrigation Canal
	Underground Water sources
	Source of drinking water
	River basins

Figure 1. Spatial Dataset for NGTL routing

Spatial and non-spatial data required in determining NGTL routes are in large volume and vary very much. Requirement of studying effectively these data has revived raster based GIS methods in the recent years [26,27]. GIS is one of the most effective tools to analyzethe data in a whole. The data which will be used as input in NGTL route determination is needed at many stages (cost, operation, maintenance, time, efficiency of the line etc.) of pipeline project [28].

The first stage in raster based route planning model is defining the factors that affect NGTL route determination. Factors affecting the route determination of NGTL project vary in themselves. The factors affecting the determination is quiet different in crossing between the land and the sea. The effect ratios of these factors must be also identified. These rates called as factor weights are able to be determined in different ways. In this study, raster based route planning model has been generated by taking account of land crossing of NGTL in a dynamic structure in which necessary factors, weights of factors and related data layers can be studied as a whole (Figure 2).

Accuracy Analysis
Factors and Data
NTGL ROUTING CONCEPTUAL MODEL
Weights
Determining of Weight Coefficients Based on Shareholders
Ensuring of Weight Coefficients Based on Professionals
Series of Spatial Analysis
MCDM Analysis
Raster Network Analysis
Determining of Alternatives Routes
Optimum Route
Checking
Factors
Data Gathering & Data Processing

Figure 2. NGTL routing conceptual model [17]

3. Factors

In raster based routing model, spatial and non-spatial data corresponding to factors affecting the route must be primarily obtained by paying attention to sufficient accuracy and precision. In Turkey context, many problems are experienced during the period from producing spatial data to sharing them [29]. Difficulties in reproduction of spatial data that are not probable, non- spatial data in a scattered structure and probable data without required quality are some of the most important reasons. Completing the process of data collection which is the most important part of the entire system in terms of time and cost, and data layers belonging to factor that will be used must be generated (Table 1).

In NGTL routing, factors, and weight of these factors vary depending on the precision of project's construction, execution and maintenance. For example, in an area in which intense landslides occur, landslide factor is amplified. Or identification of accurate obstacle is generated for flora-fauna factor while determining the route through the areas which are sensitive environmentally, such as tropical areas. In this context, factors affecting the route are generally classified in three main titles: environmental, economical and sociological factors (Table 2).

Limitation \ Layers	Land Cover	Elevation	Geology	Soil	River	Road	Flora / Fauna	Fault Line	Protected Area	Boundaries	Recreation	Railway	Ownership
NGTL length	Determined in feasibility stage												
Corridor width	Determined in feasibility stage (20 – 30m)												
Agricultural areas crossing	×			×									
Forest area crossing	×												
Rocky area crossing	×												
Pasture area crossing	×												
Wetland crossing	×												
Residential area crossing	×									×			
Low expropriate costing areas crossing													×
Active landslide area crossing													
Fixed landslide area crossing	×		×										
Potential landslide area crossing	×	×	×		×	×							
Fault line crossing			×										
River crossing	×				×								
Road and railway crossing	×					×						×	
Steepest slope crossing		×											
Unsuitable lithological area crossing			×										
Existing public corridor crossing	×					×						×	×
Existing and potential mine area crossing			×										
Archeological and historic area crossing											×		
Recreation area crossing											×		
Natural protected area crossing									×	×			
Protected area crossing							×		×				
Water resources crossing	×												
Flora / Fauna crossing							×						

Table 1. Limitations which affect NGTL route and relevant data layers

	Economical	Environmental	Sociological
Land Cover	X	X	X
Elevation	X		
Geology	X		
Soil	X		X
River	X	X	
Road	X	X	
Railway	X		
Flora / Fauna		X	
Earthquake/Fault	X		
Protected Area		X	X
Boundaries			X
Recreation		X	X
Ownership	X		X

Table 2. General classification of factors affecting NGTL Route

4. Determination of factor weights

In the process of route planning of pipelines, multiple qualitative and quantitative parameter should be evaluated in a whole and need to be decided basing on the results. When some of these parameters contradict with each other, the process of decision-making also called as the MCDM method is used to get more reliable results. In MCDM method, due to contradictory criteria, choosing the most appropriate alternative is very difficult for decision makers. After consideration of criteria, methods to remove these contradictions must be used to obtain the most accurate results [30].

The Analytic Hierarchy Process (AHP) is a method that is widely applied in decision theory; it is a paradoxical measurement method that includes measurable or abstract criteria. The Analytic Hierarchy Process (AHP), besides being a method which finds common practicing field in decision theory, is a measuring method which takes the controversial measurable and/or abstract criteria in to consideration. In AHP, if a decision is to made, the information and the experiences that are nearly in equal importance with the data are also taken into consideration. AHP is a tool which can be used in a wide scale from personal decisions to complicated management ones. The power of the theory is in both its simplicity and its applicability to varied conditions.

In AHP, specifying the scale is significant. In order to specify this scale, first, particular sets of number are taken and by using these numbers, it is decided how the modified priorities will be associated with each other. A scale consists of three elements: the mass of objects, the mass off numbers and determination of the mutual connections between the object and numbers.

In a standard scale a unit must be used to generate the values of the scale. Standard scale is used in measurement of objects or facts by using a unit developed to measure a specific feature. The numbers obtained from the scale function only in terms of human mind and do not have a value alone. The scale shown at the Table 3 is the basic scale of AHP.

Intensity of importance	Definition	Explanation
1	Equal importance	Two factors contribute equally to the objective
3	Somewhat more important	Experience and judgment slightly favor one over the other.
5	Much more important	Experience and judgment strongly favor one over the other.
7	Very much more important	Experience and judgment very strongly favor one over the other. Its importance is demonstrated in practice.
9	Absolutely more important	The evidence favoring one over the other is of the highest possible validity.
2,4,6,8	Intermediate values	When compromise is needed.

Table 3. Pair wise scale in AHP

Similar items in every stage of hierarchy are compared for the next level criteria. The results taken from these comparisons are stated as numbers above. This scale shows the meaning of values from 1 to 9. The values in the scale specify the density of connections of items. The matrix of pair wise comparison is generated as a result of pair wise comparison of all items. As the comparison of an item with itself is defined as a number,1; the values of 1 are drawn out at diagonals of matrix. A number of n(n- 1)/2 comparison is made in an item matrix. The reason of it, the values of the 1 take place on the diagonal on account of comparing the items with themselves. In matrix, evaluation as the number of items of diagonals on the upper side is needed. This situation is derived from that evaluations under diagonal is opposite of the values upper side of the diagonal. Therefore the needed number of evaluation will be {(n*n)-n}/2.

4.1. Pair wise matrix and factor weights

Pair wise comparisons are one of the AHY components. The weight of the parameters used to determine the NGTL route are calculated by taking into consideration the pair wise comparisons of the parameters and the impact relative to each other affecting the route determination they make. Normally, the relative value of two parameters is based on the preference of the decision maker. In this study, Environmental Impact Assessments (EIAs) were prepared and examined for pipeline routing, existing applications, and scientific researchers to compare factors and sub-factors. Moreover, information was obtained from interviews with masters in BOTAS (Petroleum Pipeline Corporation), which is responsible for the pipelines in Turkey, and conversations with masters or experienced people in different corporations. Furthermore, current NGTL construction works were examined, and the relative degree of importance of the factors was determined at the end of the study [31,32].

The accuracy imparted by these weight values is studied by carrying out spatial analysis, questioning and cost evaluation [31,32]. Natural gas pipelines of which routes determined by using classical methods were optimized with this raster based route determination model and the values of weight were tested by comparing the results with original data on the land [31]. The results and statistical findings have shown that the weight values determined by this method are compatible with the original land data.

The number of factors defined for NGTL route planning is 10 in this study. In addition to these factors, absolute barriers were included in the process of routing.

First, determining relationships between basic factors affecting the route of NGTL is important. The matrix of pair wise comparisons between layers were generated to determine which layers are affected and to what extent they are affected (Table 4). Weights for each layer in the NGTL route were calculated. Moreover, consistency Ratios (CRs) related to the layers were calculated to determine the importance of these works.

	A	B	C	D	E	F	G	H	I	K	Weights
A	1	1	2	3	4	7	8	5	6	9	0.256
B	1	1	1	2	3	6	7	4	5	8	0.205
C	1/2	1	1	1	2	5	6	3	4	7	0.156
D	1/3	1/2	1	1	1	4	5	2	3	6	0.116
E	1/4	1/3	1/2	1	1	3	4	1	2	5	0.084
F	1/7	1/6	1/5	1/4	1/3	1	1	1/2	1	2	0.033
G	1/8	1/7	1/6	1/5	1/4	1	1	1/3	1/2	1	0.025
H	1/5	1/4	1/3	1/2	1	2	3	1	1	4	0.061
I	1/6	1/5	1/4	1/3	1/2	1	2	1	1	3	0.044
K	1/9	1/8	1/7	1/6	1/5	1/2	1	1/4	1/3	1	0.020
CR: 0,0136 < 0,10											

A: Land Cover, B: Slope, C: Geology, D: Soil, E: Landslide, F: Stream, G: Road, H: Flora/Fauna, I: Protected Area, K: Recreation

Table 4. The matrix of pair-wise comparisons to determine the weights of factors that affect NGTL routing

The quality of the work is determined with the help of evaluation of obtained results in the AHP. Whether decisions support each other or whether they are meaningful can be determined. This work is carried out with CR in the AHP. Acceptable high point's value for CR is 0.10. If CR is higher than 0.10 decision maker has to control his comparisons again. When weights related to data layers in NGTL route plan are examined (Table 1), factors such as land cover, slope, geology, soil and landslides affect the route more than the other factors. Sub-factor weights were calculated for each factor with the matrix of pair wise comparisons (Table 5).

After factors were identified and relevant data layers were created, the weights of these factors and sub-criteria needed to be identified in the second stage. In this study, after criteria, taken into consideration in routing studies by using classical methods, the views of experienced and professional people, legal procedures and practices carried out in developed countries were examined as a whole, the process of determining the weight of factors was started. In this process, the most important processing step is interviews conducted in various institutions and organizations engaged in NGTL. As a result of determining the weights of factors supported with the interview results and literature studies, the weights of factors and sub-criteria affecting NTGL route determination were identified by AHP.

In raster based route determination model, after needed factors were identified and formed properly, also some limitations need to be introduced. These limitations do not take any weight value and define a barrier. In route planning, the lands where the passing is strictly forbidden is defined as absolute barrier, and the lands in which passing is likely despite of its difficulty are also defined as relative barrier. Absolute barriers are determined in sense of the benefits expected from the entire project by users. In generated route planning model, residential zones, landslide areas, wetlands and fault lines were defined as absolute barrier and their weights were exemplified by defining as "∞".

Factors / Sub-Factors	Weights	CR	Factors / Sub-Factors	Weights	CR
Land Cover	**0.263**	**0.0247**	**Slope**	**0.211**	**0.0108**
Forest	0.096		$<10^0$	0.031	
Cultivated Areas (Seasonal Agriculture)	0.043		$10 - 20^0$	0.060	
Agricultural Areas	0.063		$20 - 30^0$	0.081	
Wetland*	0.134	∞	$30 - 40^0$	0.124	
Rocky Areas	0.226		$40 - 50^0$	0.152	
Pasture Areas	0.028		$50 - 60^0$	0.185	
Settlement Areas*	0.411	∞	$>60^0$	0.367	
Geology	**0.162**	**0.0443**	**Stream**	**0.040**	**0.0063**
Acid-Intermediate Intrusives	0.473		River	0.444	
Basic-Ultrabasic Rocks	0.288		Stream	0.053	
Metamorphic Rocks	0.149		Canal	0.262	
Volcanic Rocks	0.054		Brook	0.153	
Sedimentary Rocks	0.036		Creek	0.089	
Protected Area	**0.049**	**0.0290**	**Recreation**	**0.023**	**0.0167**
Level I	0.407		Upland	0.039	
Level II	0.129		Tourism Center	0.262	
Level III	0.079		Historical Monument	0.492	
Urban Protected Areas	0.052		Picnic Areas	0.069	
Historical Protected Areas	0.333		Promenade Areas	0.138	
Soil	**0.130**	**0.0278**	**Road**	**0.030**	**0.0238**
I. Class soils – Excellent Agriculture	0.269		Highway	0.486	
II. Class soils	0.251		Three Lane Road	0.222	
III. Class soils	0.193		Two Lane Road	0.121	
IV. Class soils	0.104		Stabilized Road (two lane)	0.090	
V. Class soils	0.081		Stabilized Road (one lane)	0.044	
VI. Class soils	0.045		Seasonal Road	0.037	
VII. Class soils	0.037		**Landslide**	**0.092**	**0.0334**
VIII. Class soils – Non Agriculture	0.020		Active Landslide Areas*	0.633	∞
Fault Line *	∞		Potential Landslide Areas	0.260	
			Old Landslide Areas	0.106	

* absolute barrier

Table 5. Factor and Sub-factor weights affecting the NGTL route

5. Nabucco and Egypt-Turkey NGTL routing using the model

A study has been carried out to show the applicability of raster based GIS model formed for studies of NGTL route determination. For this purpose, route determining studies have been carried out for Nabucco NGTL and Egypt-Turkey NGTL projects which have made concrete progress towards implementation and for which feasibility studies have been carried out. At first, the possibility of data was examined depending on factors affecting NGTL route. Problems especially in obtaining spatial data existed based on the size of the study area. Using spatial data, source and scale of the data are shown on Table 6. In evaluations that were executed, it was seen that when the most of the factors affecting route were taken into consideration, data that was missing (depending on weight values) did not change the route direction that was identified [31].

Data	Data Type	Data Source	Date	Scale
Elevation	Line	General Command of Mapping	2008	1/100.000
Geology	Polygon	General Directorate of Mineral Research	2008	1/500.000
Fault	Line	General Directorate of Mineral Research	2008	1/500.000
River	Line	General Directorate of Mineral Research	2008	1/100.000
Road	Line	General Directorate of Highway	2008	1/100.000
Railway	Line	Turkish State Railways	2008	1/100.000
Boundaries	Point	General Directorate of Rural Services	2008	1/100.000
Lake	Polygon	General Command of Mapping	2008	1/100.000
Forest	Polygon	General Command of Forestry	2008	1/100.000

Table 6. The data used in route determination of the Nabucco NGTL Project.

In this study, the size of pixel was chosen to be 250 m based on the scale of data. Especially in these kinds of large-scale projects, routes determined by using raster based web analysis are often defined as a corridor and not a line depending on the size of the pixel. The route determined by using 20-30 meters- wideth large pixel size symbolized a corridor. Pipeline construction route of which width is 20-30 meters was able to be determined.

5.1. Generating cost surface map

After data layers were generated, NGTL cost surface map was formed by using the ArcGIS 9.3 software developed by ESRI Company and spatial analysis extension in this software. NGTL route determination query can be made by using generated cost surface map. It is easy to reach construction routes called as absolute route by using sensitive data in a corridor 250m wide.

The weighted cost surface is generated by using pixel-based arithmetic processes on raster data layers formed for each surface separately. Weights needed for each layer are shown in Table 2. The value of pixels on this cost surface describes the total transition cost that belongs to the area on the surface (Figure 3).

P_w : $P_i * W_i$

P_w : weighted layer

P_i : i^{th} data layer
W_i : i^{th} data layer weight

In this implementation, the pixel size was chosen to be 250 m. based on the scale of the data.

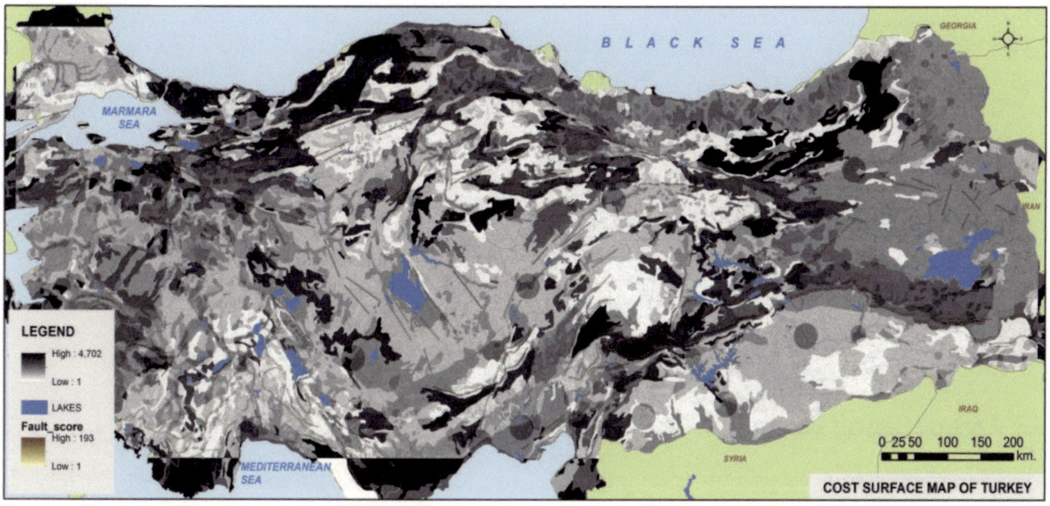

Figure 3. Cost surface map for the NGTL routing

5.2. Nabucco NGTL routing

The length of the pipeline recommended for Nabucco project has been defined as approximately 1584 km. It crosses a total of 11 cities starting from Kars to Canakkale. In terms of the provincial administrative boundary, the longest crossing 332,8 km belongs to Kars city; the shortest crossing 71,9 km belongs to Kirikkale City (Figure 4). Although the route seems like a straight line, it makes deviations in significant distances in some crossing depending on the weight of the factor. For example, in Ankara provincial boundary, it has been seen that NGTL has deviated from line direction to not passing through the dense residential area. Route has made road crossing a total of 432 times including two highways.

The Nabucco Natural Gas Transmission Line Project is a new natural gas pipeline that will begin at the eastern border of Turkey and will connect the Caspian Region and the Middle East via Turkey, Bulgaria, Romania, and Hungary with Austria and further with Central and Western Europe gas markets. The pipeline will be approximately 3300 km long, stretching from the Georgian/Turkish and Iranian/Turkish borders to Baumgarten in Austria. Additional feeder pipelines are possible for Iraqi gas. Based on technical market studies, the pipeline has been designed to transport a maximum amount of 31 bcm per year. The first aim of the project is to supply gas to the countries on the route, and then gas is to be transported to Western Europe according to the wishes of other countries in the following years.

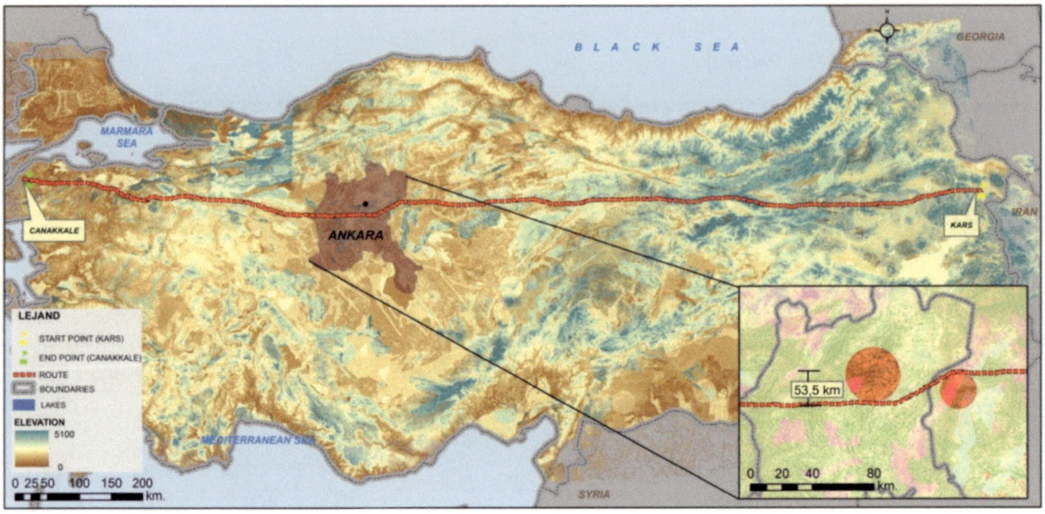

Figure 4. NGTL route on elevation data

5.3. Determining Egypt-Turkey NGTL route

The recommended length of the pipeline for the NGTL project of Egypt-Turkey is approximately stated 387 km. It passes through a total of 4 cities starting from Kilis, a county of Gaziantep to Yusufeli, a county of Sivas. In terms of the provincial administrative boundary, the longest crossing 124,1 belongs to Sivas; the shortest crossing 49 km belongs to Gaziantep. The route crosses 28 streams including a river. Turkey-Egypt NGTL, plans to deliver gas to Europe market besides Turkey, passes over Israel and Syria. The route crossing is shown on figure 5 overlaid over cost surface crossing.

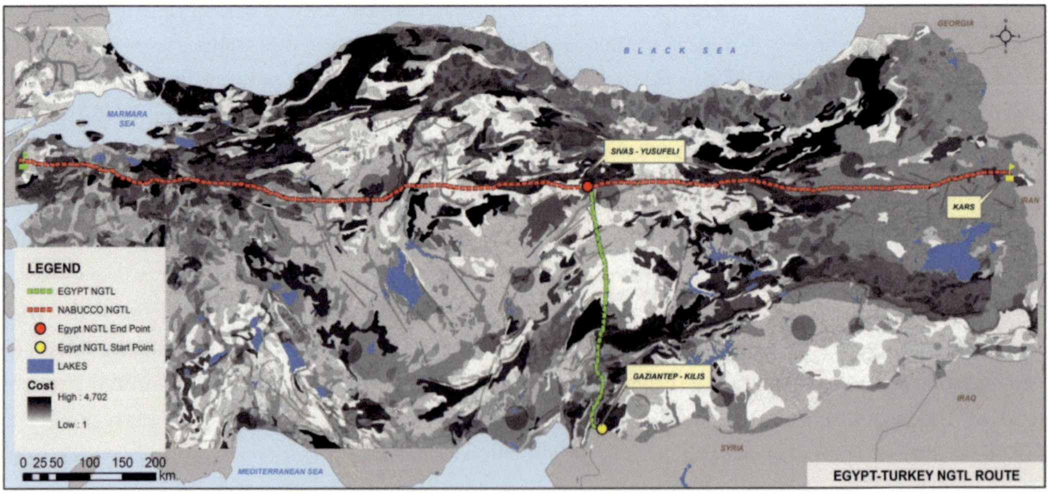

Figure 5. Egypt-Turkey NGTL route on cost surface

6. Results and findings

Some analysis and questioning have been carried out for Nabucco NGTL and Egypt NGTL route to show the efficiency of the route determining model developed within this study. Nabucco NGTL route crosses 226 streams of which 195 are brooks, 3 are rivers, and 9 are streams. It has been seen that Nabucco NGTL has crossed North Anatolia Fault Line in the province border of Erzincan (Figure 6).

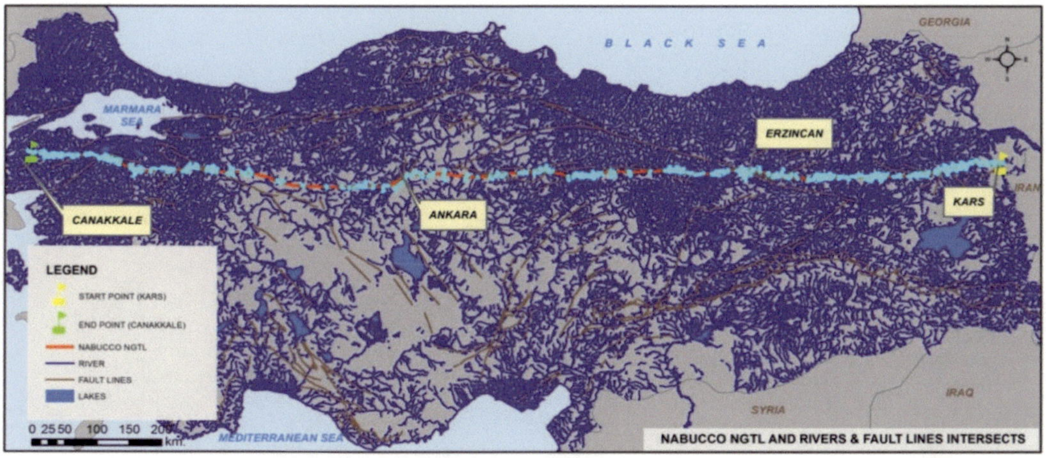

Figure 6. Crossing of rivers and fault lines of Nabucco NGTL

For Nabucco NGTL, the slope is at least 2%, a maximum of 32% and an average of 5%. While the minimum height of the pipeline is 52 meters, the height value can reach 3402 meters in some circumstances. The average height value of the pipeline is 1324 meters. Crossing of the pipeline through the areas of which slope value is more than 25%, defined to be of high slope, is a total of 108 pixels, in other words, approximately 27 km. Additionally, the length of the crossing on the high lands, 2500 m elevation is 580 pixel, i.e., approximately 145 km.

The crossing of Nabucco NGTL on the lithology units is statistically evaluated. In this stage, rocks are placed in four categories. According to this evaluation, the route crosses soft areas over 3636 pixels, which is approximately 909 km, and crosses 970 pixels of hard rock, which is approximately 242,5 km. These results show that Nabucco NGTL route on the pattern does not pass through the areas that increase the cost due to geological reasons.

It is seen that the proposed route for Nabucco NGTL has not passed through the center of dense residential areas like metropolitan, city, county, town. This result shows that accurate obstacle limitation for residential areas gives proper results. Egypt NGTL route crosses 28 streams of which 23 are brooks, 1 is a river and 4 are streams. Additionally, it crosses The South Anatolia Fault Line.

For Egypt-Turkey NGTL, the slope is at least 1%, a maximum of 30% and an average of 4%. While the minimum height of the pipeline is 350 meters, the height value can reach 2400 meters in some circumstances. The average height value of the pipeline is 1280 meters.

Also proposed route for Egypt Nabucco NGTL has not passed through the center of dense residential areas like metropolitan, city, county, town (Figure 7).

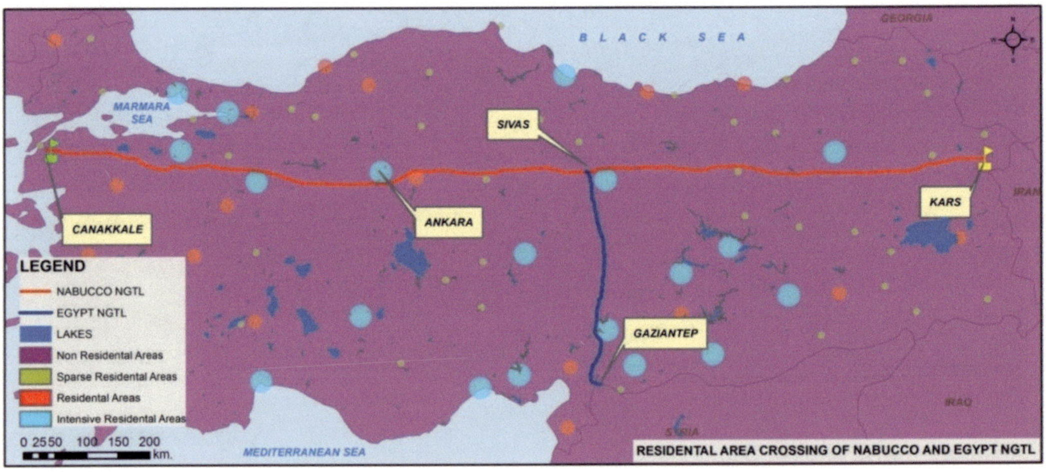

Figure 7. Residental Area Crossing of Nabucco and Egypt-Turkey NGTL

7. Conclusions

The NGTL route identification is complex and requires the analysis of large quantity of data and many parameters depending on the length of the project. GIS is one of the tools to perform this analysis effectively. GIS provides a large number of analytical functions that are capable of replacing manual and traditional methods of natural gas pipeline route planning. It is a powerful tool to integrate thematic layers in an automated environment to compute the shortest possible route with associated costs, which eventually reduce the cost and time of project execution and thus the operating expenses. The integration of GIS and the AHP provides a baseline for complex decision making in which the variant nature of criteria and stakeholders can be accounted for successfully.

Raster-based data models and raster-based network analysis are necessary to determine the surface resistance and to model the NGTL route determination appropriately. One of the basic steps of route determination is to determine the factors that affect the route and their weights. In this stage, AHP presents effective solutions.

The accuracy of the results for this model is directly proportion of the quality of the information used. Especially in Turkey, spatial data it should be generalized with the use of a satellite image with the appropriate resolution. In this study, cost surfaces map that is 250x250 m pixels in size were generated based on available data and the data quality for Turkey. For any NGTL projects, the most appropriate corridor, which is 250 meters wide, can be determined easily with this map.

This model can easily be adapted to determine the necessary factors and calculate the weights for linear engineering structures, such as pipelines, waterlines, roads, channels, railways, and energy transfer lines.

This model is designed for Nabucco and Egypt-Turkey Pipelines Turkey crossing, but the model can be applied universally. In this model surface passage criteria factor weights can be changed and alternative routes can be created. Additionally, same factor weights can be used on same surface characteristics in developed and developing countries.

Author details

V. Yildirim
Karadeniz Technical University, Department of Geomatics Engineering, Trabzon, Turkey

T. Yomralioglu
Istanbul Technical University - Department of Geomatics Engineering, Maslak - Istanbul, Turkey

R. Nisanci
Karadeniz Technical University, Department of Geomatics Engineering, Trabzon, Turkey

8. References

[1] Huang Y, Zhang G, Wang J, Huang J, (2009) Design of Long Distance Pipeline Information Management Based on GIS, International Conference on Information Management, Innovation Management and Industrial Engineering, Available: http://ieeexplore.ieee.org/stamp/stamp.jsp?arnumber=05370526. Accessed Feb 12.

[2] Yang Z S, Jin J H, Yu P S, Chen C Y, Hao Q J, Zhai, Y Z (2010) Environmental hazards and contingency plans along the proposed Chinese–Russian Oil Pipeline route, NE China, Cold Regions Science and Technology. 64,3: 271-278.

[3] Ahmad N (2006) Determining Optimal Path Using GIS, Available: www.gisdevelopment.net. Accessed Feb 10.

[4] Yomralioglu T (2009) Cografi Bilgi Sistemleri Temel Kavramlar ve Uygulamalar. Fifth Edition, Istanbul, Secil Ofset. 479 p.

[5] Rosado R I J, Fernandez J L A, Garcia G E, Zorzano S P (2005) Advanced Model for Expansion of Natural Gas Distribution Networks Based on Geographic Information Systems, Proceedings of the Fifth IASTED International Conference Power and Energy Systems, Available: http://www.actapress.com/Abstract.aspx?paperId=20783. Accessed Jan 24.

[6] Luettinger J, Clark T (2005) Geographic information system-based pipeline route selection process, Journal of Water Resources Planning and Managament, 131,3: 193-200.

[7] Wang D, Wu M, Wang W, Wang F, (2009) The Determination of Optimal Rescue Route Based on Gas and Oil Pipeline GIS, Proceedings of the International Conference on Pipelines and Trenchless Technology, Available: http://ascelibrary.org/proceedings/resource/2/ascecp/361/41073/60_1. Accessed Jan 25.

[8] Chand A, Gloven M, (2009) Using GIS to Support New Pipeline Construction and Material Procurement, Pipeline&Gas Journal, Available:http://www.petroitg.com/CMS.pdf. Accessed Jan 25.

[9] Saha A K, Arora M K, Gupta R P, Virdi M L, Csaplovics E (2005) GIS-Based Route Planning in Landslide-Prone Areas, International Journal of Geographical Information Science, 19,10: 1149-1175.

[10] Yu C, Lee J, Stasiuk M J M (2003) Extensions to Least-Cost Path Algorithms for Roadway Planning, International Journal of Geographic Information Science, 17,4: 361-376.

[11] Baban S M J, Yusof K W, Foster L D I, Ramlal B (2004) Modeling the Optimum Routes for Linking Potential Reservoir Sites to Demand Areas in Mountainous Tropical Islands, Surveying and Land Information Science, 64,3: 183-189.

[12] Cevik E, Topal T (2003) GIS-Based Landslide Susceptibility Mapping for a Problematic Segment of the Natural Gas Pipeline, Hendek (Turkey), Environmental Geology, 44: 949-962.

[13] Gutierrez Z G, Rubio O A (2004) Quantitive Assessment for Selecting the Route for a Gas Pipeline in Yucatan, Mexico, Human and Ecological Risk Assessment, 10,2: 451-460.

[14] Rylsky I A, (2004) Optimization of Pipeline Routes Using GIS-Technologies, Vestnik Moskovskogo Universiteta, Geografiya, 5,4: 34-40.

[15] Malpica J A, Pedraza J (2004) Roads Extraction Through Texture From Aerial And High-Resolution Satellite Images, The International Society for Optical Engineering, 4170: 358-366.

[16] Jafari J, Danehkar A, Khorasani N (2010) Application of Raster Images ,n Computing Environmental Risk Assessment (ERA) of Pipelines, World Applied Sciences Journal, 9,1: 91-100.

[17] Yildirim V (2009) Development of a Raster Based Dynamic Model with Geographical Information System for the Determination of Natural Gas Transmission Pipelines, PhD thesis, Karadeniz Technical University, Trabzon, Turkey.
[18] Orhan A H, Yılmazer I, (2006) Alignment Selection Criteria for Pipelines Essential, Pipeline & Gas Journal, 233:5, 43-44.
[19] Cluff S L, Page A R, Slemmons B D, Crouse B C (2003) Seismic Hazard Exposure for the Trans-Alaska Pipeline, 6th U.S. Conference and Workshop on Lifeline Earthquake Engineering, California, USA.
[20] Dey P K, Gupta S S (2001) Feasibility Analysis of Cross-Country Petroleum Pipeline Projects: A Quantiative Approach, Project Management Institute, Project Management Journal, 32:4, 50-58.
[21] Rowland A, (2005) GIS route selection analysis for Oil and Gas exploration using Multi-Criteria Decision Analysis- Case Study of the Oil and Gas Producing area of the Niger Delta in Nigeria, MSc Thesis, University of Greenwich, London.
[22] Dey P K (2001) Integrated approach to Project feasibility analysis, Impact Assessement and Project Appraisal, 19:3, 235-245.
[23] Anavberokhai I O (2008) Introducing GIS and Multi-Criteria analysis in road pah planning process in Nigeria, MsC Thesis, University of Gavle, Department of Technology and Built Environment.
[24] Berry K J, King D M, Lopez C A (2004) Web-Based Application for Identifying and Evaluating Alternative Pipeline Routes and corridors, Available: www.innovativegis.com/basis/present/gita. Accessed Jan 14.
[25] Yomralioglu T, Reis S, Nisanci R (2002) GPS ile Hareket Halindeki Araçlardan Elde Edilen Gerçek Zamanlı Verilerin Orta Ölçekli CBS Çalışmalarında Kullanılabilirliği, Selçuk Üniversitesi Jeodezi ve Fotogrametri Mühendisliği Öğretiminde 30. Yıl Sempozyumu, Selçuk Üniversitesi, Konya, 107-115.
[26] Vincent C T, Yong H (2002) Assessment of Airborne Lidar and Imaging Technology for Pipeline and Safety Applications, Pecora 15/Land Satellite Information IV/ISPRS Commission I/FIEOS Conference Proceedings, Canada.
[27] Hall S, Nelson A, Meehan D, Mukherjee J (2005) Disseminating Critical Pipeline Infrastructure Data With GIS, Esri User Conference, USA.
[28] Hausamann D, Zirnig W, Schreier G, Strobl P (2005) Monitoring of Gas Pipelines – A Civil UAV Application, Aircraft Engineering and Aerospace Technology: An International Journal, 77:5, 352-360.
[29] Reis S (2003) Çevresel Planlamalara Altlık Bir Coğrafi Bilgi Sistemi Tasarımı ve Uygulaması: Trabzon İl Bilgi Sistemi (TİBİS) Modeli, Doktora Tezi, KTÜ Fen Bilimleri Enstitüsü, Trabzon.
[30] Yalcin A (2005) Ardeşen (Rize) Yöresinin Heyelan Duyarlılığı Açısından İrdelenmesi, Doktora Tezi, KTÜ Fen Bilimleri Enstitüsü, Trabzon.

[31] Yildirim V, Nisanci R, Yomralioglu T, Uzun B (2008) Raster-based GIS data guide economic pipeline construction, Oil & Gas Journal, 106:13, 62-68.

[32] Yildirim V, Yomralioglu T (2007) GIS Based Pipeline Route Selection by ArcGIS, ESRI Users Group Conference, Accessed: http://proceedings.esri.com/library/userconf/proc07/papers/papers/pap_2015.pdf USA. Available Jan10.

Chapter 6

Natural Gas Virtual-Pipeline for Alternative Energy Distribution

Miguel Edgar Morales Udaeta, Jonathas Luiz de Oliveira Bernal,
Luiz Claudio Ribeiro Galvão and José Aquiles Baesso Grimoni

Additional information is available at the end of the chapter

http://dx.doi.org/10.5772/48711

1. Introduction

Virtual-pipeline is an alternative method of transporting natural gas to places where there are no pipeline networks available. It is based on a modular system of compression or liquefaction, transport and decompression and/or regasification of natural gas, which communities, industries, gas stations and others can use. In virtual-pipelines compressed/liquefied natural gas is transported by various modes - highway, railway, waterways and the sea.

For efficient use the end users –residential users, the gas stations, and industrial users - should be located near the pipeline; otherwise, extension branches would be cost-prohibitive. Even users with high potential for consumption could not be using this fuel on account of their distance from existing pipelines. This was found to be a great problem by gas distributors. Trying to find a solution some companies have developed a new technological option: natural gas supplied through virtual-pipelines, which has the great merit of bringing the natural gas to regions not served by conventional pipelines.

The virtual-pipeline is a system that allows the natural gas transportation in the form of compressed/ liquefied gas using modules coupled to mobile platforms, which are transported by trucks, ferry boats, boats and/or rail platforms. When the product reaches its destination, the module is connected to a decompression/regasification station for ready consumption. The great advantage is that the amount of natural gas transported varies according to the needs previously determined by the customers. In short, versatile systems are supported by natural gas compression/liquifaction modules, regulation stations and the transportation system itself.

For each region a study is performed to determine particular dimensions, where the amount of gas to be consumed is previously determined, e.g., from probabilistic analyses of variables such as: consumption, distances, compliance with the transport, environment and security norms, climate condition, in addition to the technical and economic viability.

Thus, the Virtual-Pipelines and their specific features, such as type and storage capacity, means of transportation, compression and/or liquefying, loading, unloading and supplying, offer the remote customer the possibility of using the natural gas long before a conventional pipeline is constructed. This in turn contributes tremendously to the region's economic development.

In places where there is no natural gas, the industry is limited to work or develop products and services using the energy sources available in the region, and does not keep up with new technologies that can be more economical, with lesser emission levels and greater quality and reliability. The entrepreneurs can - having the possibility to use natural gas in the local market of any region, which is a cleaner energy source compared to the oil byproducts – develop more competitive goods and services in relation to internal and external competitors, thus stimulating their penetration into new markets.

The Virtual-Pipeline generates the flexibility for a more appropriate industrial location, and the possibility to have better utilization of geographical resources resulting in better cost/benefit relation. Natural gas facilitates the region to dispose goods that would not normally be developed in the region. This new technology also allows increasing number of GNV (vehicular natural gas) service stations, with a relatively smaller investment, emphasizing that this technology is also employed to other productive and service sectors, of which scattering demand disables the supplying through ducts (condos, commercial centers and small industries). This way, the users of vehicles provided with GNV have more flexibility to drive through the country side of the country, with natural gas available even in places where there are no conventional pipelines.

The utilization of the virtual-pipeline must not be treated as a steady and definite alternative, but as a faster way to take the natural gas to places where there is no technical or economical viability for the conventional pipeline. Virtual-Pipeline is a passport to new frontiers, consolidating the natural gas consumption and prepares the region for the future utilization of conventional gas supply. Given this case, the system of Virtual-Pipeline can be transferred to a new region that needs to be developed.

Due to increasing Green house gas concerns, the use of natural gas is encouraged. Additionally, with recent advances natural gas extraction from shale gas has become extremely popular. As sources of NG remain remote from locations of use viable technologies remain: Compressed Natural Gas (CNG), Liquefied Natural Gas (LNG), Gas-to-Liquid (GTL) and Gas-to-Wire (GTW). These four technologies promote one way or another, the foundations of a virtual-pipeline. In this chapter, emphasis will be on the first two, and the third is a complementary one.

The last technology, GTW, is not discussed at length in this paper. The GTW technology consists of the production of electrical energy using turbines of open or combined cycles, close to the place of gas production, and the energy so produced is transmitted to the consumer market through transmission cables.

The GTL (Gas-To-Liquids) technology on the other hand, comprises basically of conversion of natural gas into synthesis gas through a reformation reaction using steam (auto-thermal or partial oxidation) and subsequent conversion of the synthesis gas (syngas) into synthetic fuel through a Fischer-Tropsch reaction. The synthetic fuel so produced can be viewed as efficient use of natural gas itself, and further can displace oil consumption. Even considering the high costs for the construction of the plant, the value of the utilization of natural gas in the absence of pipeline network becomes attractive considering the low-cost of NG as compared to oil.

The benefit of synthetic fuel from GTL is somewhat similar to that from LNG. The LNG, through the cooling results in volume reduction for NG, facilitates its transportation by tank-ships, and subsequently, allows it to be regasified and distributed through pipelines to its final destination. The GTL, in comparison, transforms the NG into a synthetic fuel, which will be supplied as a liquid and used to fuel cars, buses and, possibly, jet engines. However, it is necessary to know that the capital investment for the construction and installation of a GTL plant are extremely high. It is estimated that U$ 2,5 bi is necessary to generate only 100 thousand barrels per day; and the thermal efficiency of a GTL plant can only reach 60% (i.e., lots of energy is lost along the process before the fuel can reach the end user).

It is important to highlight that the largest consumer of natural gas in the world, the United States, by 2012, has reverted its position regarding the use of new technologies for unconventional NG and has started accelerated exploitation of shale gas. Also, it is said that the brute production of NG in the EEUU is the biggest one in all times, and it has reached 2.500 billion cubic feet on October 2011. It is further estimated that this value will keep growing, mainly due to the abundance of this unconventional gas (shale gas). And also due to the fact that in the case of the shale gas, the oil tanker knows exactly where the shale is the gas is derived economically.

This paper primarily discusses the natural gas virtual- pipeline as alternative to distribution of energy in the form of CNG and/or LNG. Additionally, though to a lesser extent, the production of synthetic liquid fuel is discussed as an alternative.

2. Virtual-pipeline

The concept of virtual-pipeline aims to distribute the natural gas to places where the physical or economic conditions deem the installation of a real pipeline unfeasible.

The so called Virtual-Pipeline consists of the transportation of the Compressed Natural Gas by pressure or liquefaction in the natural form at low temperatures. Once the mean or state of the natural gas is defined (compressed or liquid) for the achievement of the virtual-pipeline, it will be transported into modular containers to the unloading plants. This allows the gas to reach the consumer centers in an economically feasible way, competitive or not, in general far removed from the physical pipeline network. It is important to highlight that the economic coefficient is relative, being necessary to consider several logistical factors, such as effective cost of the diesel transported to these remote places.

Natural gas may be distributed in three forms: compressed natural gas – CNG (Natural Gas in its physical compressed state); liquefied natural gas – LNG (Natural Gas in its physical Liquefied state); and, the natural gas hydrates –GNH (Natural Gas in its physical "solid" state). Where the CNG is more commercial and known, it is normally used as direct fuel to internal combustion engines or for heating of processes in industries. The LNG is more expensive and multi-faceted technologically, it is also considered as a vector of distribution to great distances. LNG often proves economically feasible for maritime transportation to great distances. Although, the NGH, i.e., Natural Gas Hydrates, are technologically more complex, they are already past the testing phase, and a few plants are already under construction on industrial scale.

2.1. Relevant aspects of the distribution via virtual-pipeline

The virtual-pipeline, as already defined, can be identified as the natural gas distribution through containers, independently of the natural gas physical state. In the event of the virtual-pipeline the compressed state is the most dominant one, i.e., the CNG. In this sense, the CNG is the natural gas processed and conditioned for transport, commonly, e.g., into ampoules and cylinders, at environmental temperature, pressure 70 to 250 bar, with a volume reduction from 60 to 225 times the volume occupied under standard conditions.

The virtual-pipeline allows meeting the demand in regions where there is no infrastructure of gas pipelines distribution nets. It also aims to encourage the development of new markets for the natural gas, once it is highly flexible. It must be said that the transportation of CNG is complex and it involves risks of accidents due to the high pressure operation. Basically, the natural gas is taken out first, next the gas is compressed, and later the gas is transported to the purchasing point.

As time passed, the demand for gas has increased in all regions, however, its supply through conventional pipelines has not kept up with those needs. As a result a Compressed Natural Gas (CNG) system that can meet the cities' needs for electricity generators, automotive sector, industries, industrial utilization, etc., is very much the need of the hour.

In Russia, for instance, there are large swaths of populations that do not have access to the gas distribution network. As a result, they developed LNG (Liquefied Natural Gas) technology as an alternative transport which allows them to provide these places with natural gas by liquefying the gas, transporting it into trucks or vessels of convoy and subsequently regasifying it for its distribution, obtaining great results for system efficiency.

Nowadays there are enterprises working with the CNG or LNG technology. In South America, for instance, companies specialized at working with modular CNG achieve a distribution of more than 4,500,000 m^3/month to cities in Argentina. Such companies are intending to enter the Brazilian market together with local distributors in order to supply nearly 600,000 m^3/month of natural gas.

There are other technological options available in the South American natural gas market as an alternative of transport: the cryogenic system. The gas is liquefied at -162°C and stored into special tanks for its transportation to the consumer place. The cryogenic alternative of LNG is practical and great volumes of gas converted into liquid can be transported.

3. Technological aspects of the virtual-pipeline

During the technical execution of a virtual-pipeline, the CNG can be nominated gas into transitory state, due to the volume reduction and storage into pressure vessels. The supply of CNG in bulk consists of acquisition, receiving, compression, storage, delivery, commercialization and quality control.

The virtual-pipeline system is based on modular technology, which in turn allows scaling up of the system proportional to the demand. The Virtual-Pipeline described here (see fig. 1), is based on 3 technological developments, such as:

- Modular stations of CNG compression.
- Modular plants of pressure regulation.
- Modular system of Natural Gas storage and transport.

Figure 1. Technology developed in Argentina

First, a compressor is installed. Later, the CNG loading platforms are set at the place in a dispositive named STM - Storage and Transport Module – Tanks measuring 1500 m^3; 2.2m x 3.6m x 2.6m. The STM are carried by trucks and transported by land to the unload platforms. Together with the unload platforms a controller plant which will reduce the outlet pressure (200 bar) to the pressure at the distribution net (4 bar).

In figure 2, the process of Loading of Galileo Technology is shown.

Following is the process of loading and unloading:

Head Station: A MICROBOX/MICROSKID (a complete and compact system of intrinsically safe compression), connected to a natural gas source, it can be an existing pipeline, compresses the gas within the STM modules of transport. These are on platforms, called PA-C, which allow the fulfillment and interchange between the module and the Transport System (step 1, see figure 3).

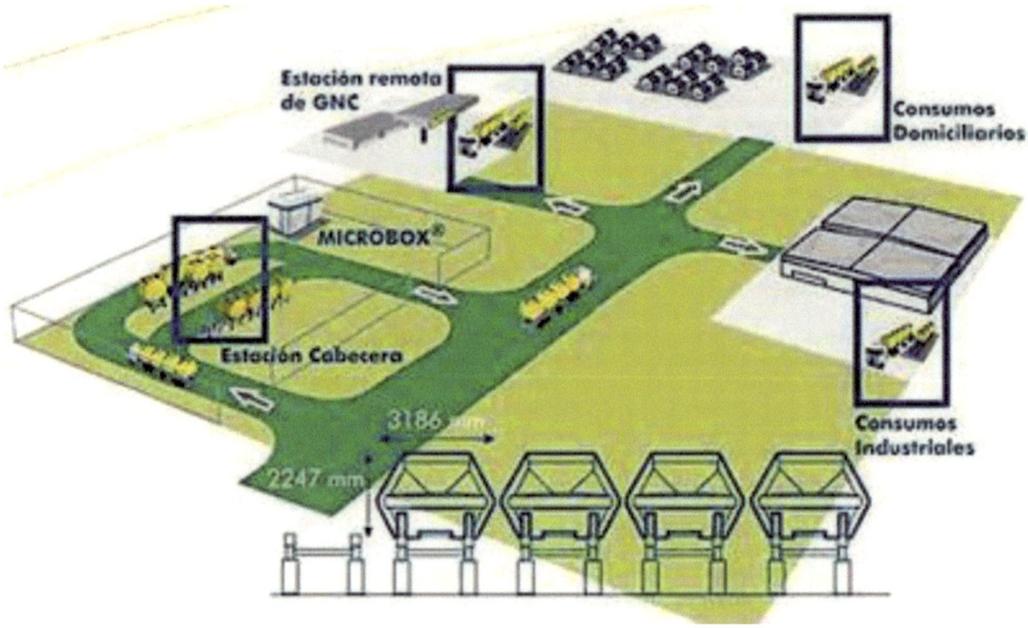

Figure 2. Loading System for the Galileo Technology in Argentina

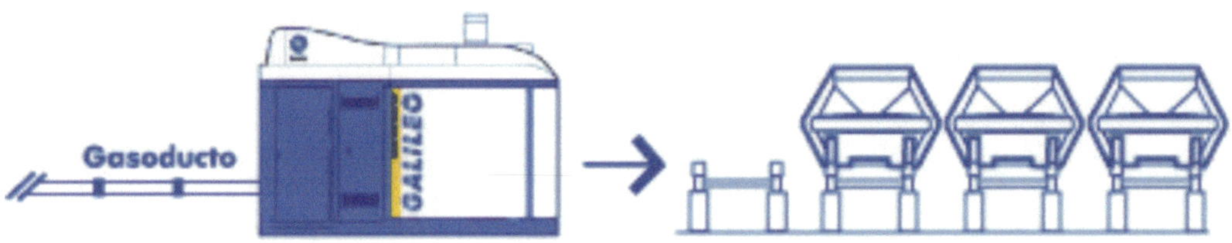

Figure 3. Step 1

Loading and Unloading: At its arrival, the transport trailer interchanges the empty STM from the consumer point by the filled ones. This interchange is executed by means of TS machines (CNG Transport System) that are on the trailer (step 2, see figure 4)

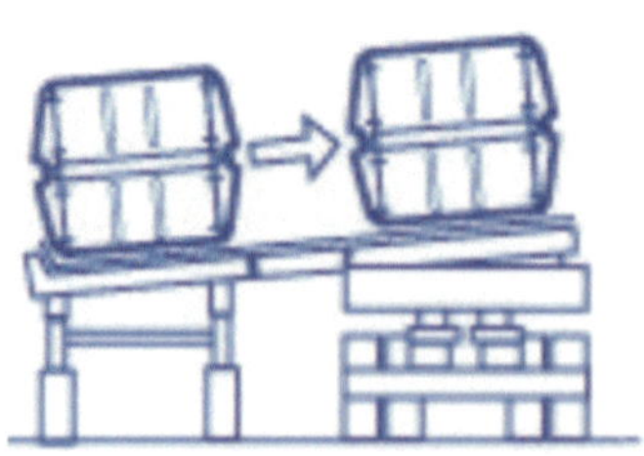

Figure 4. Step 2

Transportation: The vehicle transports the STM modules by land at the same speed as other loading transports, with no need for additional security system. The STM modules are linked to the trailer by an anchorage system. (step 3, see figure 5).

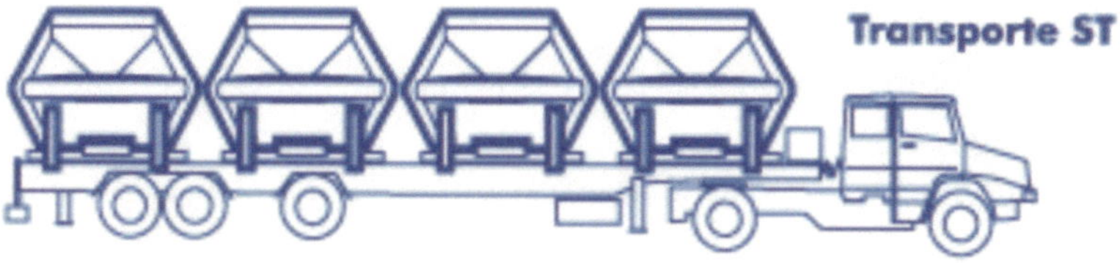

Figure 5. Step 3

When the consumer points are reached, the STM modules are unloaded on platforms, called PA-D (Plataforma de Abastecimento-Descarregamento) [SU-P Supplying – Unloading Platform]. In a previous step by a regulation plant, the STM are connected to the residential supplying net (step 4, see figure 6).

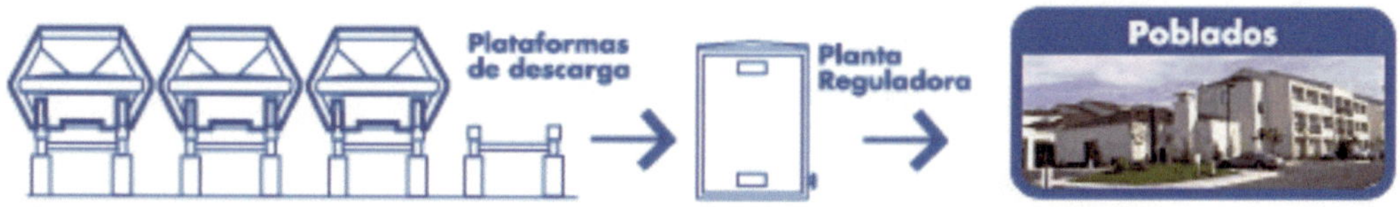

Figure 6. Step 4

At industrial applications, depending on the kind of industry, the platforms are connected to regulation plants. These have a catalytic heating system and the capacity to produce 10,000 m3/hour (step 5, industrial supplying, see figure 7).

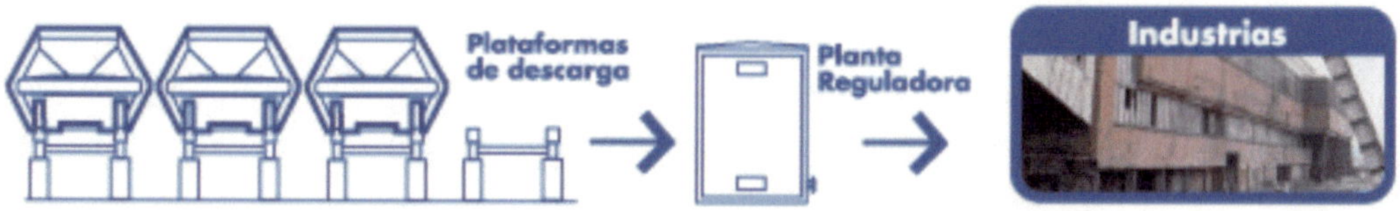

Figure 7. Step 5

Remote stations of CNG: The STM modules are connected to the filling station through an unload BOOSTER (small compressor and dispenser designated to recompression and supplying, respectively) and are managed by the Rotating Cascade (step 6, see figure 8).

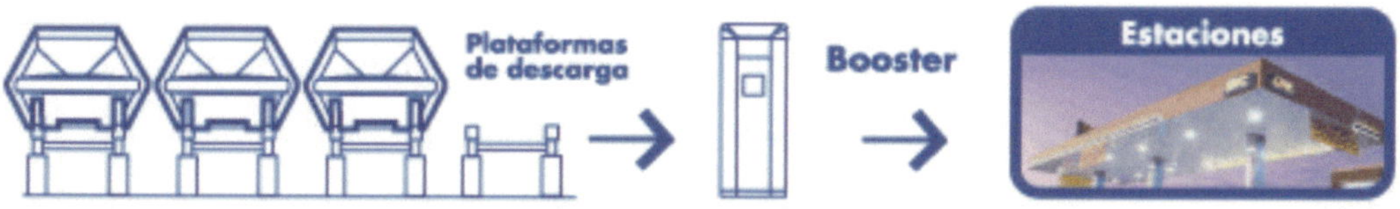

Figure 8. Step 6.

3.1. Cryogen technology

This system consists of a gas caption station at a certain point, mechanisms for gas compression, injection into special tubes disposed on trucks, and decompression stations at the consumer end. The system developed by Cryogen makes it possible to use the dragging of the tubes as a supplying source for direct consumption, i.e., the supplying source in the industrial line.

To avoid the interruption of supplying the empty trailer is substituted by the filled one and taken to the caption point for its reloading. The most important limitation consists of the distance between the caption point and the consuming point. Longer distances with more than 200 km make the alternative difficult because of cost. Distances between 150 and 200 km are considered ideal.

The Cryogen system presents two alternatives for the utilization of CNG: the first one related to the trucks parked at the consumer point that decompress the gas with an independent energy equipment, using the environmental conditions to decompress; the second one uses compressor to keep the necessary pressure at the filling of the vehicular cylinders.

The first system – compression, transportation, decompression – for industrial consumption, includes the setting up of the compression system at 250 bar for the filling of the transport unit, the decompression system for industrial use between 2 and 30 bar. The standard Cryogen system consists of a transport system equipped with 150 cylinders of 125 liters capacity each, summing up 5.155 m3 of NG each journey.

The second system – compression, transportation, decompression, compression – is used directly for the automotive sector. This system is composed by a valve at the caption point, a compression system at 250 bar, filling mechanism, equipment for loading and unloading of cylinders and decompression and compression systems for the vehicular natural gas (VNG) supplying stations. Figure 8 shows the manner in which the modules of Criogen technology are adapted to the trucks used for transportation.

4. Technological aspects of LNG

The LNG (Liquefied Natural Gas) is liquefied by the reduction of its temperature to -162 ºC and regular atmospheric pressure and it occupies around 1/600 the volume of NG in gaseous state.

At first, the LNG relative to virtual-pipeline presents basic distribution conditions similar to the CNG, however, distances, volume and financial-economic analysis are different.

Technologically, in the case of its final use in transportation vehicles, the LNG is stored at low pressure into thermally isolated tanks and with capacities that can vary from 174 to 511 liters. The isolation of the reservoirs, even though very effective, cannot keep the temperature low.

The LNG is stored as a cryogenic product, i.e., in liquid state at its evaporation temperature of -162°C. However, due to thermal losses into the storage tank, there is usually some boil off vapor that must be vented to maintain constant pressure.

The LNG is transported in towing trucks that carry more than 40 thousand liters, into small cistern and towing trucks, railway wagons and methane ships for LNG with capacity reaching 114 million liters. The towing trucks for LNG are often used to resupply the LNG stations, such as the delivery of gasoil or gasoline.

The system consists of deposits of LNG (liquefied natural gas) at the extremities of the "line", and a series of vehicles making the transportation of it. The same way as gasoline transportation is performed between the supplying stations.

Nevertheless, the components are a little more complex considering that the LNG must be kept at low temperatures in order to avoid its evaporation. The containers must have a double covering layer and be isolated by an insulating material developed by NASA, called super isolating "SI". Just one inch SI is more efficient than 30 inch sponge or glass fiber. Besides, the stationary containers have SI, isolating dust and "vacuum".

The transferring equipment between containers consists of a centrifugal pump that can be submersed or be apart between the terminals. Separated pumps usually need a long time of cooling to reach operable temperatures.

The transportation tanks are available in several sizes, normally projected to minimize the cost per weight, respecting the local laws of maximum weight and conditions of the highways. This means aluminum tanks with no transferring pumps. In the United States, for example, a vehicle used for the LNG transportation has its total capacity of 21,500 kg. It is important to emphasize that this technology means great investment costs.

Small vehicles are used for remote areas, high pressures and pumps on board, e.g., in the United States special tanks are used to fuel 400 liters vehicles in 5 minutes or less.

There are also LNG wagons, but the supplying stations are limited to places with railway access.

In the event of the virtual-pipeline, now, the considerate and known technologies are those that allow the delivery of natural gas in both dominant physical states of the natural gas: liquefied and compressed. In this sense the supplying stations for both, LNG and LCNG where, certainly, the LCNG is the compressed natural gas (CNG) produced from the liquefied natural gas (LNG). For those regions distant from the transportation and distribution system of natural gas, the installation for this kind of stations is advantageous. The LNG, for example, can reach the supplying station by train or highway, in tank-trucks. Finally, the components of a station or substation of LNG and CNG (i.e., LNG and LCNG), are basically:

- Cryogenic system: it concerns a tank for LNG storage (normally, with a capacity between 50 to 100 thousand liters);

- Transferring system: it is the tubage (it can be isolated by a vacuum system), the pump, the heating changer, the supplier and the control panel.

Thus, a LCNG station (GNLC in Portuguese), compared to the LNG station has one more pump, a vaporizer and a supplier. In the event of the cryogenic and compressed system, the LNG at cryogenic temperature (-162ºC) and low pressure is converted into CNG, having more energy efficiency (less energy consumption) than the compression station. Although, the required power by the system pump/vaporizer is 1/10 to 1/20 of a traditional compression system.

It is interesting to mention that for determined transport vehicles, when the necessary autonomy is not reached with the CNG, the use of LNG has its advantages. That is why a LCNG station can fuel both LNG and CNG vehicles, as a conventional supplying station. The time of supplying with LNG is relatively equivalent to that of regular fuel. This is the reason why the fuel system, for a vehicle provided with LNG has such features: a reservoir; a vaporizer where the secondary fluid is the engine cooling liquid; an indicator of the level at the reservoir; and, to provide the engine with LNG at gaseous form.

5. Synthetic fuel from natural gas

During the World Wars, the natural resources became scarce, leading to a lack of liquid fuel and ammonia base required for explosive production. This resulted in a search for artificial processes to obtain from the well known synthesis gas (syngas), primarily derived from charcoal. More specifically, during the 1920 decade, in the last Century, the production of liquid hydrocarbons has been initiated from the synthesis gas using iron based catalysts developed by the scientists Franz Fischer and Hans Tropsch.

The Fischer-Tropsch method – FT converts the synthesis gas, compound by hydrogen and carbon monoxide into a wide range of hydrocarbons. The synthesis gas can be produced by a great variety of sources: from sources derived from oil, such as the natural gas, to sources as biomass or even residential organic garbage. Through the utilization of an Iron-Cobalt catalyst, combinations of these liquid hydrocarbons of great commercial use can be obtained, such as: diesel, naphtha, Dimethylether, methanol and others.

Some time ago, the investments in this technology did not stabilize due to the volatile cost of oil in the last years, causing a restriction in the supply for other priority markets of natural gas.

Another process to obtain liquid fuel from the gas is known as Davy Process Technology, that consists the production of a pumping product and sulphur-free from the natural gas. Because it is a compact technology, it is quite indicated for off shore installations of NG extraction.

5.1. Diesel from natural gas

The product of GTL is a liquid, somewhat similar to that of diesel, and further eliminates sulphur completely. As a result, GTL has the potential to serve as a replacement for diesel, and thereby can exploit an existing market.

The use of GTL derived fuel worldwide, mainly in the developed countries, would help to improve the air quality in the cities due to the absence of sulphur. Still, concerning a much more articulated position, it can be used as precursor for efficient use of energy and reducing emissions, while meeting the fuel property requirements for the transport sector.

GTL diesel, in addition to not having sulphur, does not have long chain aromatic compounds. Features that turn this fuel into an ideal mixture for intermediary distillated market, very attractive to be used as product for fuel with very low emissions, having cleaner and ecological characteristics. Also, use of GTL derived diesel bolsters a country's energy security through diversification of fuel sources.

5.2. Methanol from natural gas

Also known as Methyl Alcohol, Wood Alcohol, Methyl Carbinol, Colonial Spirit, Wood Spirit, Methyl Hydroxyl, Methyl Hydrate (CH3OH), it is produced through a three step process:

- Syngas, or Synthesis Gas production
- Syngas conversion to brute methane
- Purification through distillation

There are six companies that own the patent over this process: Haldor Topsoe, M.W. Kellog, Lurgi, Mitsubishi Gas Chemicals, Synetix and Krupp Udhe. Synetix (ICI) that is majority in the market followed by Lurgi and Mitsubishi Gas Chemicals (AL-Shalchi, 2006).

In 1999 the global demand for methane was 27 million ton and in 2009 was almost 53 million of tons.

5.3. Dimethyl ether

Also known as dimethylic ether, methylethylic ether, methoxymethane or DME, it is the simplest of ethers presenting its chemical formula (CH3OCH3). Its physical-chemical properties are similar to butane and propane, facilitating both storage and handling.

The current production technology is the methane dehydration, with new ways of productive processes being developed, mainly used as an aerosol propellant.

5.4. Other sub products from GTL

The FT process produces a great range of hydrocarbons, from methane to molecules with more than 100 carbon atoms. In spite of commercial processes tries to break the great molecules in a search of bigger proportions of diesel aiming at greater market for its products, there is the possibility to produce the specialty products, such as:

- Linear normal paraffin (especially used in surfactants)
- Intermediate and combined paraffins (solvents)
- Wax

Given the specialty products, such as the ones mentioned above offer attractive profit margins related to fuel products, the possibility of producing polymers from its base becomes uninteresting, as it would affect the price balance and the GTL conversion in large scale.

6. Geo-energetic analysis for a virtual-pipeline in Bolivia

The Bolivian Government has made it a national priority to provide all the regions in the country with natural gas and to change the energy matrix based on natural gas. The Hydrocarbons Ministry has been analyzing better natural gas supplying options to all regions, through both pipelines or virtual-pipelines. It is important to emphasize that in Bolivian territory there are large natural gas reserves, however the productive places currently find themselves geographically distant from the north of the country (Bolivian Amazonian region).

One of the greatest challenges to Bolivia is to change its energy matrix on the North of the country, which is geographically distant from the natural gas producing regions. The virtual-pipeline was evaluated to supply natural gas as an alternative to diesel that was used for power generation, and also to supply natural gas for other uses.

The incorporation of natural gas into the energy matrix of these regions would not only contribute to a cheaper energy supply, but it could also stimulate its use in the vehicular, commercial and residential sectors, improving life quality of its inhabitants and generating a cleaner local development.

It is observed then that there is a great need to supply natural gas to the North of the country. There is the VP (Virtual-Pipeline) technology to transport gas in an unconventional way, but an important aspect for the implementation of virtual pipeline is the analysis of access routes to the local consumer relevance, such as the cities of Guayaramerín and Riberalta. In both cases there is the issue of land access, which during the rainy season is practically impossible. Basically the 4 months, on average, the rainy season lasts, it is not possible to transport by truck, since the unpaved roads become impassable in the rain. The river is the only means of access, and therefore the virtual - pipeline implementation must be based in these areas after assessing the feasibility of transporting the gas in tanker-trucks from the river from Cochabamba to Guayaramerín. Geographically Cochabamba is at the center in Bolivia and Guayaramerím to the North right on the border with the Brazilian city of Guajará Mirim (see Figure 9).

6.1. Bimodal corridor aiming at virtual-pipeline

It all starts as the thesis of the use of the Bolivian territory to bi-oceanic interconnection (Pacific - Atlantic or vice versa). Such facility offers a range of alternatives to reach distant markets, primarily the United States and Asian countries. These alternatives are not competing with each other, but are complementary, tending to maximize the benefits.

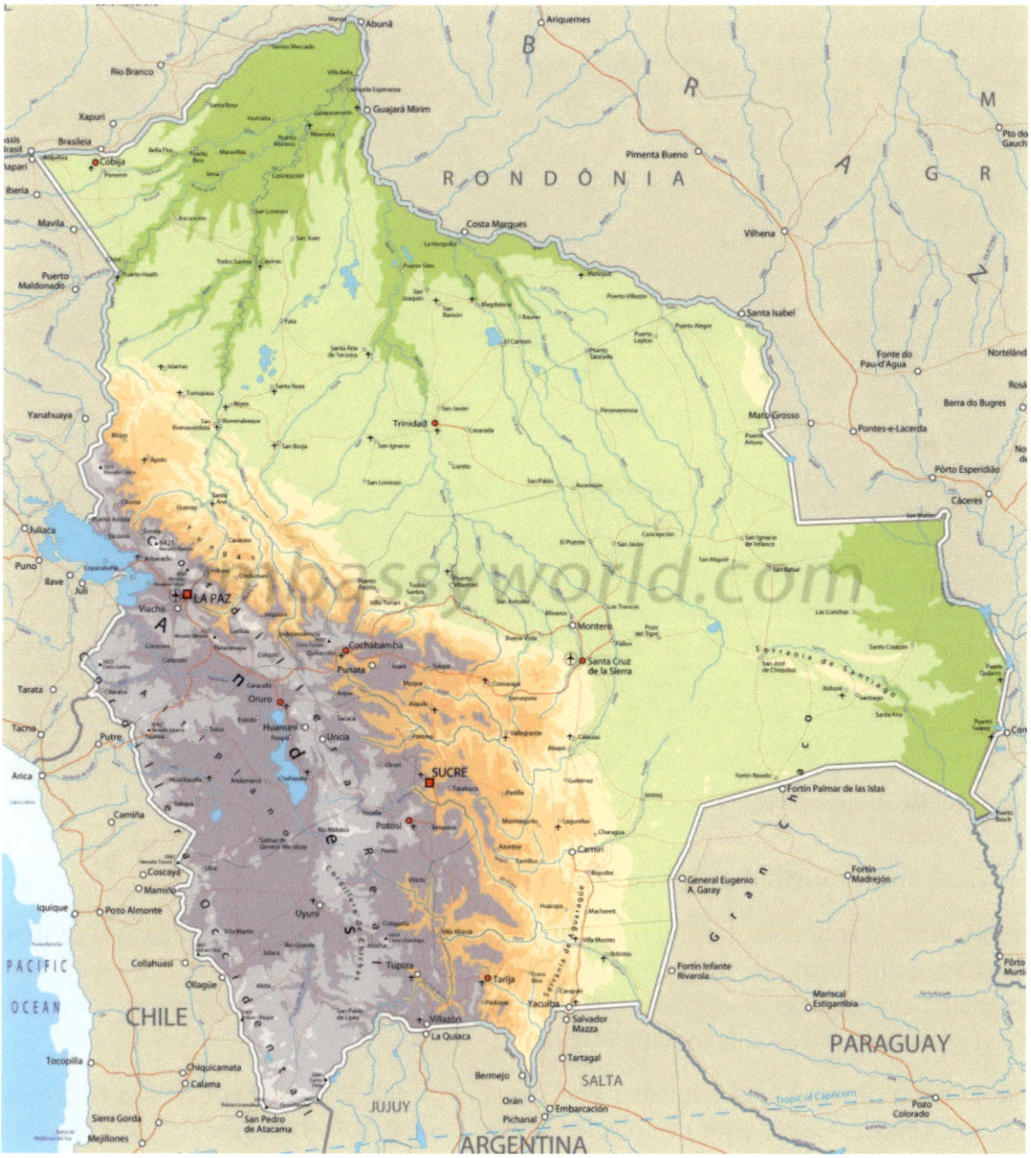

Figure 9. Bolivia Political Map

The alternative of an export corridor allows communicating with the Pacific Ocean, in the case of Brazil, the central area of Rondônia State, part of the State of Acre, and part of the State of Mato Grosso. In the central point and place of departure is the city of Porto Velho. The corridor would allow Bolivia to get to America on the Pacific coast(see figure 10).

Figure 10. Geographical map of Bolivia from Pacific to Brasil

6.2. Case study: Virtual pipeline in Bolivia

In this item, it is not the intention to parse or analyze scientific basis of the VP (virtual-pipeline), but search, in the context of scientific research in engineering, establishing pragmatic benchmarks of applicability in Bolivia for more remote regions such as north Bolivia (Amazon region shared with Brazil).

Bolivia has a physical environment varied (see Figure 11). It is located between the Andes and the Amazon Basin, with the altitude points around the 7,000 meters of sea level, for example, 6542 in the Nevado Sajama (the Andes) and negative altitudes in the Chaco region, but stand out formally 90 meters of sea level in the Rio Paraguay(Paraguay River).

Bolivia is then divided into three geographic regions: the Andean Region, formed by the Bolivian Altiplano and by the Andes, with altitudes of over 4000 meters, the Andean Sub-Region, consisting of valleys and mountain rainforests of the Yungas, with a average altitude of 2000 meters of sea level., and the Plains Region, composed of subregions: Amazon, Platense and the Gran Chaco, with an average elevation of 400 meters of sea level.

It is crossed from north to south by the Andes (Figure 11), divided into three systems: the Cordillera Occidental, the Cordillera Central and Cordillera Oriental. For this reason it can be deduced that the construction of pipelines across the Bolivian territory is little more than improbable, since they would cross, virgin forests, rivers, mountains and snowy areas. As a result one needs to consider other ways of supplying more remote regions of the capitals of each Department (equivalent to State in the Brazilian case).

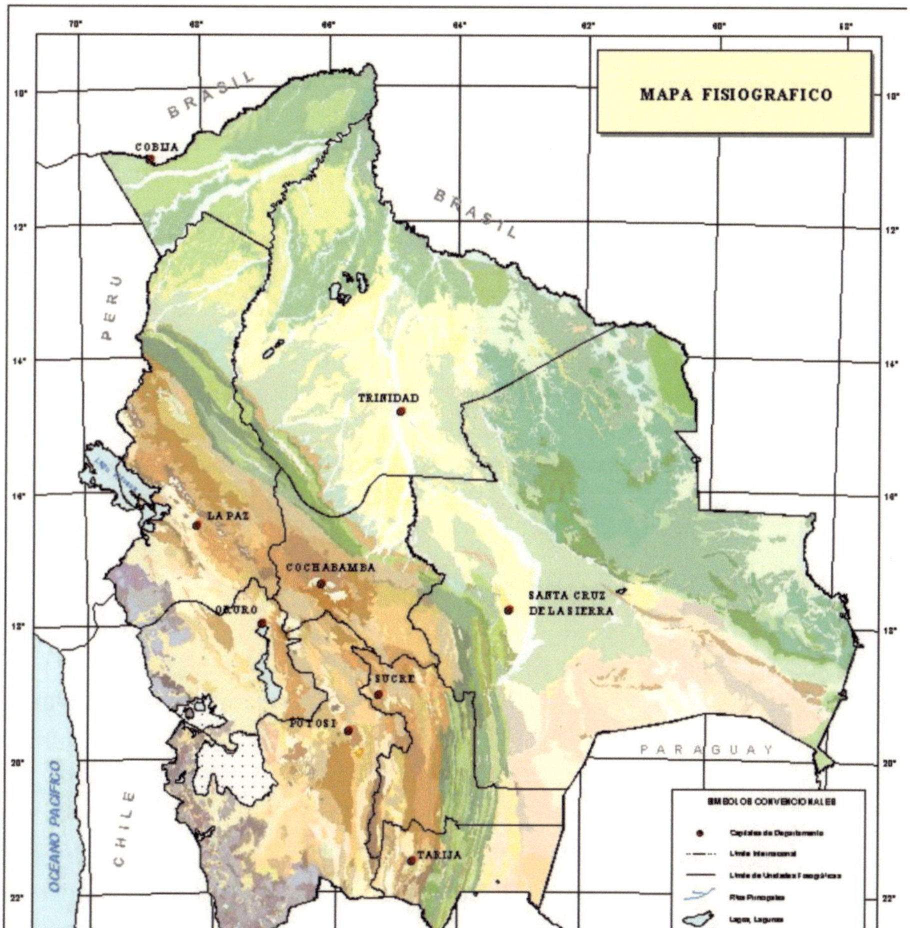

Figure 11. Topographic map of Bolivia.

An alternative to this problem is the so-called Virtual-Pipeline which consists in transporting the Compressed Natural Gas (CNG) under pressure or liquefied at low temperatures, transported in containers to the modular plant unloading, which will allow gas to reach the people away from the pipeline network cost-effectively. It must be emphasized here that the economics are relative, e.g., the effective cost of diesel transported to these remote sites.

The construction of physical pipelines, particularly in complex terrain and urban areas can mean high investment costs, such as the Bolivian pipelines - Brazil, which required an investment of two billion dollars. The relatively high costs are compensated by higher volumes of gas transported through pipelines. In many cases the supply of natural gas to the pipeline in remote areas, and access costs are high compared to the volumes demanded. Under these conditions, the use of the VP (Virtual Pipeline) appears as a reasonable alternative to reduce the investment costs.

The choice between a physical and a virtual pipeline is a matter of capital costs, losses due to time spent, transportation distances, volumes of demand, etc. According to industry estimates, the economy of using the VP oscillates around 25% of costs. This percentage depends very much on the circumstances of each particular project.

The advantages of the virtual - pipeline where conventional systems for natural gas distribution are not economically viable or nonexistant, are clear because VP

- Allows supply of natural gas to all locations where the pipelines are not enough;
- Facilitates the "Flexibility" to enjoy the natural gas in its entirety - industrial use, commercial use and residential sector;
- Allows customers of the remote stations of CNG or LCNG to use natural gas up to 200 bar;
- Offers very low operating costs per m^3 of gas transported, that are generally smaller than any other system.

6.3. Analysis geo-energy for a virtual pipeline in Bolivia

According to research conducted by NRECA (National Rural Electric Cooperative Association) and DAI (Development Alternatives, Inc.), we expect high demand for natural gas in the Chapare region, located in the Department of Cochabamba (Figure 12) in the short term. From information consolidated by 2006, currently there are 2281 potential users in the residential category, 266 small businesses, large businesses 9, 6 small industries, two large industries and 600 vehicles wishing to be adapted to CNG.

Given the need mentioned above, the development of gas distribution in the Chapare (Department of Cochabamba as shown in Figure 12), it appears that from an investment of 3.8 million dollars there will be Compressed Natural Gas (CNG) for the benefit 21,000 inhabitants of the communities of Entre Rios, Ivirgarzama, Chimore, shinahota and Villa Tunari. The objective is to eliminate government subsidies for LPG and diesel, and establish a utility company with local participation in the Chapare.

But to change the energy matrix in the north, are being studied alternatives to the use of diesel for electricity generation by natural gas. The incorporation of natural gas in the energy matrix of these regions would not only contribute to providing a cheaper energy service, but can also, as already mentioned, its use to be driven sectors of small industry, rural, transport, commercial and residential use in the search for a universal, rational and reliable energy in Bolivia.

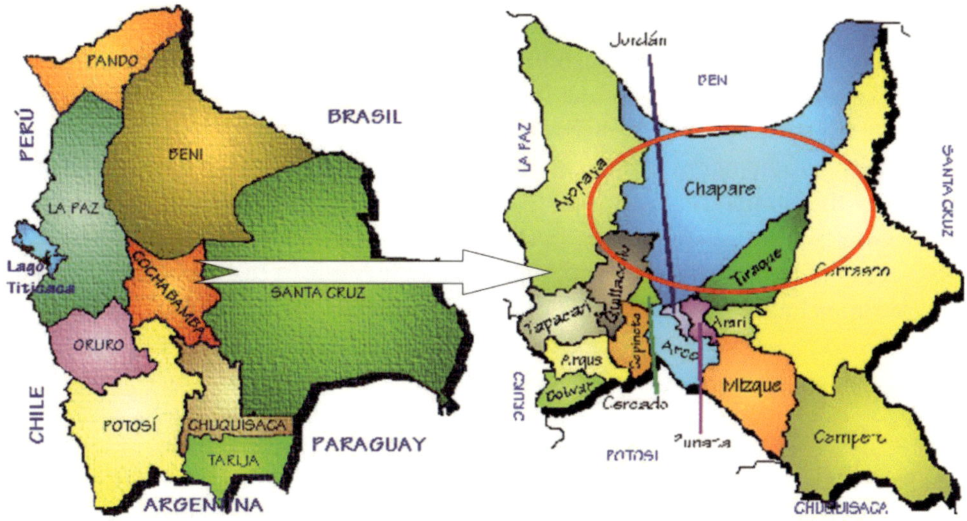

Figure 12. Location of Chapare in Cochabamba – Bolivia

The Departments of Beni and Pando (see Figures 10, 11 and 12) there are numerous power plants not integrated into the National Interconnected System (SIN). These plants use diesel for electricity generation, which absorbs much of the subsidy that the State grants to the fuel. According to Petrobras Bolivia in 2006 are required approximately 3,500 cubic meters of diesel fuel monthly to the departments of Beni and Pando.

The Government of the Beni Department is very interested in virtual pipeline project to supply the cities of Guayaramerín and Riberalta (see Figure 10), since the capital Trinidad is running the project interconnection Caranavi-Trinidad. It should be pointed out that the first two cities are the fastest-growing regions, besides the capital, Trinidad.

From the information posted above, it is observed that there is a great need to supply natural gas to the north. There is VP technology to transport the gas in an unconventional way, but an important aspect for the implementation of the VP (virtual-pipeline) is the analysis of access routes to the aforementioned locations Guayaramerín and Riberalta, within the Department of Beni. In both cases there is the issue of land access to rainy season it is practically impossible. For with the aforementioned, there are at least four months or more to the north of Bolivia (Amazon and countless rivers) is under heavy rains and it is not possible to transport by truck, since the unpavedroads become impassable in the rain. Thus, both pragmatically and in terms of sustainable river transport mode should be considered as a form of transportation, even as the river is the only means of access and in some cases throughout the year. These factors and others relating to modern development, should be considered as motivation to consider the pipeline as a tool for virtual supply. Still, the VP must support its implementation in these areas after assessing the feasibility of transporting the gas in tanker ships from the river up Guayaramerín Cochabamba (see Figure 13).

The route along the river faces the problem of navigation during the dry months, as the navigation is possible only between Trinidad and Guayaramerín, while in the rainy season is made from Puerto Villaroel (Cochabamba) to Guayaramerín.

One of the most important aspects to define the viability of the virtual pipeline project in northern Bolivia is to analyze the possibility of making a navigable passage defined between the rivers Mamore and Ichilo called Ichilo-Mamore route (see Figure 14) during the whole year. In fact, this route when stabilized, would enable the virtual pipeline through a tank lighter, and can supply natural gas to all regions of the north. Moreover, to convert at once in a waterway that stretch between the 2 rivers, above, is a type targeted by soy producers in Brazil, since it would establish transport corridor between the bi-oceanic Atlantic and Pacific. Since bi-oceanic corridor that would allow an increase in the trade of soybeans, and, and also facilitate import/ export of all kinds of merchandise.

6.3.1. Basics of bimodal corridor

To demonstrate the viability of the virtual pipeline to northern Bolivia, aimed at diversifying the energy matrix and the massive use of natural gas thus reducing costs to the grant of traditional fuels. And also to consolidate and demonstrate that use of the waterway via the intermodal system reduces costs and facilitates a cleaner socioeconomic development. This item seeks to present some aspects of the bi-oceanic corridor. At the same time it demonstrates a synergistic way of harnessing a bimodal corridor (waterway and highway).

As already mentioned the basic reference is the proposal, based on comments by the Japanese of using the Bolivian territory to an interconnection between Pacific and Atlantic oceans or vice versa, which is called bi-oceanic. Thus the Brazilian soybean markets, for example, as China would be more than a thousand nautical miles closer. One is as easy to trade in goods of a high strategic value as it is soy, represents an opportunity to facilitate internal transport of goods in Bolivia, that opportunity may be entitled to the relevant secondary parents. One should also mention other distant markets, such as the United States for the pacific coast and other Asian countries. Since the alternatives do not compete with each other, are complementary, tending to maximize the benefits.

Accordingly, the basic aspects are identified (technical and economic) related to the bi-oceanic corridor, from studies of international research funds, the Department of Beni in Bolivia. These studies resulted in the so called "Study of the master plan for development of river transport in the Amazon region." Everything for the consolidation of the export corridor. The study allowed the definition of an alternative route for the implementation of the export corridor with an inter-modal transportation system that allows communication with the Pacific coast from the city of Porto Velho by land to the city of Guajará-Mirim in Brazil , passing through the waterway to Puerto Villarroel (starting at Guyaramirin through Tinidad) in Bolivia, and again by land to Arica in Chile. As shown in Figure 13.

The alternative export corridor allows communicating the central area of Rondônia State, the State of Acre, and the State of Mato Grosso. At the central point, and the place to start is the city of Porto Velho. The corridor would allow Bolivia to get to Arica (Chile) on the Pacific coast. A description of the distances, the state of the stretches of road and the travel corridor are shown in Table 1 and Figure 14.

Figure 13. Bi-oceanic corridor

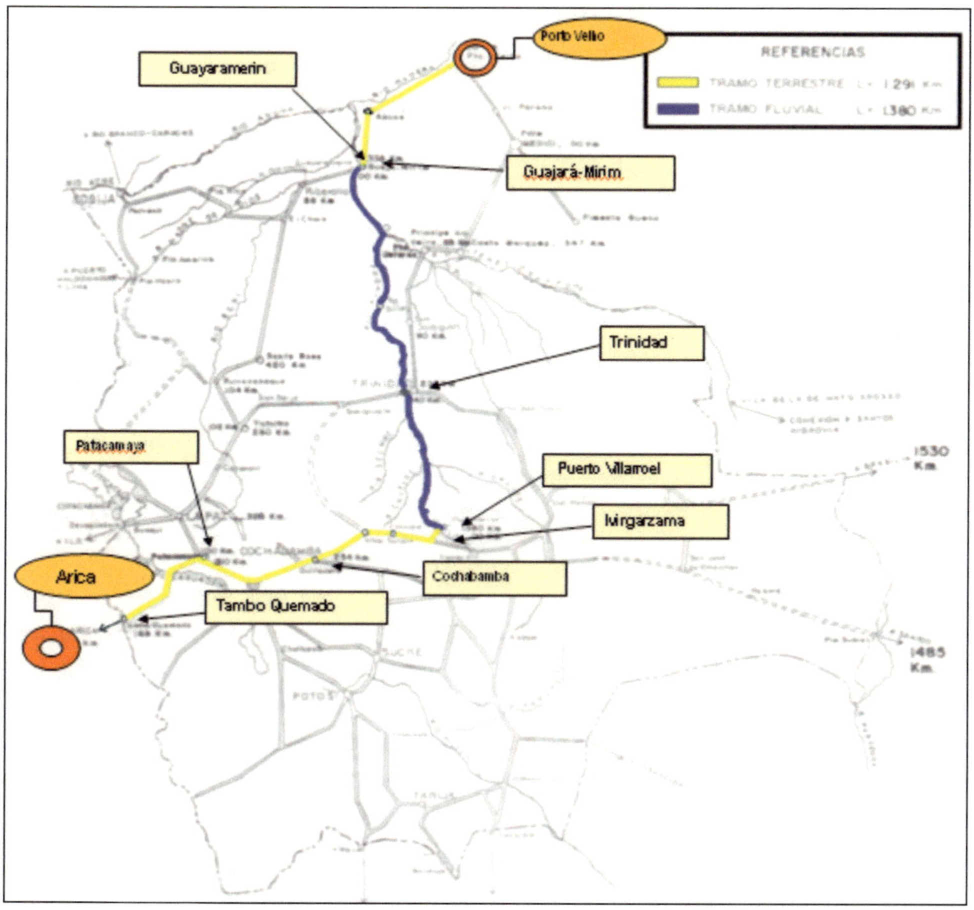

Figure 14. Bi-modal corridor design

Bimodal Export Corridor			
From	**To**	**Distances (KM)**	**Situation**
Porto Velho	Guayaramerin	338	By land (pavement)
Guayaramerin	Puerto Villarroel	1.380	By river (navigable)
Puerto Villarroel	Ivirgarzama	24	By land (pavement)
Ivirgarzama	Cochabamba	230	By land (pavement)
Cochabamba	Patacamaya	310	By land (pavement)
Patacamaya	Tambo Quemado	188	By land (pavement)
Tambo Quemado	Arica	201	By land (pavement)
	Total	**2.671**	

Table 1. Bimodal Corridor Distances

Currently there is infrastructure already existing to support this proposal. The bi-modal corridor would not have to start from scratch.

7. Some aspects of the synthetic fuels NG

Finally, GTL will be discussed mainly to show that it is an alternative to the virtual-pipeline which uses physical states (compressed and liquid) of natural gas to diversify this energy source. Natural gas also gains value through its chemical transformation, with the final product as synthetic fuels, of which the greatest interest is the synthetic diesel (clean diesel). GTL diesel also relies on the use of logistics and distribution system of traditional petroleum-based fuels for end use.

In this sense, the question to be examined a little more and it also refers to the industrialization of natural gas in Bolivia, and production of synthetic fuel (in the case of diesel it would be cleaner than regular diesel). This is the case of the GTL. The case of the GTL diesel has the advantage of the possibility of exporting synthetic diesel and would not have to import it. The GTL would generate productive jobs due to the secondary economy including other synthetic derivatives (already mentioned). Moreover, the use of GTL diesel would provide a cleaner environment in urban areas locally, and an overall effect in the case of greenhouse gases.

Energy resources play a key strategic role in the development of a country, primarily from its economic activity. The worldwide trend of increased energy demand, especially when it comes to a developing country, making the energy resources and energy demand, both in becoming a critical issue in terms of socioeconomic status. Ensuring supply and demand in balance in order to maintain rational over to fear an economic development, a sustainable environment and welfare of society as a whole is a constant challenge.

Therefore, it must be clear and it is interesting that the synthetic fuels produced from natural gas industrialization turn gains even when assessed in light of the virtual pipeline. Certainly it would be showing the actual multiple use of this resource, in this case, transforming it through the GTL process in synthetic derivates. From a technical standpoint, it is important to identify economic investments in technology and production costs for both projects, both GTL and VP or, in the sense competitive and / or synergistic effect of both. In commercial terms, the goal should be to identify current and future situations and VP GTL market and what it represents. From the viewpoint of economics, the intent is to determine relevant issues intrinsic to the GTL project and / or VP. Finally, but not least, legal and political questions relating to regulatory issues of conversions from natural gas, both chemical (GTL) and physical (VP), must be mastered, mainly relating to the site of end use.

7.1. Prices of goods

In the event that is being dealt here, i.e., a GTL project, the sales prices on the open market of derivatives stemming from natural gas, there is a significant degree of importance in order to be commercially viable. The real economic opportunities and financial feasibility of projects like GTL, in practice have always been influenced by the relative prices of natural gas.

In the specific case of Bolivia, even as this country has a current structure of state-controlled prices of natural gas in the mouth of the well has a forecast for the next year, from 4.0 to 5.0 U.S. $ / MMBTU. Therefore, the GTL project in Bolivia would be limited to a large scale. It would also require the government to demonstrate the willingness in the item relating to taxes, even more so that companies have in the future natural gas reserves, have the power to make the decision to directly monetize its reserves, thus eliminating the need (determined by the current rules) that there is another company that operates a GTL plant.

However, the current state of affairs with respect to oil prices allows the development of type GTL ventures. This is because that when the value of a barrel of oil is over U.S. $ 30 / Bbl, a GTL plant could reach an internal rate of return (IRR) of at least 15% for the worst case. Therefore, the GTL products (based on Fischer-Tropsch technology), due to its superior quality and could be sold at a competitive price compared to products from conventional oil refinery.

7.2. Feasibility of GTL

When it comes to evaluate the feasibility of an enterprise GTL it must necessarily be applied the 1:10 rule, this means that 2 TPC (trillion cubic feet) of natural gas reserves, should produce during 20 years, 20 000 BBL / d of synthetic diesel on average (ultra-clean fuel).

In the Bolivian case, it is initially planned to produce 15,000 BBL / d between the third and fifth years, and so on increasing to maximum capacity. Thus, the operation of the GTL plant to reach full capacity in the seventh or eighth year. Verified bibliographic information consulted, units, such as those considered for the Bolivian case, range in capacities from 10,000 BBL / day to 160,000 bbl / day.

Moreover, according to feasibility studies, investment depends on the size of the plant. Where also it must be taken into account that such endeavors have returns every time the price of a barrel of oil is above thirty dollars. From the standpoint of investment, production of synthesis gas (commonly known as syngas), a key component in the GTL process corresponds to about 50% of the plant cost, when the separator unit including air. Since the synthesis of the Fischer-Tropsch (FT) requires about 15% of the costs. However, the step of improving the product requires 10% of the capital investment. Since about 25% remaining is for the additional systems such as power generation and infrastructure.

For this reason and to define the feasibility of implementation and the size of a GTL plant, it is necessary to take into account various aspects such as supply and demand for FT products worldwide, demand for the domestic market of fuels, the availability of proven reserves of natural gas, costs of investment and technological trends. All this highlights the fact that when it comes to the physical layout of the plant is rather complex.

Author details

Miguel Edgar Morales Udaeta*, Jonathas Luiz de Oliveira Bernal,
Luiz Claudio Ribeiro Galvão and José Aquiles Baesso Grimoni
USP, GEPEA/EPUSP (Grupo de Energia do Departamento de Engenharia de Energia e Automação Elétricas da Escola Politécnica da Universidade de São Paulo), São Paulo, SP, Brasil

Acknowledgement

The authors thank FAPESP - São Paulo Research Foundation, because this work is the result and was developed based on the research support received from FAPESP, through research projects: FAPESP 2005/03059-0 "*Modeling the Procedure for Systemic, Viable and Sustainable Use of Natural Gas in Bolivia Including the Development in Brazil*", and FAPESP 2009/03168-4 "*Background and Introduction To Supply Chain of Natural Gas*".

8. References

Anuario Estadístico, Ministerio de Servicios y Obras Públicas, Viceministerio de Electricidad, Energías Alternativas y Telecomunicaciones, 2004.

Associação Portuguesa do Veículo a Gás Natural. Introdução ao GNL-GNC, <http://www.apvgn.pt>. Lisboa, APVGN, 04. nov. 2007.

California Energy Commission, April, 2010.

Carrera G. A. Zamalloa "Avaliação de Alternativas Tecnológicas (GNL e GTL) para a Viabilização de Jazidas de Gás Natural Remotas em Países em Desenvolvimento" – Estudo de Caso: Jazida de Camisea no Peru" 2004.

Carroll, J. Natural Gas Hydrates: A Guide For Engineers. Elsevier Science, ISBN: 9780080570020. eBook, 2009.

Centro de Documentação e Informação de Bolívia, 2006; www.cedib.org, 04/2006.

Centro de Tecnologia do Gas. Gás Natural, <http://www.ctgas.com.br>. Rio Grande do Norte, CTGas, 24 out. 2007.

Código Petroleiro Davenport de 1995-2000.

* Corresponding Author

Department of Energy – National Energy Technology Laboratory. Energy, <http://wwwe.doe.gov>. Washington, DC. DOE, 21 set. 2006.

Doria, M.S., 2003. "Gas Bolivia"; La Paz - Bolivia.

Duke Energy Gas Transmission Canadá. Natural Gas, <http://www.duke-energy.com>. Canadá. 04 nov. 2007.

Enbridge. The Northern Pipeline Story, <http://www.enbridge.com>. Ontario, 12 jul. 2002.

Energy Information Administration. International Energy Annual – Natural Gas Tables, <http://www.eia.doe.gov>. Washington, DC. EIA, 25 set. 2006.

Energy Press-Bolivia, www.energypress.com/Bolivia, 03/2006.

Foster Wheeler, Gas to Market Technology, 2002.

Galileo, S.A. website in order to inform, educate and communicate. www.galileoar.com; Buenos Aires – Argentina. 2012.

Galvão, L.C.R.; Udaeta, M.E.M. "Aspectos Relevantes do Gás Natural Visando o Planejamento Energético". In: III CBPE, São Paulo -SP. Junho de 1998. SBPE, Anais.

Geology College, November, 2009.

Grimoni, J.A.B. Galvão, L.C.R., Udaeta, M.E.M. Iniciação a Conceitos de Sistemas Energéticos para o Desenvolvimento Limpo. São Paulo, Edusp, 2004.

Institute of Americas, El Gasoducto Sudamericano "Mesa Redonda Ejecutiva Sobre La Integración Regional Energética". Presentada por: Ministerio de Industria, Energía y Minería del Uruguay y el Instituto de las Américas. Montevideo, Uruguay, August of 2005.

Instituto Nacional de Estatísticas de Bolívia; www.ine.gov.bo, 02/2006.

International Association For Natural Gas Vehicles. Natural Gas Vehicles, <http://www.iangv.org/>. Auckland New Zealand, IANGV, 12 ago. 2007.

International Forum: "Industrialización Del Gás Boliviano: Sueño o Realidad?" La Paz, December, 2003.

Kinn, L.C., 2004. "Política Energética Integral"; Santa Cruz - Bolivia.

Mde – Ministerio de Desarrollo Económico, Comisión Política de Estado sobre el Gas Natural, 2002. "Política de Estado sobre la Utilización del Gas Natural"; Bolivia.

Naturalgas. Natural Gas, <http://www.naturalgas.org>. Washington, DC, 28 out. 2007.

Oliva, R.C.R., "Exeqüibilidade da industrialização do gás natural na Bolívia e a sustentabilidade de abastecimento a mercados além das suas fronteiras", São Paulo – Brasil, 2006.

Petrobrás. Programas Tecnológicos, <http://www.petrobras.com.br>. Rio de Janeiro, 04 nov. 2007.

Semena, Estudio de Factibilidad Socioeconómica, Financiera e Impacto Ambiental del Corredor de Exportación Porto Velho – Guayaramerin – Trinidad – Puerto Villarroel – Arica o Iquique o Matarile o Ilo, 2006.

The Center for LNG. LNG Research, <http://www.lngfacts.org>. Washington, DC. CLNG, 28 out.2007.

Udaeta, M. E. M. et al. Pesquisa Tendências Tecnológicas do Setor Petróleo e Gás. Programa de Recursos Humanos para o Setor de Petróleo e Gás. São Paulo, Instituto de Eletrotécnica e Energia da Universidade de São Paulo. São Paulo, IEE, 2004

Udaeta, M.E.M. Modelamento do procedimento para o uso sistêmico, sustentável e viável do gás natural na Bolívia incluindo o aproveitamento no Brasil. FAPESP, IEE/USP, GEPEA/EPUSP, CESU/UMSS. São Paulo – Brasil. 2007.

Udaeta, M.E.M.; Galvão, L.C.R.; Lafuente, R.J.O. Capitulo Bolivia. In: PAULA, Ericson de. (Org.). Energía para el Desarrollo de América del Sur. Mackenzie. São Paulo, 2002, p. 69-110.

Udaeta, M.E.M.; Grimoni, J.A.B.; Burani, G.F.; Rigolin, P.H.C.; Massara, V.N. Fundamentos e Introdução à Cadeia Produtiva do Gás Natural. São Paulo - Brasil, Edusp, 2010

Udaeta, M.E.M.; Lafuente, R.J.O. "Perspectiva del rubro energético en Bolivia y gas natural". Cochabamba – Bolivia. Acta Nova Revista Semestral de ciencias y tecnología de la UCB. Vol. 2 Nº 2, Junho de 2003, p. . ISSN – 1683 – 0768.

Udaeta, M.E.M.; Zurita, R.O.R.; Lafuente, R.J.O.; Galvão, L.C.R. "La Industria Energética en Bolivia y su Vocación Integradora Através del Gas Natural" In. IV Encontro Brasileiro dos Profissionais do Gás, São Paulo, Gasbrasil, 2003, CD-ROM, Anais.

Udaeta, M.E.M; Reis, L.B; Lafuente, R.J.O; Zurita, R.O.R; Burani, G.F. "Análisis de Ia Industria Energética en Bolivia en el Marco del Mercado Competitivo". Rio de Janeiro - Brasil. Periódico "Revista Brasileira de Energia" -Vol. 8 No 1- 2001, SBPE. ISSN OlO4-303X.

Vega, F.F. de.; "El Gas Natural en América Latina y el Caribe". Buenos Aires – Argentina, 2004. PennWell. American Petroleum Institute. Programs and Services, <http://www.api.org>. Washington, DC, API, 04 nov. 2007. Gasvirtual. Informe, <http://www.gasvirtual.com.br>. Rio de Janeiro, 05 nov. 2007.

Villanueva, Luz Z. D. Uso de Gás Natural em Veículos Leves e Mecanismo de Desenvolvimento Limpo no Contexto Brasileiro. São Paulo, Tese de Doutorado do Programa Interunidades de Pós-Graduação em Energia da Universidade de São Paulo, 2002.

Yacimientos Petrolíferos Fiscales Bolivianos; www.ypfb.gov.bo, 01/2006.

Chapter 7

Reliability Analysis of the Regasification System on Board of a FSRU Using Bayesian Networks

Marcelo Ramos Martins and Adriana Miralles Schleder

Additional information is available at the end of the chapter

http://dx.doi.org/10.5772/45803

1. Introduction

The global economic downturn has reduced the demand for all types of energy while the world's capacity to produce natural gas is surging. As a result, LNG is becoming an important energy source option. Large reserves of natural gas exist around the world, often in areas where there is no market or where the resources exceed the demand. Therefore, this natural gas must be shipped to areas where there is demand, and, to reduce costs, the gas is liquefied, reducing its volume by about 600 times. Thus, the storage and regasification system usually occurs in onshore plants. The LNG is stored in a double-walled storage tank at atmospheric pressure until needed. Then, the LNG is pumped at a higher pressure and warmed until it turns into gas again.

From this viewpoint, the FSRU (Floating and Storage Regasification Unit) is becoming a new economic and flexible alternative for the storage and regasification system. The FSRU costs less than an onshore facility of similar capacity; it provides a faster return to the capital invested because time is saved by not having an extensive planning and permitting process as usually occurs with onshore developments. Moreover, construction time is reduced, assuming the conversion of an existing LNG carrier. Additionally, as the FSRU can be moved from one demand area to another, paperit is a flexible and attractive feature in countries with seasonal demand or where there is an unstable market.

It is also worth noting two more FSRUs attractive features. First, an accident in one onshore plant might produce considerable impact on neighboring areas and on the local population (this risk may be even worse due to the possibility of a terrorist attack), as reported by [1, 2]. Second, LNG provides clean energy as compared with traditional fuels, and it is a significant alternative to diversify the national energy matrix.

Because the regasification system usually occurs in onshore plants, the processing in vessels is pioneering. These vessels were formerly used for transporting liquefied gas and were transformed into FSRUs to gasify the LNG. Therefore, the development of efficient technologies for LNG exploration and distribution is essential. Risk and reliability analysis are vital to the development of these technologies.

As reported by [3], an increasing number of recent studies have applied BN in reliability analysis, risk analysis and maintenance due to the benefits that BN provides in contrast with traditional tools, such as its capacity for representing limited or incomplete knowledge, local dependencies and multi-state variables and its ability to deal with any probability density function. Its practicality to model common cause failures and the capability to update the state of knowledge according to new evidence are also important advantages that have been explored in reliability analysis using BN.

This chapter proposes the construction of a BN to evaluate the reliability of the regasification system of a FSRU and presents additional developments over a previous published study [4], in which the regasification system BN, converted from a fault tree, was evaluated to model the probability of an undesired event. The first part of this chapter provides an overview of the BNs, followed by explanations of the Regasification System as well as the Regasification System BN. The model shown herein was built and executed using the commercial tool, AgenaRisk, available in [5]. In the second part of this chapter, the reliability of the regasification system is evaluated and information is obtained such as critical components and subsystems and other conditions that affect the system reliability.

2. Bayesians networks

As defined by [6], BN is a graphical structure for representing the probabilistic relationships among a large number of variables and for making probabilistic inferences with those variables. A BN is a direct acyclic graph (DAG) with the nodes representing the variables and the arcs, their conditional dependencies. The BN qualitative analysis provides the relationships between the nodes while the quantitative analysis may be performed in two ways: a predictive analysis or a diagnostic analysis. The first one calculates the probability of any node based on its parent nodes and the conditional dependencies. The second one calculates the probability of any set of variables given some evidence.

The nodes and arcs are the qualitative components of the networks and provide a set of conditional independence assumptions that may be represented through a graph notion called d-separation, which means that each arc built from variable X to variable Y is a direct dependence, such as a cause-effect relationship.

If the variables are discrete, the probabilistic relationship of each node X with their respective parents pa(X) is defined by its Conditional Probability Table (CPT) while for continuous variables, this probabilistic relationship is defined by its Conditional Probability Distribution (CPD), which represents conditional probability density functions. The quantitative analysis is based on the conditional independence assumption. Considering three random variables X, Y and Z, X is said to be conditionally independent of Y given Z, if $P(X,Y \mid Z)=P(X \mid Z)P(Y \mid Z)$. The joint probability distribution of set of variables, based on their conditional independence, can be factorized as shown in Eq.1:

$$P[x_1, x_2, \dots, x_n] = \prod_{i=1}^{n} P[x_i | \text{Parent}x_i] \quad (1)$$

The graphical representation is the bridging of the gap between the (high level) conditional independence statements that must be encoded in the model and the (low level) constraints, which enforce the CPD [7].

Given some evidence, beliefs are recalculated to indicate their impact on the network. The possibility of using evidences of the system to reassess the probabilities of network events is another important feature of the BNs; it is interesting to determine critical points in the system. Classical methods of inference of a BN for this purpose involve computation of the posterior marginal probability distribution of each component, computation of the posterior joint probability distribution of subsets of components and computation of the posterior joint probability distribution of the set of all nodes. The analysis and propagation of evidences allowed by BN are useful to explore or forecast some system behavior that is unknown or requires more attention as in [8]. In addition, the propagation of evidences offers the possibility to check the influence of redundant systems or critical equipment, such as equipment that requires a long time to repair.

In the last years, the number of studies that presented the use of BN in reliability analysis have been increased [9-11]; traditional models, such as fault trees and block diagrams, have been replaced by discrete BN. However, to perform an efficient application of BNs in reliability assessment, such network models must be hybrid models formed by discrete and continuous variables. The evaluation of hybrid networks poses a challenge; there are limitations of inference algorithms, such as dealing with state space explosion and finding an appropriate discretization. However, a new and efficient dynamic discretization of the domain and an iterative approximation method which produce finer discretization in the regions that contribute more to the structure of the density functions associated with a robust propagation algorithm were proposed by [9,10]. This approach is implemented in the commercial BN software package, AgenaRisk [5], which is used in this study. There are other commercial tools for the calculation of BNs, such as [12] and [13]. [14] is an excellent reference for the theoretical aspects of BNs and algorithms. This approach allows the BN to deal with any probability distribution function, unlike the traditional tools which are capable of dealing only with exponential distributions.

Another relevant BN feature is its practicality to model common cause failures; the Common Cause Failures (CCFs) are the failure of more than one component due to the same cause, which can render the redundancy protection useless significantly affecting the system reliability. More details about CCFs and proposed methods to deal with are reported in [16,17]. Including CCFs in a BN is not a complex process. If the BN is a network composed of only discrete variables, the CCF probabilities are included directly to the CPT, as [10] reports, and it is not necessary to use additional constructs such as when using fault trees. If the BN is a network composed of continuous variables, the CCFs are represented by additional nodes; one node for each group of similar components is included in the BN. The frequency rate of these failures is estimated according to the redundant components; [16, 17] explain the methods for obtaining these rates.

It is also worth noting that BNs are efficient to model multi-state variables, local dependencies and limited or incomplete knowledge. Multi-state variables and local dependencies are important to build a more realistic model; traditional tools usually use just binary variables and are unable to represent local dependencies, for example to represent how a malfunction of the equipment affects other equipment. Occasionally, there are not enough or satisfactory statistical data about the system to perform the reliability analysis; in this situation, BN builders ask relevant questions to a group of specialists and explain the assumptions that are encoded in the model, and the domain experts supply their knowledge to the BN builders ([18]and [19] demonstrated this process).

The reliability analysis presented in this chapter explores the benefits of BN use, such as the inclusion of CCFs, modeling with continuous variables, propagation of evidence and local dependences representation.

3. Methodology

The proposed methodology is a combination of different techniques already used. Proposals of different authors and several techniques were combined to compose the methodology, which resulted in the formation of a four-step methodology: familiarization, qualitative analysis, quantitative analysis and complementary analysis.

In the first step, familiarization, all the information available about the system and the operation must be collected. The second step, qualitative analysis, is the step at which the relationship among the system components must be identified and, as a result, a BN is built to represent the system. Next, in the quantitative analysis, the priori probabilities of root nodes and the conditional probabilities tables for non-root nodes are defined allowing the evaluation of the joint probability of a set of variables. Finally, the complementary analyses must be performed by evaluating the posterior probabilities: criticality analysis, the analysis of different scenarios of interest and the conditional reliability analysis. These analyses allow improving the reliability analysis through an evaluation that is not possible through traditional tools. The criticality analysis means to find the set of components or subsystems that have greater influence in the system behavior; the analysis of different scenarios can be used to model any situation of interest, such as the impact of including redundancies, the impact of a component fault or any other condition that affects the system reliability; and the conditional reliability analysis provides information about the system behavior over time. Figure 1 presents an overview of the methodology steps; the figure is divided into two parts: the first in which all the tasks to be performed at each step are listed and the second which lists the means suggested for these tasks.

4. Application

In this section, the reliability analysis of the regasification system is performed by using the methodology of the previous section. First, the information collected about the system is presented. Then, the qualitative analysis is performed. Subsequently, a quantitative analysis

	Step 1 *Familiarization*	Step 2 *Qualitative Analysis*	Step 3 *Quantitative Analysis*	Step 4 *Complementary Analyses*
Tasks	Understanding the system and identifying possible scenarios that the system will be submitted to.	Representing the system physically and functionally; Representing the relationships between system elements;	Completing the construction of BN with quantitative data (priori probabilities, CPT and density functions); Estimating the probability density of system failure and reliability for a given mission time.	Analyzing the criticality; Analyzing different scenarios; Analyzing the conditional reliability.
Means	Data review; Interviews with experts.	Functional tree; Unifilar diagram; Block diagram; Fault tree; Bayesian Networks.	Bayesian Network (Inference)	Bayesian Network (posterior probability)

Figure 1. Methodology

will be conducted, in which the failure probability density and the system reliability are estimated for a given mission time of the system. And finally, the complementary analysis is presented. The AgenaRisk (Desktop Agena Risk, 2011)[5] was used to build the Bayesian Network and to make the inferences about the system.

4.1. Familiarization - The regasification system

Usually, vessels are used for LNG transportation; however, in the last years, these vessels also began to participate in gas regasification and directly supply net pipes. The regasification process onboard adds new hazards to the operations of LNG vessels, because in addition to LNG, there is now compressed gas in the process. Accidents in this process may reach the storage tanks, causing very severe consequences.

In the vessel studied, a cascade system was used to regasify the LNG. In this system, the LNG is heated in two stages. In the first, it is heated by a propane compact heat exchanger (HE1), and its temperature increases from -162ºC to -10ºC; at this stage, the natural gas is already vaporized, but this temperature is too low for delivery to the pipeline, where the heating process would continue. In the next stage, the gas is heated by seawater in a shell-and-tube heat exchanger (HE2), and the temperature reaches 15ºC. The first stage uses no water due to the possibility of water freezing in direct contact with LNG. The propane used in the first phase works in a closed loop. When the propane leaves the LNG heat exchanger HE1, its temperature is approximately -5ºC, and it is liquefied (propane at 4.7 bar liquefies at approximately -5ºC); hence, it is pumped into a titanium heat exchanger (HE3) and heated, by sea water, up to 0ºC at 4.7 bar and vaporizes. It then returns to the LNG exchanger HE1. This system must have an effective thermal insulation to avoid an unexpected heat gain of the LNG or propane inside the tubing, which could result in a gas expansion and possibly cause a tubing rupture. A more detailed description of the regasification system is given in [4]. A diagram of the system is shown in Figure 1, and the nomenclature used is presented in Table 1. The thermal insulation for each heat exchanger will also be considered (I1, I2 and I3).

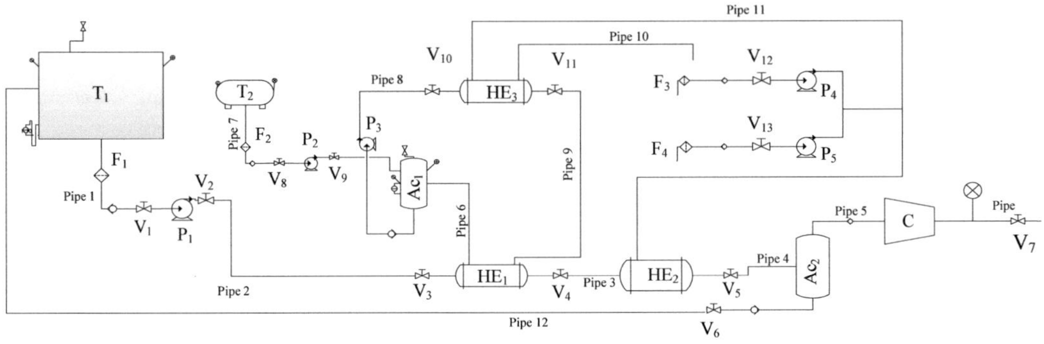

Figure 2. Regasification system

Item	Description	Item	Description
T1	LNG tank	**P2, P3**	Propane pumps
T2	Propane tank	**P4, P5**	Water pumps
V1, V2, V3, V6, V8, V9, V10, V11, V12, V13	Gate valves for liquid	**Pipes 1, 2, 3, 4, 5, 6, 7, 8, 9, 10, 11, 12**	Pipes with less than 10"
V4, V5, V7	Gate valves for gas	**Delivery Pipe**	Pipe with 12"
P1	LNG supply pump	**Ac1**	Propane accumulator
F1	LNG filter	**F2**	Propane filter
Ac2	LNG accumulator	**HE2**	Sea water/LNG shell&tube heat exchanger
HE1	Propane/LNG compact heat exchanger	**HE3**	Propane/sea water titanium heat exchanger
C	Compressor	**F3 e F4**	Water filter

Table 1. Nomenclatures

4.2. Qualitative analysis

Qualitative analysis should provide a clear view of the system and the relationships between system elements; this representation may be produced by building a block diagram or a fault tree and then converting it into BN, as presented by [4] and [11] or may be directly produced from the system analysis.

The regasification system was represented in a hybrid BN. Continuous nodes were built to represent the time to failure (TTF) of the basic components and subsystems, and discrete nodes to represent the state of the system or subsystem.

Another important BN feature is its capability of modeling local dependences. The regasification system has local dependences between heat exchangers and insulators; the failure probability distributions of the nodes "Heat Exchanger LNG/Propane" (HE1), "Heat Exchanger LNG/Water" (HE2) and "Heat Exchanger Propane/Water" (HE3) change if the insulation fails. If the insulation fails, the heat exchanger failure probability increases. This variable has a conditional dependence that is not possible to address with traditional approaches such as Fault Trees (FT). In FT analysis, it is not possible represent local dependence, but it can be modeled in a simple way with BNs. To include this dependence in the model, an arc was built between these nodes (Figure 3). With this approach, it is possible to model how the malfunction of any equipment affects other equipment.

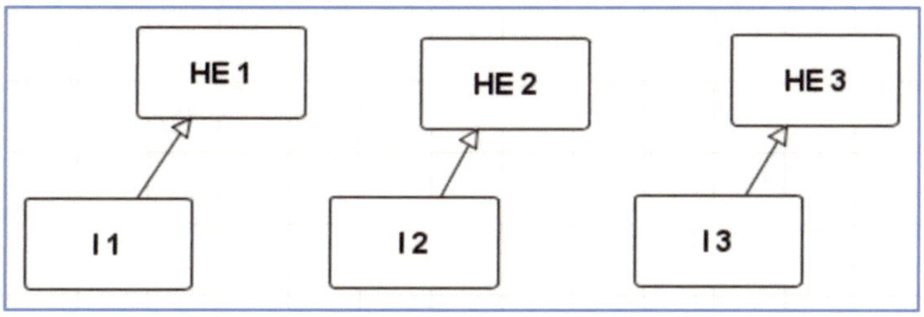

Figure 3. Conditional dependences

Finally, to complete the qualitative analysis, the CCF must be included. This system has a redundant subsystem in which a CCF may occur; a parallel system provides water for heat exchangers HE2 and HE3. The CCFs are important contributors to system unreliability and typically exist among redundant units. A CCF in this subsystem directly affects the reliability of the whole system. In the BN, one node is included for each group of redundant components to verify the CCF effects; each node is a representation of the CCF associated with groups of similar equipment: CCF1 (valves), CCF2 (filters) and CCF3 (pumps).

The Regasification System BN is illustrated in Figure 4; the top node represents the whole regasification system. Below are the subsystems and even the basic components and the CCFs are highlighted. Also, there is node "R" which is the reliability node; it will be used to evaluate the system reliability for a specific mission time.

4.3. Quantitative analysis

Quantitative analysis begins with the inclusion in the BN of the priori probabilities of root nodes; these probabilities can be provided by statistical data or be estimated by experts. Next, the relationships between nodes must be specified. And finally the joint probability of the network is obtained, which, in the case study will serve to obtain the system reliability for a given mission time. The root nodes that represent the basic components are completed by probability density functions representing the TTF of each basic component.

The relationships between components are represented by basic constructs, such as the AND and OR gates, used in fault trees. The AND gate, where the output will fail when all input components fail, has a probability of failure of its output in the time interval [0,t], given by:

$$P(\zeta_{And} \leq t) = P(\zeta_1 \leq t, \dots, \zeta_n \leq t) = P(\max\{\zeta_i\} \leq t)$$

Where ζ_{And}: time to failure of AND gate
ζ_i: time to failure of component i

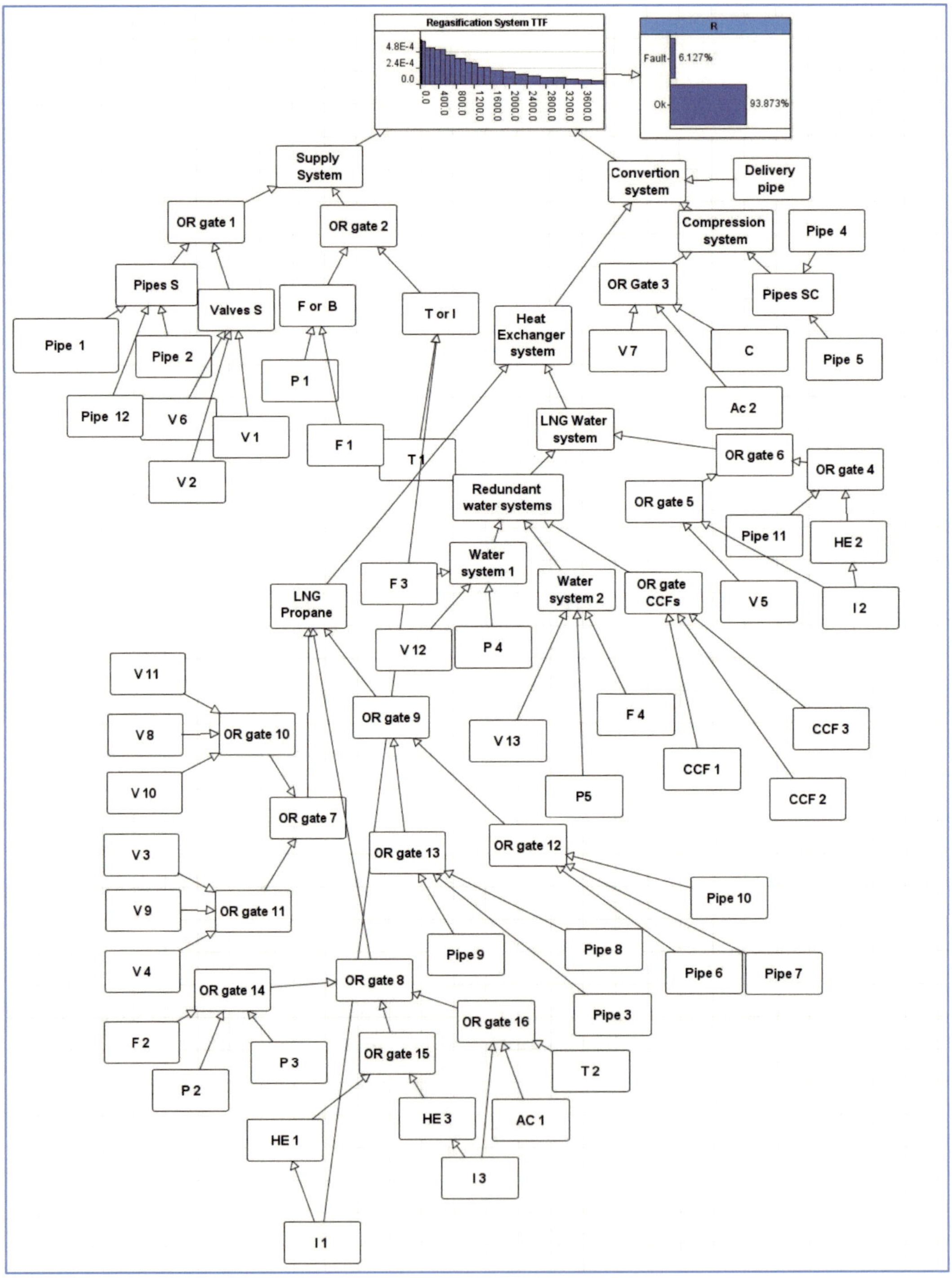

Figure 4. Regasification system BN

The OR gate, where the output will fail if at least one input component fails, has a probability of failure of its output, in the time interval [0,t], given by:

$$P(\zeta_{OR} \leq t) = 1 - P(\zeta_1 > t, \ldots, \zeta_n > t) = P(\min\{\zeta_i\} \leq t)\,)$$

where

ζ_{OR}: time to failure of OR gate

ζ_i: time to failure of component i

Although BN is able to deal with any kind of prior distribution, the components were considered to have constant failure rates (λ) which means that the time-to-failure distributions were assumed to be exponential. Thus, the probability of a component to fail at time T within a given mission time t is calculated as $P(T<t)) = 1-e^{-\lambda t}$, except for the insulation. Statistical data about the probability of the insulation failure were not found along this investigation, but these distributions may be estimated by expert judgment. BN builders ask relevant questions to a group of specialists and explain the assumptions that are encoded in the model, and the domain experts supply their knowledge to the BN builders. [18,19] demonstrated this process. In the current study, for the insulator node, the failure distribution was assumed to be a Weibull distribution with shape factor s = 6 and inverse scale β = 1/10000. It is worth noting that any distribution can be used in the BN, which is another benefit that BNs offer over those of traditional tools. The failure rates (provided by [20] and [21] are listed in Table 2).

Component	**Failure rate (λ) (hr^{-1})**	**Component**	**Failure rate (λ) (hr^{-1})**
C	$1.709\text{x}10^{-4}$	T2	$2.883\text{x}10^{-5}$
P1	$4.801\text{x}10^{-5}$	Ac1	$2.883\text{x}10^{-5}$
P2, P3	$4.801\text{x}10^{-5}$	Ac2	$2.883\text{x}10^{-5}$
HE1	$3.857\text{x}10^{-5}$	P4, P5	$5.120\text{x}10^{-6}$
HE3	$2.175\text{x}10^{-5}$	V1, V2, V3, V6, V8, V9, V10, V11, V12, V13	$6.610\text{X}10^{-6}$
Pipes 1, 2, 3, 4, 5, 6, 7, 8, 9, 10, 11, 12	$6.700\text{x}10^{-9}$	HE2	$2.739\text{X}10^{-5}$
Delivery pipe	$6.300\text{x}10^{-9}$	T1	$2.523\text{X}10^{-5}$
V4, V5, V7	$1.517\text{x}10^{-5}$	Filters 1,2,3,4	$4.155\text{X}10^{-7}$

Table 2. Failure rates

As mentioned in the previous section, there are local dependencies between the heat exchangers and insulation. In order to model these dependencies, the failure rates of the heat exchangers were adjusted: the first considers the failure rate of the exchanger in the case the insulation works and the second considers a failure rate higher for the heat exchanger if the insulation fails. Thus, for heat exchanger HE1, the failure rate increases from $3{,}857\text{x}10^{-5}$ to $4{,}000\text{x}10^{-5}$ when the insulation fails, and similarly for the other two heat exchangers, these failure rates are in Table 3.

State of insulation	Failure rate (λ) (hr^{-1})
I_1	HE_1
Ok	3,857x10^{-5}
Fault	4,000x10^{-5}
I_2	HE_2
Ok	2,739x10^{-5}
Fault	3,700x10^{-5}
I_3	HE_3
Ok	2,175 x10^{-5}
Fault	3,000x10^{-5}

Table 3. Conditional dependences

Finally, to complete, the BN are included the CCF frequency rate; this was calculated using the Beta Factor Model, which was presented by [16] , and the factor beta was assumed to be β = 0.1, as recommended by [17]. These frequency rates are in Table 4.

The use of BN allowed the inclusion of CCFs in the model despite the use of continuous variables, which is not possible with traditional tools such as fault trees and diagram blocks.

	CCF Groups	λ group $\lambda_g = \beta\lambda_c$
1	Valves	6.610 x10^{-7}
2	Water pumps	5.120 x10^{-7}
3	Water filters	0.416x10^{-7}

Table 4. CCF Groups

The BN was evaluated at a mission time t = 96 hours, which is the time required to regasify all of the stored gas in the vessel. The inference of this BN, with all parameters, allows obtaining the prior probability for node R (which represents the Regasification System Reliability at a mission time), and, for a mission time 96 h, R = 0.93873. The prior reliability system is the first information provided by the BN; however, the BN may provide many more data, and it allows several analyses concerning the system behavior.

4.4. Complementary analyses

4.4.1. Criticality analysis

The criticality analysis allows verifying which component or subsystem causes the most impact at the reliability system by evaluating the posterior probabilities. In this chapter, the first criticality analysis performed concerned the impact of each subsystem on the system reliability.

The graph in Figure 5 is a visual perspective, where the length of the bars is a measure of the impact of each node on the target node (the TTF of the Regasification System). The first bar indicates the range between the lowest and the highest value for the expected TTF to the regasification system given the LNG/Propane subsystem. The initial point of the bar (equal to 0.5) is the expected value for the regasification system TTF given the LNG/Propane subsystem's TTF being nearly zero. The end point (equal to 2569) is the expected value of the regasification system TTF given the LNG/Propane subsystem's TTF significantly exceeds the mission time, which influences the regasification system TTF. After this point, the regasification system TTF stabilizes. Even if the LNG/Propane subsystem's TTF increases, the regasification system TTF does not change. The second bar represents the expected values for the regasification system's TTF conditioned on the Compression subsystem's TTF. The Compression subsystem TTF begins to influence the regasification system TTF when the expected TTF of the regasification system is around 0.5 and this influence ends at the point that the expected regasification system´s TTF is 2427. The next bars are plotted using the same concept.

The graph shows that the most critical subsystem is the LNG/Propane subsystem, which has the most significant influence on the regasification system TTF, and, as expected, the subsystems that have minor influence on reliability are those that are redundant: the water subsystem.

This analysis may also be performed to find critical components; it is an effective tool to search critical points in the system, which contributes to map equipment and subsystems that require improvement and special attention in the maintenance plan. Once seen that the critical subsystem is the LNG/Propane subsystem, the criticality analysis of this subsystem can be performed; this analysis shows that the critical components of the LNG/Propane subsystem are pumps P2 and P3.

In this study, a diagnosis analysis was also performed: given the evidence that the regasification system failed, which component is likely to cause the fault. The failure evidence is included in BN and the beliefs are recalculated. The graph in Figure 6 is the visual perspective where the length of the bars is a measure of the impact of each component on the Regasification system TTF given the evidence "fault of the system". As can be seen, the compressor is the component that has the most influence on the Regasification System TTF given the fault system.

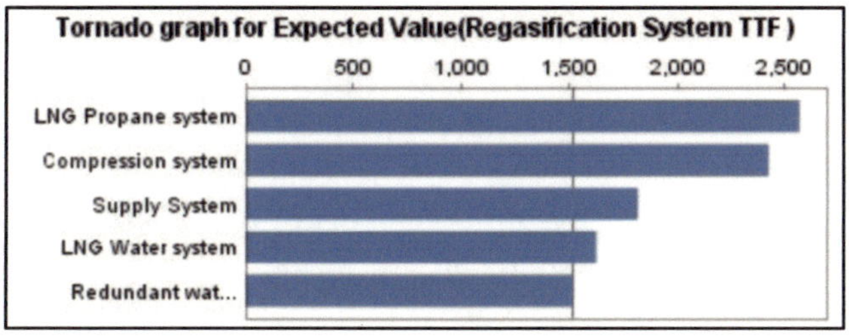

Figure 5. Criticality of the subsystems

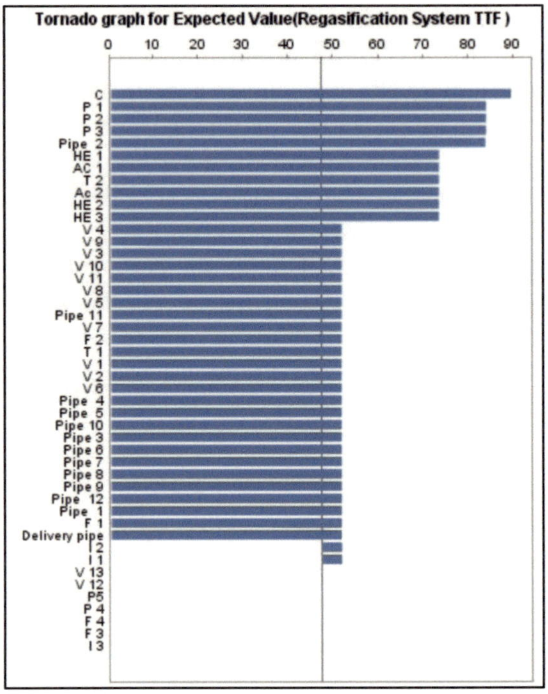

Figure 6. Impact of each component on the Regasification system TTF given the fault system

4.4.2. Analysis under different scenarios

BN allows the study of the evidence propagation and this way the study of different scenarios by evaluating the posterior marginal probability distribution.

Assuming that there is interest in improving the reliability of the regasification system, it is reasonable to consider improving the reliability of the most critical subsystem, but there is generally an associated cost. Thus, the BN can be used to verify the change in the system reliability due to the improvement of the critical component or subsystem. It is only necessary to include the evidence in the model to perform this analysis. In other words, a scenario is created by assuming that the LNG/Propane subsystem does not fail during the mission time, and then the beliefs are recalculated given this evidence. Thus, it is possible to see the reliability of the system improve from R = 0.93873 to R= 0.96143. This is the greatest possible improvement in the system reliability acting on this subsystem (LNG/Propane subsystem). As mentioned in item 4.4.1, the critical components of this subsystem are pumps P2 and P3; thus, a scenario is created in which the evidence that these two pumps will not fail at mission time is included; the beliefs are recalculated to verify the improvement that can be reached by improvements in the reliability pumps. The greatest possible improvement in the system reliability acting on pumps P2 and P3 made the reliability system increase from 0.93873 to 0.94735.

Another issue is to improve the system reliability by adding a redundancy; in our case, for example, a redundant compressor, once this is the critical component given the fault system.

However, this action also incurs costs; thus, it is necessary to know the real effect of adding this redundancy. A BN is an effective way to model the appropriateness of adding a redundancy. A new node that represents the redundant compressor was included in the BN, and then the beliefs were recalculated to indicate the impact on the system reliability. For the system presented here, the reliability increased from R = 0.93873 to R = 0.95721. The same analysis may be performed to calculate the effects of a redundant compression subsystem in addition to a redundant component.

The evaluation of the real effects of a redundancy inclusion on the system is essential in decision making process about resources and tasks to improve the reliability system. A BN allows performing this analysis using a simple method.

4.4.3. Conditional reliability

The failure rates of the equipment with exponential failure distributions are constant; however, the failure rates of the equipment that have a different distribution modify over time; subsequently, the reliability is modified. The BN allows including a different previous operation time for each component by the inclusion of a node that indicates if a components has a previous operation time. Then, the beliefs are recalculated and the posterior probabilities are evaluated given this evidence.

As an application, a previous operation time of 1000 hours was included as evidence to the insulators, which have a Weibull distribution. Figure 5 shows a partial view of the system BN, in which the extra node which indicates if there is a previous operation time can be seen. In blue, the posterior probabilities given a previous operation time are illustrated, and in green, a new component is illustrated, without a previous operation time. It is possible to simulate different previous operation times for each equipment. For a mission time 96h, the failure probability of the insulator increases from 9.499×10^{-13} to 5.094×10^{-5} and the reliability system decreases from 0.93873 to 0.93859. This analysis allows checking the conditional reliability for any component operation time and, therefore, to know the reliability over time.

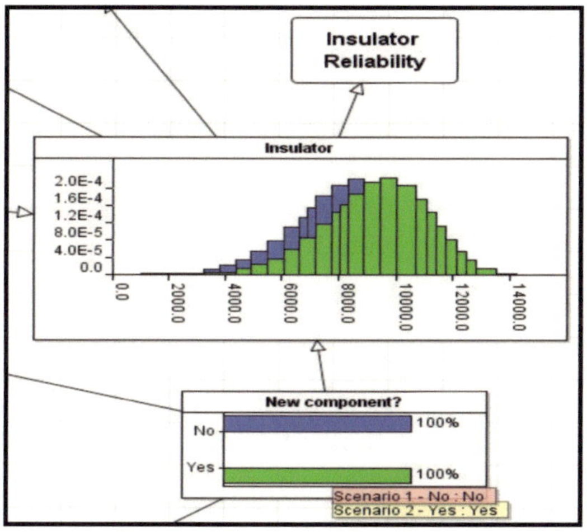

Figure 7. Conditional Reliability

5. Conclusions

It is essential to get more information about the system to make the analysis more efficient; however, based on the available data, the BN analysis shows that the LNG/Propane subsystem is the subsystem that caused the greatest impact on the reliability of the regasification system. Thus, the maintenance plan should pay special attention to this subsystem and consider the inclusion of a redundant subsystem, once that subsystem reliability may increase the regasification system reliability by more than 2%; a redundant subsystem is reasonable and not only the improvement of the critical components of this subsystem (P2 and P3), once they may improve the reliability system by less than 1%. The maintenance plan must pay special attention to the compressor, too, once this is the most critical component of the system given the evidence: system fault. The scenarios analyzed here are only a few examples; many others scenarios can be simulated using the BN capability to model several scenarios, according to the evidences.

The conditional reliability analysis shows that a significant change may occur in the failure probabilities components which do not have exponential distributions. Thus, it is recommendable to perform a detailed study about the failure distribution of each equipment, and then assess the conditional reliability behavior.

It is possible to model different scenarios, such as improvements in critical components, the impact of common cause failures, the impact of including redundancies and any other condition that affects the system reliability.

This chapter described the basic configuration of the regasification system on board a FSRU, evaluated the assessment of its reliability using the proposed methodology allowing to obtain detailed results for the mission time, in addition to permitting to consider some possible scenarios that led to the identification of critical points and considering possible improvements to the system reliability. Besides the evaluation system regasification, this chapter also highlights the potential of BN for the improvement of reliability studies.

In future work, the reliability analysis applied in this study may be expanded to the entire FSRU. In addition, a study about the system dependability may be performed; yet, to perform this study, the assessment of the maintenance process will be necessary. Also, the impact of a failure in this system may reach other areas of the vessel, such as the LNG tanks, which can lead to severe consequences. Therefore, future work may also focus on performing a risk analysis of the FSRU.

Author details

Marcelo Ramos Martins and Adriana Miralles Schleder
Naval Architecture and Ocean Engineering Department,
University of Sao Paulo, Brazil

6. References

[1] Ramos, M. A.; Martins, M. R.; Droguett, E. L.; Souza, E. P. "Quantitative risk analysis and comparison for onshore and offshore LNG terminals: the port of Suape – Brazil Case". In: International Conference on Ocean, Offshore and Arctic Engineering, 2011, Rotterdam, Netherlands. Proceedings of the ASME 2011 30th International Conference on Ocean, Offshore and Arctic Engineering, 2011. p. 1-8.

[2] Natacci, F. B.; Ikeda, N. H.; Martins, M. R. (2010) —Consequence analysis of a liquefied gas leakage||. Proceedings of the ASME, 29th International Conference on Ocean, Offshore and Artic Enginnering, OMAE 2010.Shanghai, China.

[3] Weber, P., Oliva, G. M., Simon, C., Iung, B., 2010, "Overview on Bayesian Networks applications for dependability, risk analysis and maintenance areas", Engineering Applications of Artificial Intelligence, doi: 10.1016/j.engappai.2010.06.002.

[4] Schleder, A. M. ; Martins, M. R. ; Souza, G. F. M. (2011) —Bayesian networks on risk analysis of a regasification system on an offshore unit.|| In: International Conference on Ocean, Offshore and Arctic Engineering, 2011, Rotterdam, Netherlands. Proceedings of the ASME 2011 30th International Conference on Ocean, Offshore and Arctic Engineering, 2011. p. 1-8.

[5] Agena risk desktop (2011). Available at: http://www.agenarisk.com/ products/desktop.shtml, accessed at 05/04/ 2011.

[6] Neapolitan, R. E. (2004) —Learning Bayesian Networks||. New Jersey, PearsonPrentice Hall, 2004.

[7] Langseth, H., Portinali, L. (2005) —Bayesian networks in reliability||, Reliability Enginnering and System Safety 92, 2007, p.p. 92-108.

[8] Jones. B.; Jenkinson, I.; Yang, Z.; Wang, J. (2009) —The use of Bayesian networking for maintenance planning in a manufacturing industry||. Reliability Engineering and System Safety 95, 2010, p.p. 267-277.

[9] Marquez, d., Neil, M., Fenton, N. (2010) —Improved reliability modeling using Bayesian Networks and dynamic discretization||. Reliability Enginnering and System Safety 95, 2010, p.p. 412-425.

[10] Marquez, D., Neil, M., Fenton, N. (2011) —A new Bayesian Networks approach to Reliability modeling||. Available at: http://www.agenarisk.com/studies.shtml, accessed at 05/04/ 2011 .

[11] Bobbio, A.; Portinale, L.; Minichino, M.; Ciancamerla, E. (2001) —Improving the analysis of dependable systems by mapping fault trees into Bayesian networks||. Reliability Enginnering and System Safety 71, 2001, p.p. 249-260.

[12] Netica Software, available at: http://www.norsys.com/, accessed on November, 10th, 2010.

[13] Hugin software, available at: http://www.hugin.com/serene accessed on November, 10th, 2010.

[14] Jensen, F. V., 2001, "Bayesian Networks and decision graphs", Springer, New York, NY.

[15] Schleder, A. M. ; Martins, M. R. ; Souza, G. F. M. (2011) Preliminary risk analysis of a liquefied natural gas regasification system in an offshore unit.|| In: International

Conference on Vulnerability and Risk Analysis and Management, College Park, Maryland, US. Proceedings of the International Conference on Vulnerability and Risk Analysis and Management, 2011. p. 1-8.

[16] Smith, D. J. (2001) —Reliability, Maintainability and Risk||. Woburn,UKA. 6ª edição. Butterworth Heinemann.

[17] Schuller, J. C. H (2005) —Methods for determining and processing probabilities – CPR 12E||. The Hague, Netherlands. 2ª edition. Gervarlijke Stoffen Publicattiereeks.

[18] Charniak, E. (1992) —Bayesian Networks without tears||, AI Magazine, 1991,12-4, p.p. 50-63.

[19] Whiiaker, J. (1990) —Graphical Models en apllied multivariate statistics||, Chichester, UK, Wiley.

[20] OREDA (2002) – Offshore Reliability Data Handbook||, Det Norske Veritas - DNV, 2002, 4th Edition.

[21] HSE - Health and safety executive (2001) – Offshore Hydrocarbon Releases Statistics, 2001 - Hid Statistics Report HSR 2001 002||. Available at: www.hse.gov.uk/hid/osd/hsr2001/contents.htm

Chapter 8

Innovative Method of LNG Storage in Underground Lined Rock Caverns

Eui-Seob Park, So-Keul Chung, Dae-Hyuk Lee and Taek-Gon Kim

Additional information is available at the end of the chapter

http://dx.doi.org/10.5772/45771

1. Introduction

Today, the demand for LNG (Liquefied Natural Gas) in Korea is growing rapidly. However, it is difficult to solve problems regarding the adjustment of the demand and supply of LNG due to factors such as seasonal variations of the domestic demand for LNG, discordance among import patterns, limits of storage facilities, and so on. The supply of LNG can become unstable due to accidents at LNG producing areas. Furthermore, it is very important to secure large LNG storage facilities and to stabilize LNG supply management on a long-term basis (Chung et al, 2006).

Since natural gas is imported in the liquid phase by ships, the refrigerated storage of LNG is more economical than the conventional storage in the form of gas. Furthermore, all the storage of natural gas in Korea needs to cater for the storage of LNG in the condition of LNG from ships. Therefore, a different storage concept is required for the country's imported gas, in order to store it in the same condition as that of LNG from ships (Cha et al, 2006). At LNG terminals, conventional aboveground tank or in-ground tank type has been successfully used worldwide to store LNG. Nevertheless, these types of the storage tanks are less attractive as they require large land area for storage. Large reclamation of seaside and industrial area has already progressed and free remaining area is very small and expensive.

Based on experience of underground storage for crude oil and various types of hydrocarbons, underground storage system was thought to be more economical way to store LNG regardless of the above-mentioned conditions. Many attempts have therefore been made in the past to store LNG underground in unlined containment, though without success (Anderson, 1989; Dahlström and Evans, 2002). One of the most significant problems related to the underground storage of cryogenic material is the need to prevent leakage of liquid and gas from the containment system to the rock mass caused by tensile failures due to shrinkage of the rock mass around the caverns (Monsen and Barton, 2001). The failures of underground storage caverns were due to thermal stresses generating cracks in the host rock mass. The thermal cracks induced by extremely low temperatures in the rock mass contributed to the deterioration of the operational efficiency. Gas leakage and increased heat flux between the ground and storage occurred (Dahlström, 1992; Glamheden and Lindblom, 2002).

The way to prevent a hard rock mass from cracking at LNG boiling temperatures (–162°C) is to locate the unlined storage cavern deep enough below ground level so that the geostatic stresses counterbalance the tensile stresses caused by the cooling. The necessary depth varies with rock type from 500 to 1000m, which renders this unlined cavern storage concept very expensive (Amantini and Chanfreau, 2004; Amantini et al., 2005).

New concept of storing LNG in a lined rock cavern (LRC) with containment system which can overcome these problems has been developed by Geostock, SKEC and SN Technigaz with the help of KIGAM (Korea Institute of Geoscience and Mineral Resources). To demonstrate the technical feasibility of this concept, a pilot plant was constructed at the KIGAM in Daejeon Science Complex in 2003, which was operated for storing liquid nitrogen (LN2, Boiling temperature: -196°C) from January 2004 to the end of 2004, and now been decommissioned.

2. Concept of a lined rock cavern for LNG storage

The basic concept for the development of LNG storage in membrane lined rock cavern is a combination of well-proven technologies and a new concept named as "formation of ice ring (Fig. 1):

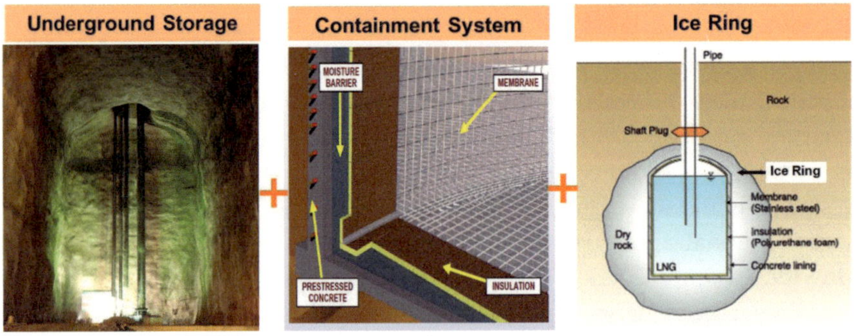

Figure 1. Three basic concept of the lined rock cavern system for storing LNG

2.1. Underground storage cavern

The underground storage technology, which proved reliable, experienced a strong development related to its intrinsic qualities but also on the innovations and technical progress from which it regularly profited. Currently, large capacity of storage cavern units with more than several hundreds of thousands of cubic meter, for a broad range type of hydrocarbon products and crude oil, are constructed and operated successfully especially in Korea.

2.2. Membrane containment system

The membrane containment system, which was introduced in 1962 on a prototype ship, has since been successfully used in several LNG carriers and storage tanks. This membrane system provides the proper thermal protection to the surrounding rock mass, preventing excessive stresses and crack formation and reducing boil-off to level comparable with conventional LNG storage tanks.

2.3. Ice ring and drainage system

Formation of ice ring as an impervious second barrier against the leakage of contained LNG can be explained by the each stage of construction of lined rock cavern (Fig. 2).

Groundwater is temporarily removed from the rock surrounding the cavern during the first phase of construction. This preliminary de-saturation of the host rock mass aims at preventing unacceptable hydrostatic pore pressure and ice formation behind the cavern lining. After excavation of cavern and installation of containment system are completed, LNG will be stored in the cavern and then the cold front starts propagation from the cavern at the same time. When the cold front has advanced far enough from the cavern wall, drainage can be stopped to allow groundwater progressively to rise up and quickly form a thick ring of ice around the cavern. After ice ring is completely formed, operation of the drainage system is stopped. The drainage period during LNG storage will last several months or years depending on the thermal properties of rock masses and hydro-geological characteristics on the site.

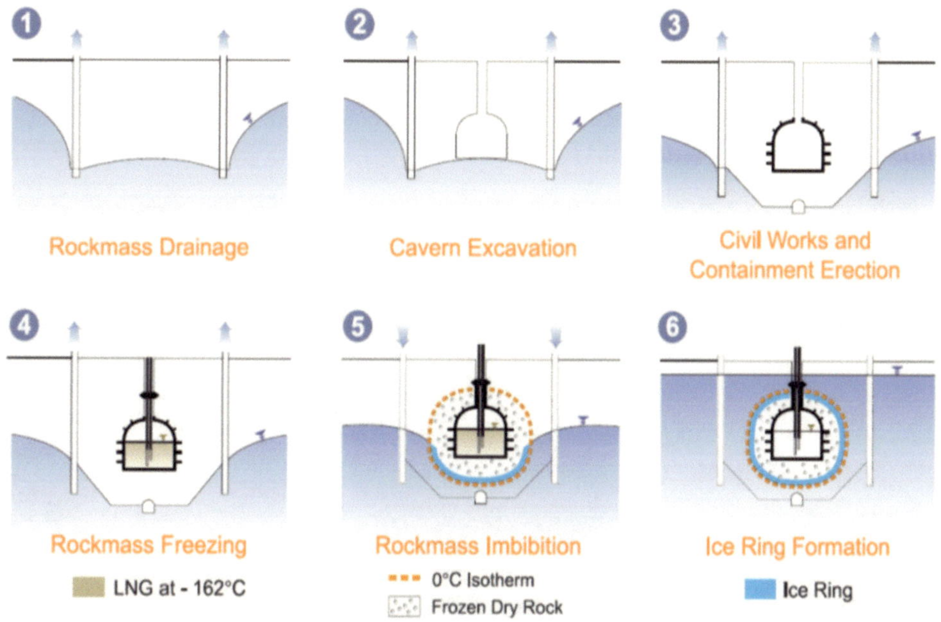

Figure 2. Schematic diagram of formation of ice ring

3. LNG pilot plant

The objectives of this pilot project are as follows:

- To demonstrate the feasibility of the lined rock cavern concept,
- To check adequacy between the results of previous computations and laboratory tests regarding in-situ measurements,
- To test the overall performance of the storage and
- To gain sufficient experience to improve design and construction of industrial scale projects

3.1. Underground pilot cavern

The LNG pilot plant is located in Daejeon, about 200 km south of Seoul, in an existing cavern implemented within the KIGAM research facilities. The previous room for cold food storage was enlarged to test the overall performance of a lined rock cavern for LNG storage. The base rock consists of Jurassic biotitic granite intruding Pre-Cambrian gneiss, which has an RQD of 80-86 and the most frequent Q-value of 12.5. The Q-value requires no supports or minor unreinforced shotcrete (about 40mm) and bolt according to support categories of Q-system. Therefore, it is proper to adapt rock bolting to stabilize main cracks of existing cavern and ensure stability of possible crack position.

Access to the pilot cavern was provided through an existing horizontal tunnel, and the experimental cavern roof lies at a depth of about 20 m below the ground surface. In order to have the containment system completed, a concrete wall closes the entrance of the cavern. The south concrete wall of the cavern is exposed to the entrance tunnel, which is not a typical case for full-scale facilities. Additionally, a platform above the entrance of the cavern is made to install instruments, manhole and piping. The internal dimensions of the completed pilot plant have a sectional dimension of 3.5 m x 3.5 m, a length of 10 m, amounting to a working volume of 110 m^3 (Fig. 3).

Figure 3. A bird's-eye-view and cross section of the pilot cavern

3.2. Containment system

The containment system, which is used for underground lined rock caverns, is similar to the one used and improved by SN Technigaz since 1962 for the membrane type LNG storage tanks and LNG carriers. The modular structure of the containment system makes it very flexible, improving construction and adaptation to cavern geometry. The thickness of the insulating panels can be chosen as per the requirement of the thermal efficiency and the nearly unstressed membrane permits its usage in very large scale future projects.

The containment system is composed of several layers, from rock to LNG as follows (Fig. 4).

For the Daejeon LNG pilot cavern, insulating panels made of foam are sandwiched between plywood sheets and are bonded on the concrete using load-bearing mastic. The insulation panel thickness is 300 mm to ensure that the rock temperature will not fall below -50°C after 30 years and the boil-off rate will stay at an acceptable limit. Finally, a 1.2 mm thick stainless steel corrugated membrane attached to the insulation panel provides gas tightness at a low temperature. All the surfaces (e.g. bottom, walls, chambers and vault) are covered with concrete lining, insulating panels and the stainless steel membrane sheets.

Figure 4. Containment system used for the pilot test

3.3. Drainage system

The purpose of the drainage system is to reduce water entry in the cavern during concrete casting, to reduce humidity percolating through the concrete during the installation of the containment system, to drain the rock mass and maintain a low water saturation degree in the surrounding rock mass during the cooling phase, and to control the re-saturation of the host rock mass once the frozen area around the cavern reaches a suitable thickness.

Before the cavern was excavated, 21 boreholes from the tunnel drifts were drilled for drainage near the cavern walls (Fig. 5), and the water table was lowered to 8 m below the cavern floor. Also, an acceptance test for the cavern was performed by an interference test after full installation of the concrete structure. The drainage system must be operated during the construction and operation of the LNG storage cavern.

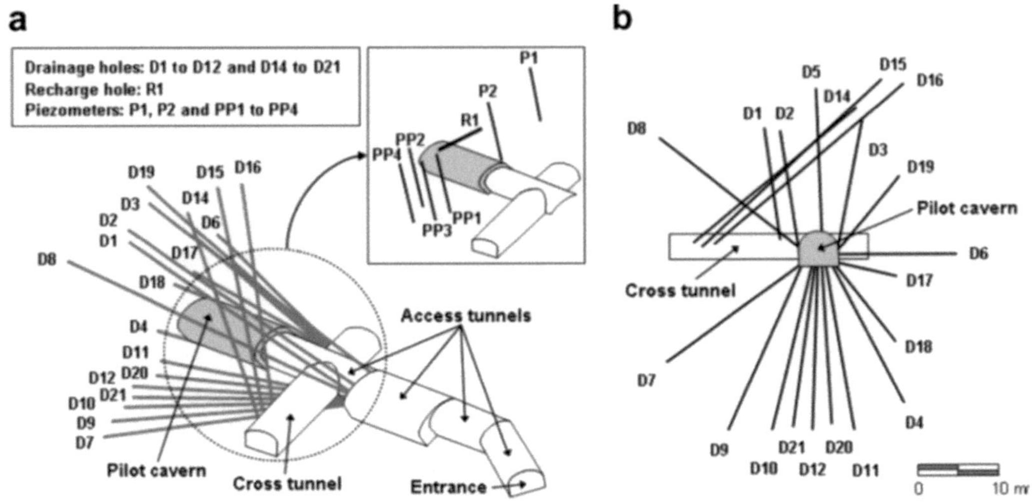

Figure 5. Schematic diagram of the pilot cavern, (a) Oblique view of drainage holes (D series), recharge hole (R1), and piezometers (P and PP series) & (b) front view of drainage holes (Cha et al., 2007)

3.4. Construction works

The transition structure between the rock and the containment system consists of reinforced concrete cast in place between a formwork and the rock, with injection holes for contact grouting. The interface cavern/access tunnel is constructed of reinforced concrete cast in place between two formworks including a closing wall and a platform used for access, piping and instrumentation. Thickness of the polyurethane insulation panel was 10 cm, and reinforced concrete barriers of 20 cm thickness were formed between the rock and the containment system.

4. Pilot test

4.1. Operation of the pilot plant

For safety and practical reasons, LN2 is used instead of LNG. The cryogenic pilot is operated from a laboratory room located in the access tunnel. It houses the Data Control System (DCS) for the process of the cryogenic pilot and for the rock monitoring. An LN2 plant is installed outside on a fenced site, containing one LN2 storage tank, a vaporizer and associated control systems in order to inert, cool down the cryogenic pilot, fill it completely with LN2 and make up to compensate for the gaseous N2 boil-off.

These operations use two insulated pipes from this dedicated site to the cryogenic pilot through the access tunnel and the special local enlargement in the cavern roof to go over the plug and penetrate inside the cavern. The additional line is necessary to release overpressures if they happen. Some instruments installed in the containment system and cavern permit to follow the main parameters (inner pressure and temperature, LN2 level, containment deformation) and to operate the pilot (Fig. 6).

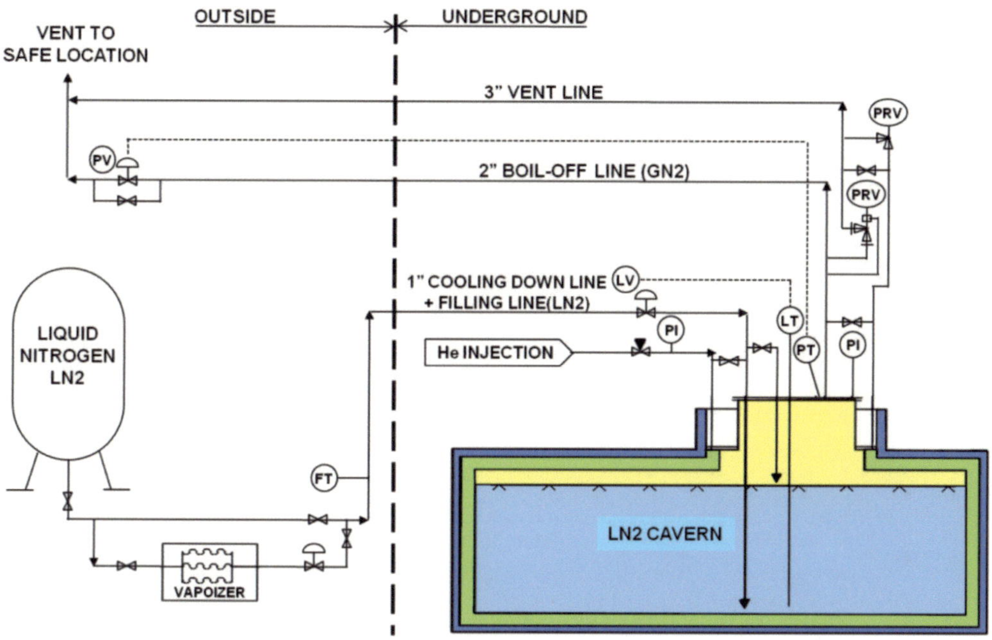

Figure 6. Schematic process flow diagram of the pilot plant

After several commissioning tests, the operation started from 10 January 2004 with a 2.6 m level of LN2 and the drainage system was stopped gradually from 10 June 2004 to 8 July 2004. The filling pipeline was closed on July 10th, and the storage has been completely empty since 12th of August, 2004.

4.2. Geotechnical monitoring system

A comprehensive monitoring system was provided for measuring temperature, thermo-mechanical and groundwater responses of the rock and concrete during pilot test. The pilot cavern and its surrounding rock mass are equipped with a comprehensive set of geotechnical instruments, which allows monitoring of temperature profiles and temperature induced displacements in rock around the cavern, the opening of rock joints on the cavern surface, the load on the installed rock bolts, settlement of ground surface, pore pressure distribution in the rock mass and the variation of ground water level. All instruments are equipped with thermal sensors to measure the temperature at the installed depth. Fig. 7 presents the general arrangement of the instruments in a cross section and a plan view (Chung et al., 2007a).

Moreover, numerous parameters such as level, temperatures, pressure and boil off rate for the containment system were monitored during the operation. The behavior of surrounding rocks and the containment system was recorded during three successive phases of operation: a) first six months duration during which a full level is maintained by filling with LN2 in order to compensate for loss of boil-off, b) second six months during which no more filling is performed allowing the cavern to empty naturally, and c) third six months during which the empty cavern is heated up till the ambient temperature is reached and the pilot cavern is decommissioned (Chung et al., 2004).

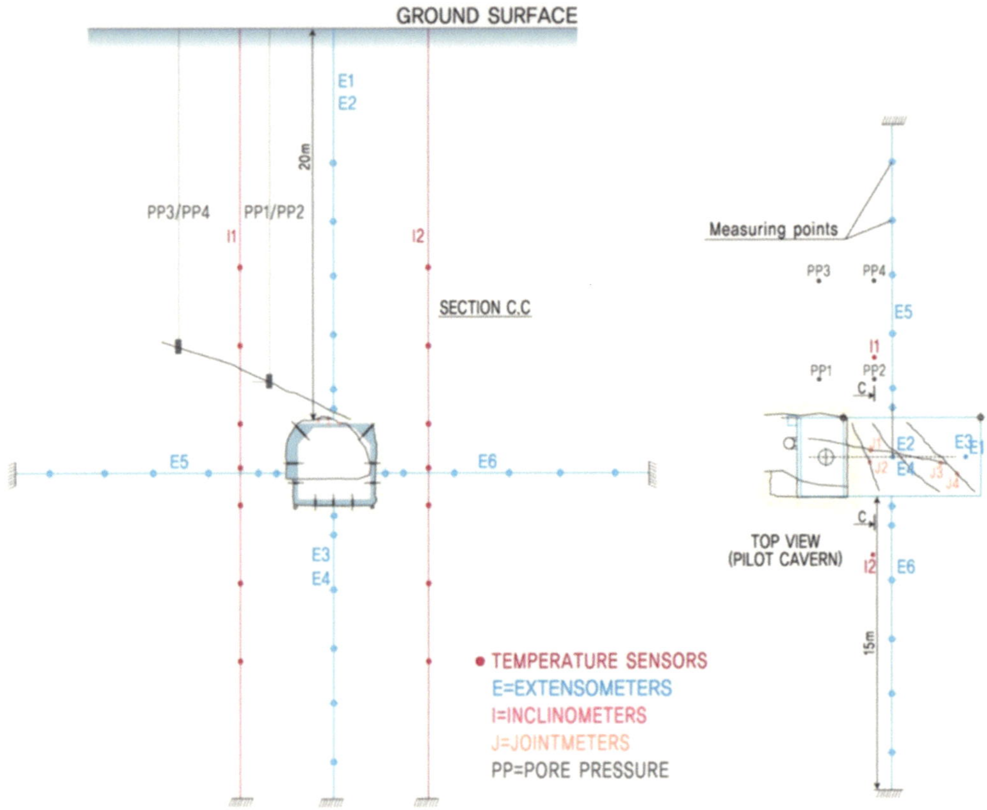

Figure 7. Arrangement of geotechnical instruments around the pilot cavern

4.3. Major results of the pilot test

The behavior of surrounding rock mass and the containment system measured during the operation of the pilot plant are analyzed and the following conclusions are made:

- The 21 boreholes drainage system has proved its efficiency during the successive construction and operation periods and allowed maintaining the rock mass de-saturated during the cooling down period. Efficiency tests performed at this stage showed that almost 99% of the natural water flows towards the pilot cavern could be drained back by the drainage system. The re-saturation of the rock mass was performed by stopping progressively the drainage. Finally, the drainage system allowed draining again the host rockmass during the thawing progress of the rockmass (Fig. 8).
- Thermal responses of the rock mass under a very low temperature of about -30°C around the rock cavern, with an absence of water, could be well predicted by numerical models such as FLAC2D code (Fig. 9).

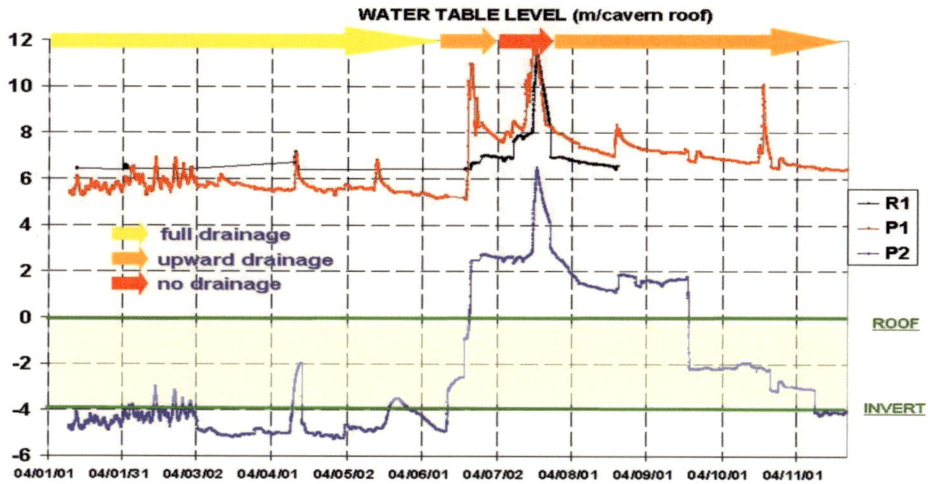

Figure 8. Piezometric variations by operation of the drainage system

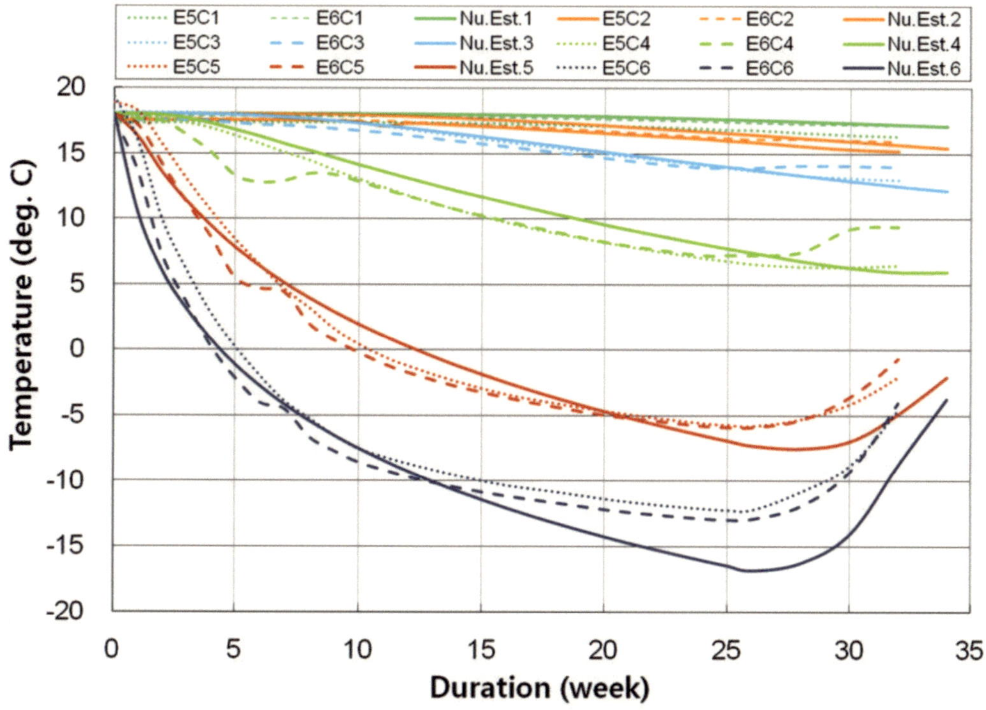

Figure 9. Comparison between measured and predicted temperatures of rock mass around the side wall of the cavern (dotted line: measured temperature, solid line: predicted temperature).

- Thermal stress-induced displacements occurred toward inner rock mass, which is favorable to stability aspects of the cavern. Displacements are of relatively low amplitude within 3~5 mm corresponding to about 0.2 % of cavern radius. Data recorded at the end of thawing period shows that small displacement (about 1 mm) is expected due to local rearrangement of the rockmass. And conventional rock reinforcements such as rock bolt and shotcrete remain effective at very low temperature during the pilot operation (Fig. 10).
- By comparing boil off gas ratio (BOR) occurred during operation of the pilot with estimated ones, BOR from underground LNG storage cavern can be estimated by numerical and theoretical methods (Fig. 11).

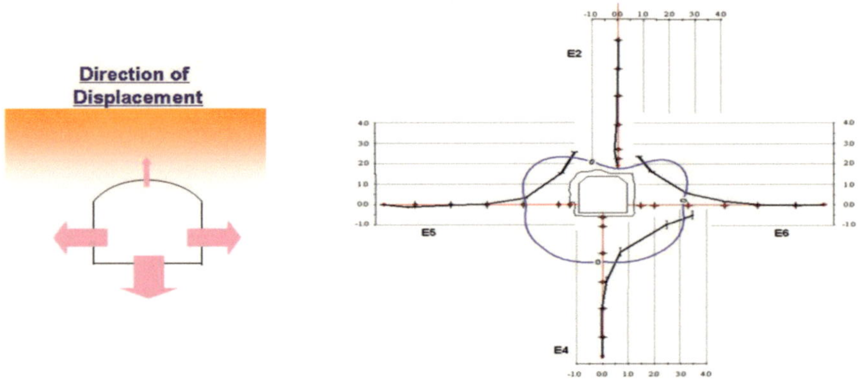

Figure 10. Displacement of rock mass around the pilot cavern

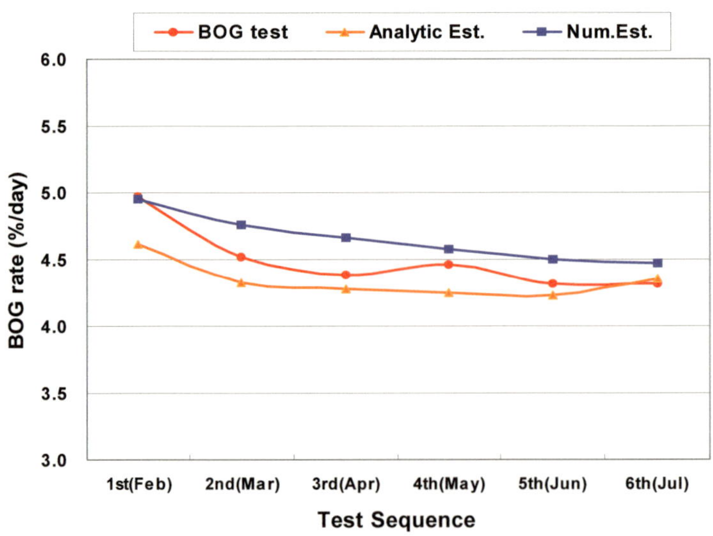

Figure 11. Boil-off gas ratio occurred during operation of the pilot plant

- After completion of the scheduled operations, the cavern and containment system were dismantled in order to judge the validity and safety of the proposed concept. Successfully, no remarkable thermal cracks were detected in the rock mass and concrete lining. There was no thermal shock by abrupt cooling near the cavern wall due to insulation barriers (Fig. 12).
- Overall, the results from the pilot test confirm that both construction and operation of underground LNG storage in a lined rock cavern is technically feasible.

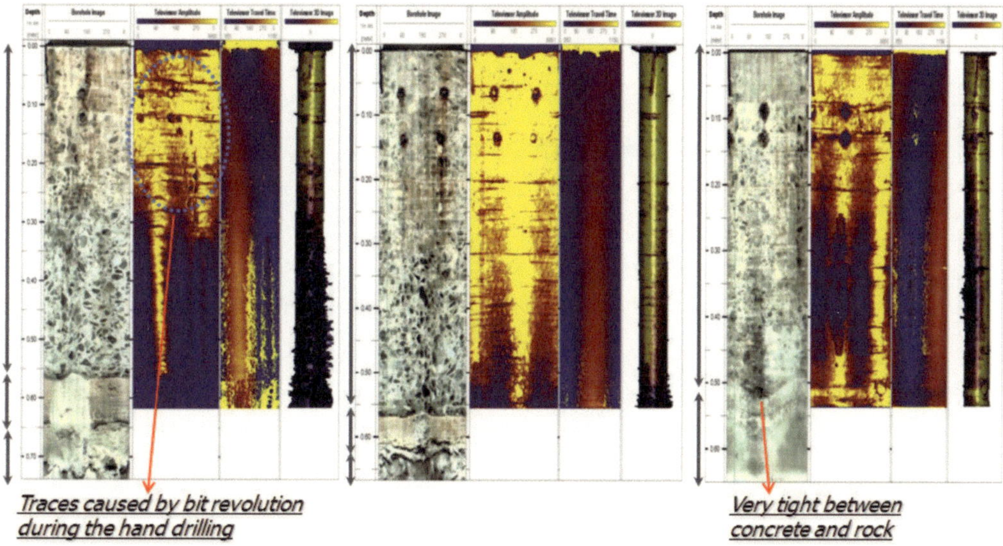

Figure 12. Results of crack inspection after decommissioning of the pilot plant

4.4. Hydro-thermal coupled analyses for formation of ice ring

A new groundwater control system was introduced to create an ice ring (ice barrier) around the LRC cavern to protect it from groundwater intrusion into the containment system. However, it can be used to balance the migration of zero degree isotherms by applying water with a specific temperature for water infiltration. In addition, the ice ring is supposed to play an important part as a secondary barrier against leakage of stored material. In the case of groundwater intrusion into the containment system, the integrity of the containment system could be destroyed because of the volume expansion of the groundwater during the freezing process. Therefore, the location and thickness of the ice ring are important factors for the stable storage of LNG in a lined rock cavern.

Combined hydro-thermal numerical models were adapted and used for investigating relevant mechanisms such as propagation of the cold front, and migration of water and ice formation in the host rock mass. Processes of ice ring formation with effective porosity of 3% are shown in Fig. 13 together with the temperature distribution of groundwater respectively. Ice ring in rock mass below the cavern is thicker than other regions. Dry zone inside of ice ring is obtained with thickness of maximum 2m. Until August 10th, ice ring is fully formed in surrounding rock. However, ice ring contacts partially the corner between wall and floor. It is found from comparison with the early interpretation of the geophysical campaign data that the real process of ice ring formation is very similar to the result from two dimensional simulations (Fig. 14).

10 June 20 June 30 June 10 July 20 July 30 July 10 August

Figure 13. Formation of ice ring and groundwater temperature around the pilot cavern (2-D simulation)

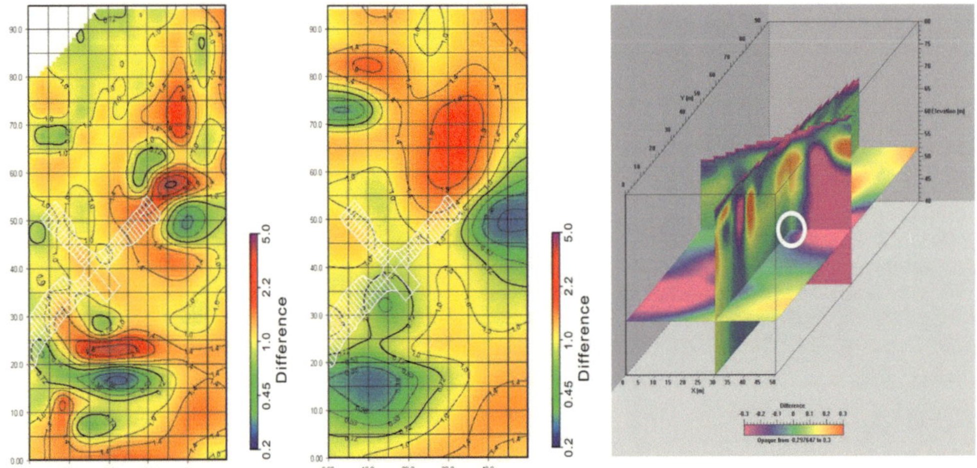

Figure 14. Change in resistivity distribution between Phase III and Phase II. Horizontal slice was made at the elevation (a) above and (b) at the level of the storage cavern. (c) 3-D fence diagram of resistivity change between two phases (Yi et al, 2005)

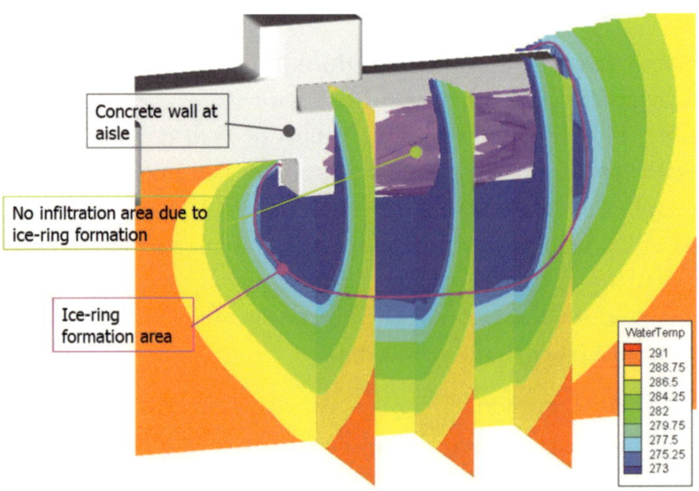

Figure 15. Shows the processes of ice ring formation by three dimensional simulations when the access tunnel is considered. Due to concrete wall exposed to the open air, ice ring was not formed and groundwater penetrated to a portion of the cavern.

From the hydro-thermal coupled analyses for simulating ice ring formation in the LNG pilot cavern, the following conclusions were drawn:

- The convective heat transfer coefficient through the roof membrane containing gaseous sky was obtained as 3 $W/m^2.K$ in the containment system with 10 cm thick insulation panel. When the containment system with more thick non-reinforced PU foam such as 30 cm or 40cm is considered, the coefficient would be reduced down to about 0.1 or 0.01 $W/m^2.K$. Because consideration of this coefficient is very important in rock mass above the cavern, probable range of the coefficient in full-scale cavern model with standard insulation specification should be obtained by appropriate numerical scheme for design purpose.
- The temperature dependency of the input properties must be considered for appropriate modeling of the cryogenic environment. Particularly when the temperature range of the simulation is wide, the variation of properties has to be taken into account.
- Numerical simulation with thermal properties of dry rock obtained by laboratory could effectively estimate the real temperature profiles in rock mass around a cavern.
- The capability of hydro-thermal modeling of ice ring formation has been verified by a comparison of numerical results with in-situ measurement data. Therefore, a similar approach could be extended to the simulation of ice ring formation in a full-scale LNG storage cavern.
- By controlling groundwater drainage system, the ice ring cab be formed easily based on the assumption that average distance of zero degree isotherm reaches 3 or 4 meter from the cavern wall in rock mass with the hydraulic conductivity of 10^{-7} to 10^{-6} m/s.
- In order to keep the continuous propagation of zero degree isotherms, groundwater close to cavern end faces should be effectively drained.

5. Typical underground storage system

This new concept of LNG storage in membrane lined rock cavern is valid for any type of export or receiving terminal, but also for peak-shaving or large capacity stockpile storage. Fig. 16 shows a full scale model for underground LNG storage system including above-ground facilities. Layout of the underground storage system and the number of storage caverns can be varied with storage capacity of LNG.

The drainage system of the lined underground LNG storage cavern is composed of drainage tunnel excavated beneath the cavern and drain holes drilled on rock surface of the drainage tunnel (Fig. 17). After the access tunnel for constructing caverns reaches to depth of cavern bottom, drainage access tunnel is excavated by depth of drainage tunnel bottom. And then drainage tunnels are excavated in parallel to the cavern with a little longer length than that. On the surface of the drainage tunnel, several drain holes should be drilled upward and arranged with same spacing. Drain holes outside the cavern boundary are inclined with a steep slope and those within the cavern boundary have a gentle slope.

The reason why drain holes are upward is that the drainage system is designed to let the groundwater be drained only by gravitational forces. That is, the groundwater within rock

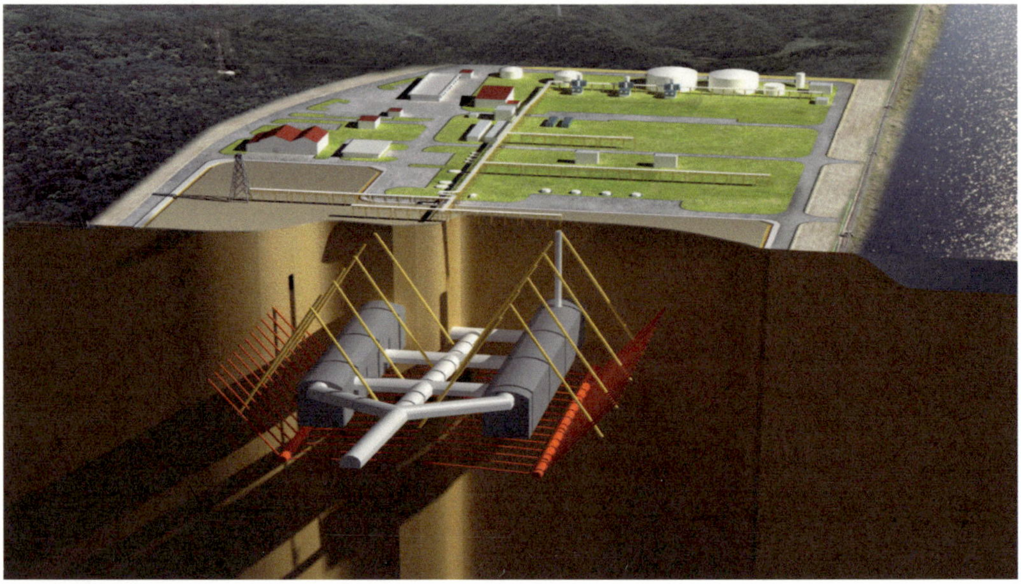

Figure 16. Bird's-eye view of a full scale model for underground LNG storage (Storage capacity of 140,000 m³)

mass flows into the drain holes along the joint-fracture channels connected those holes and is collected into the drainage tunnel only by gravitational force without pumping. In order to de-saturate sufficiently rock mass around the cavern, the position and horizontal spacing of drain holes should be designed efficiently.

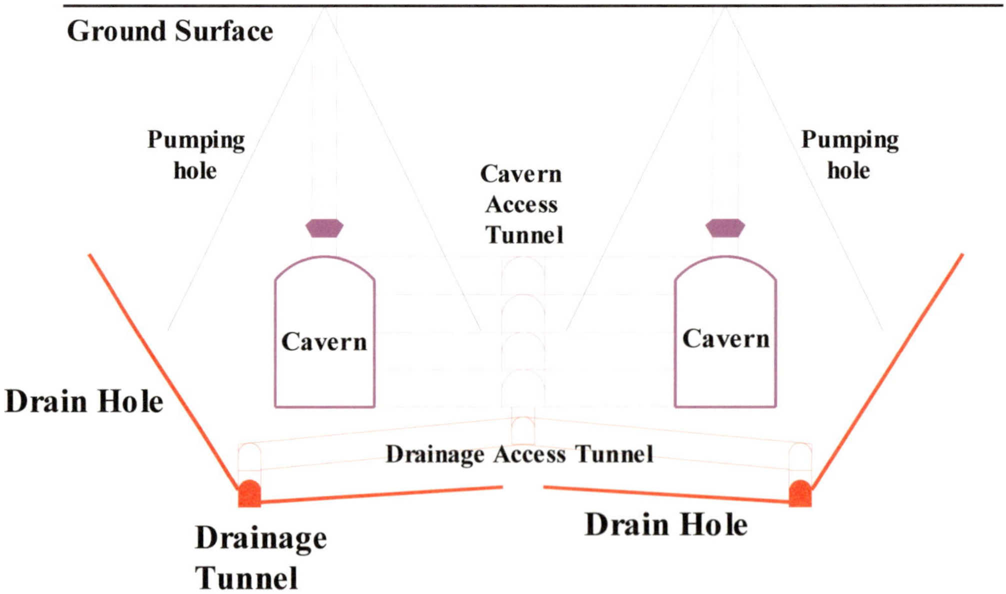

Figure 17. Typical cross section of the underground LNG storage system

5.1. Shape of storage cavern and tunnel

Geometry of the storage cavern should be adapted to local geological and hydro-geological conditions of the LNG terminal facilities. A wide variety of different geometries can be considered.

The general cross section of storage gallery is comparable to unlined underground storage technology, with horse shape galleries. Dimension of the gallery is typical 30 m height by 20 m width, which can be adapted to rock conditions. The length of cavern can be from 150 m up to 270 m, which depends upon the unit capacity per gallery. Fig. 18 shows typical sections of underground works and galleries arrangement that can be considered in good rock mass conditions.

5.2. Equipment, shaft and pipework arrangement

The equipment of storage cavern is similar to the ones of a conventional LNG terminal, allowing it to operate in exactly the same way as an aboveground tank (Fig. 19). Internal piping arrangement is also similar to the one of a membrane LNG storage tank. All pipes are embedded in a concrete structure near the bottom of the operation shaft. Inside the cavern the pipes are joined together by braces forming a pipe tower designed to permit differential pipe contraction.

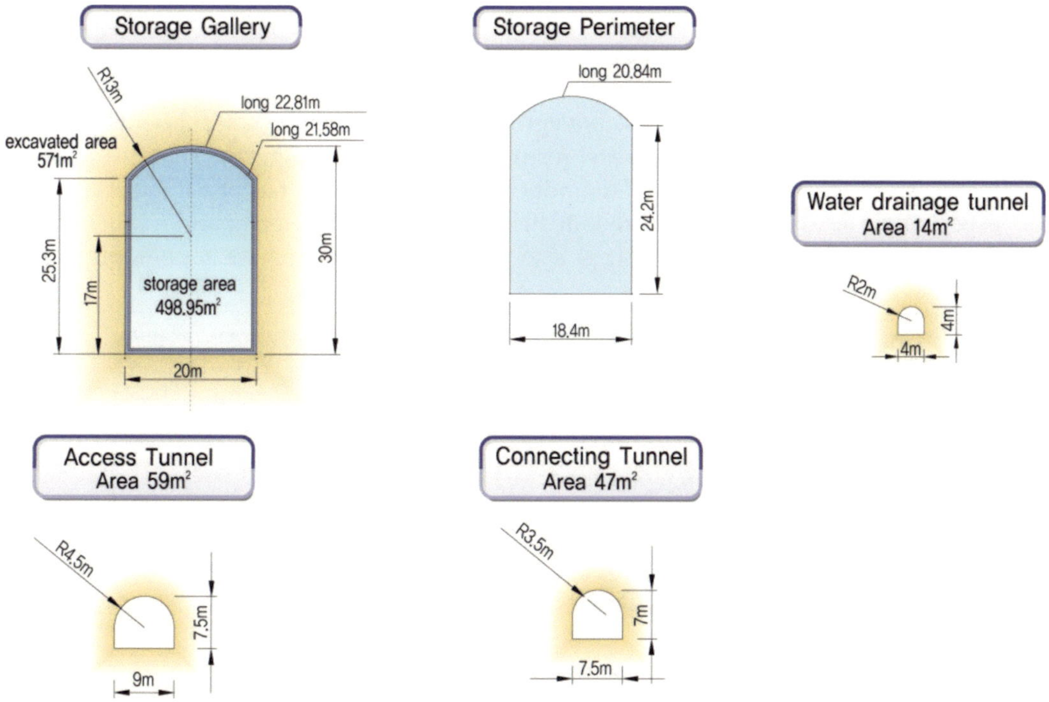

Figure 18. Typical cross section of underground galleries

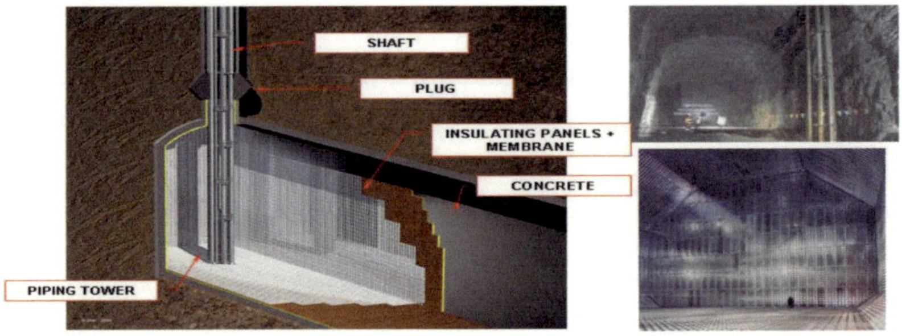

Figure 19. Typical cavern equipment arrangement (with pictures of LPG cavern storage and membrane tank of LNG carrier)

6. Advantages and economics of underground LNG storage

There are many advantages of underground storage in terms of safety, security and environmental acceptability compared with aboveground tanks and in-ground tanks:

- **Safety:** Underground storage is much safer in the case of fire on plant premises or decreased potential damages in the case of nearby industrial accidents, on account of its multi-component barrier with liner and ice ring, and lower vulnerability against earthquake and typhoon.
- **Security:** Enhances security, as underground storage can easily prevent sabotage or terrorism acts.
- **Environmental impact:** As there is no need of large reclaimed areas and less earthwork at ground level, underground storage is environmentally more friendly, and is readily accepted by people located nearby.
- **Use of land space:** Land of aboveground can be used better because of the minimization of total space required for the LNG terminal (Fig. 20). This is due to the fact that LNG storage is about 50 m underground, this can represent a huge cost saving especially in seashore areas where industries are already developed and free remaining areas are small and expensive. It is also the case in areas whose topography needs expensive reclaimed land.

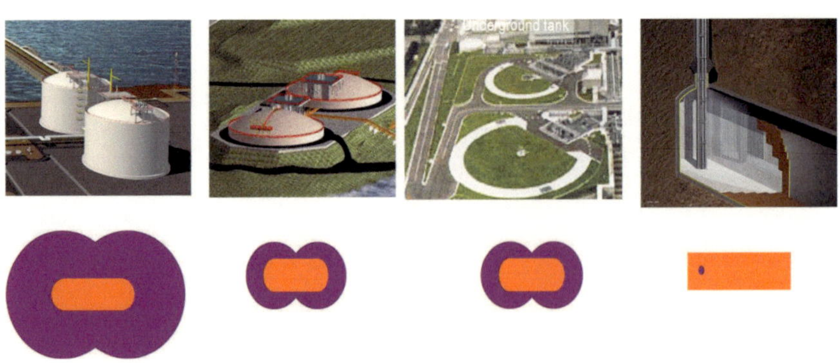

Figure 20. Required area at grade depending on storage type for 320,000 m³ (Orange colour means a storage space, purple colour means the required aboveground space)

Small galleries should be avoided wherever possible by reason of their poorer capacity/area ratio. Underground storage in the form of a gallery of around 20m width by 30m height cross section (Fig. 18) is the most favorable which has been studied geometry in cost terms versus rock behavior.

Moreover, mining technologies and membrane containment system have such flexibility that unit storage capacity has no limit. Crude oil caverns are up to 4,500,000 m^3 in Korea and LNG lined caverns can be increased till such value.

This comparative cost estimate between aboveground and cavern storage concerns only the storage itself and its equipment. It does not take into account the substantial cost saving which could be made, in the case of the cavern storage, for the safety equipment (impounding basin, peripheral retention wall, firefighting systems, etc.) and possibly for the piping length and terminal surface area reduction. Moreover, operation costs, including maintenance are very attractive.

In 2008, a national forum for cost comparison among conventional aboveground, in-ground LNG tank and underground storage was held in Korea where consisted of MKE(Ministry of Knowledge and Economics), KOGAS(Korea Gas Corp.), KNOC(Korea National Oil Corp.), KIGAM(Korea Institute of Geoscience and Mineral Resources), experts of engineering consultants and prestigious professors steered by Congress committee. Construction costs of varying stored volume from 200,000 kl to 1,000,000 kl with increment of 200,000 kl were evaluated and compared with reference price as of March 2006 in Korea (Fig. 21).

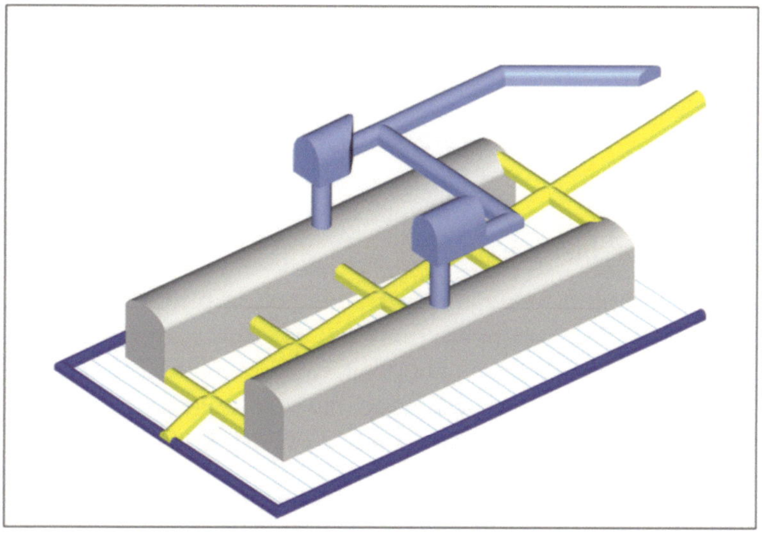

Figure 21. Typical layout of underground LNG storage system with capacity of 200,000 kl

In all cases, the underground LNG storage system is the most economic over the stored volume of 300,000 kl, and at 400,000 kl, the construction cost for underground LNG storage system can be economical by 8% and 34% compared to aboveground and in-ground tank respectively (Fig. 22).

Moreover, operation cost for underground storage units are highly competitive as compared to aboveground and in-ground tanks as systems like slab heating or fire water are not necessary or can be tremendously reduced. Based on Korean reference which has been implemented on crude oil storage by Korea National Oil Company, operation cost of underground storage is 63% less than that of aboveground one.

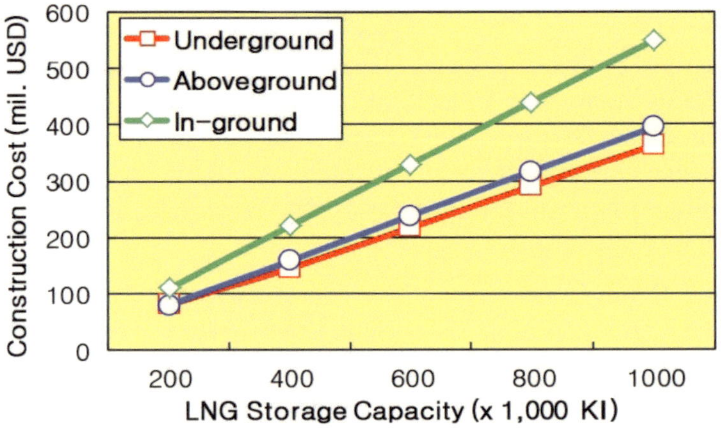

Figure 22. Capital expenditure (CAPEX) evaluation and comparison (South Korean context)

7. Conclusion

Based on experience of underground storage for crude oil and various types of hydrocarbons, underground storage system was thought to be more economical way to store LNG regardless of the above-mentioned conditions. An innovative method of LNG storage in lined rock caverns has been developed to provide a safe and cost-effective solution. It consists of protecting the host rock against the extremely low temperature and providing a liquid and gas tight liner.

To demonstrate the technical feasibility of this method, a pilot plant was constructed at KIGAM area and had been operated from January 2004 to the end of 2004. From construction and operation of Daejeon pilot plant, technical feasibility of underground lined rock storage system can be well proved. And the real scale applicability of it has been evaluated by the results from a successful operation of the pilot plant.

As compared with the conventional aboveground and in-ground storage tanks, the use of lined rock cavern LNG storage system at the LNG terminals can be more economical way in the aspects of capital expenditures and operating expenditures. In addition, it has also the advantage of safety, security and environmental acceptability against the conventional tanks.

Underground LNG storage system in lined rock caverns can be realized in due course at some countries which have suffered from the shortage of storage capacity of LNG and seasonal extreme variation of domestic demand, and of which industries are already developed and free remaining areas are small and expensive.

Author details

Eui-Seob Park
Korea Institue of Geoscience & Mineral Resources, Korea

So-Keul Chung
Korea Institue of Geoscience & Mineral Resources, Korea

Dae-Hyuk Lee
SK E&C Co. Ltd., Korea

Taek-Gon Kim
SK E&C Co. Ltd., Korea

Acknowledgement

This project was performed in cooperation with SK Engineering and Construction Co. Ltd., Korea, GEOSTOCK, and SN TECHNOGAZ, France. The authors are very grateful for permission to publish this work. Many relevant staffs of SKEC, Geostock, and SN Technigaz are also appreciated. This study is partly supported by the KIGAM's Basic R&D Project "Development of Core Technology for Underground Thermal Energy Storage in Rock Cavern".

8. References

Anderson, U.H. (1989). Steel lined rock caverns. In: *Storage of Gases in Rock Caverns*, Nilsen, Olsen(Eds.), pp. 1-10, Balkema, Rotterdam.

Amantini, E. & Chanfreau, E. (2004). Development and construction of a pilot lined cavern for LNG underground storage. *14th International Conferences and Exhibition on Liquefied Natural Gas*, Doha, Quatar, PO-33.

Amantini, E.; Chanfreau, E. & Kim, H.Y. (2005). The LNG storage in lined rock cavern: pilot cavern project in Daejeon, South Korea. *Proceedings, GASTECH 2005: 21st International Conference and Exhibition for the LNG, LPG and Natural Gas Industries*, pp.1-16, Bilbao, Spain.

Cha, S.S.; Lee J.Y.; Lee, D.H.; Amatini, E. & Lee K.K. (2006). Engineering characterization of hydraulic properties in a pilot rock cavern for underground LNG storage, *Engineering Geology* 84, pp. 229-243.

Cha, S.S.; Bae, G.O.; Lee, K.K.; Lee, D.H. & Bodin, J.L. (2007). Evaluation of drainage system around a lined pilot cavern for underground cryogenic LNG storage, *Tunnelling and Underground Space Technology* 23, pp. 360-372.

Cha, S.S., Lee, K.K., Bae, G.O., Lee, D.H., Gatelier, 2007. Analysis of rock drainage and cooling experiments for underground cryogenic LNG storage. *Engineering Geology* 93, pp. 117–129.

Chanfreau, E.; Amantini, E. & Lee, D.H. (2006). Assesing the use of underground storage tanks at LNG terminals, *LNG Journal*, June 2006.

Chung, S.K. (2006). Thermo-mechanical behavior of rock masses around underground LNG storage cavern, *Keynote lectures in Rock Mechanics in Underground Construction – 4th Asian Rock Mechanics Symposium*, pp. 19-28, Singapore.

Chung, S.K; Park, E.S. & Han K.C. (2006). Feasibility study of underground LNG storage system in rock cavern, *Tunnel and Underground Space* 16 (4), pp. 296-306 (In Korean).

Chung, S.K.; Park, E.S.; Kim, H.Y.; Lee, H.S. & Lee, D.H. (2007a). Geotechnical monitoring of a pilot cavern for underground LNG storage, In: *11th International Congress on Rock Mechanics*, Sousa, Olalla & Grossman(Eds), pp.1245-1248, Taylor & Francis, Lisbon.

Chung, S.K.; Park, E.S. & Han, K.C. (2007b). Feasibility Study of Underground LNG Storage System in Rock Cavern, In: *11th ACUUS International Conference*, Kaliampakos, D. & Benardos, A.(Eds), pp.501-506, NTUA Press, Athene.

Dalström, L.O. (1992). Rock mechanical consequences of refrigeration – a study based on a pilot scale rock cavern. PhD thesis, Chalmers University of Technology, Gothenburg, Sweden.

Dalström, L.O. & Evans, J. (2002). Underground storage of petroleum and natural gas. In: *17th WPC*, pp. 128-129, London: Portland Press Ltd.

Glamheden, R. (2001). Thermo-mechanical Behavior of Refrigerated Caverns in Hard Rock. PhD thesis, Chalmers University of Technology, Goteborg, Sweden

Glamheden, R. & Lindblom, U. (2002). Thermal and mechanical behaviour of refrigerated caverns in hard rock. *Tunnelling and Underground Space Technology* 17 (4), pp.341-353.

Jeon, Y.S.; Park, E.S.; Chung, S.K.; Lee, D.H. & Kim, H.Y. (2006) Numerical simulation of fracture mechanisms for rock masses under low temperature conditions, *Tunnelling and Underground Space Technology* 21(3-4), pp.470-471.

Jeong, W.C.; Woo, S.W.; Lee, H.S.; Lee, D.H.; Lee, J.M.; Choi, Y.T.; Jung, Y.B.; Park, E.S. & Jeong, S.K. (2006). Coupled Heat Transfer and Water Flow Analysis on Ice Ring Formation around a Underground LNG Storage Cavern, In: *Proceedings of the GEOPROC 2006 International Symposium*, Weiya Xu (ed), pp. 602-625, Nanjing China.

Jung, Y.B.; Park, E.S.; Chung, S.K. & kim, H.Y. (2011). Coupled hydro-thermal modeling of ice ring formation around a pilot LNG cavern in rock, *Engineering Geology* 118, pp.122-133.

KIGAM. (2003). Development of Base Technology for Underground LNG Storage and the Analysis on the Operation Results of Pilot Plant. *Research report by KIGAM submitted for SKEC*, 147p (in Korean).

KIGAM. (2004). Development of Base Technology for Underground LNG Storage and the Analysis on the Operation Results of Pilot Plant—2004. *Research report by KIGAM submitted for SKEC*, 144p (in Korean).

KIGAM. (2005). Development of Base Technology for Underground LNG Storage and the Analysis on the Operation Results of Pilot Plant—2005. *Research report by KIGAM submitted for SKEC*, 123p (in Korean).

KIGAM. (2006). Development of Base Technology for Underground LNG Storage and the Analysis on the Operation Results of Pilot Plant—2006. *Research report by KIGAM submitted for SKEC*, 177p (in Korean).

Kim, J.H.; Park, S.G.; Yi, M.J.; Son J.S. & Cho, S.J. (2007), Borehole radar investigations for locating ice ring formed by cryogenic condition in an underground cavern, *Journal of Applied Geophysics* 62, pp. 204-214.

Lee, D.H. (2004). Thermo-geomechanical monitoring & results, In: *Proceedings of International Symposium on LNG Storage in Lined Rock Cavern*, Energies, Saint Quentin en Yvelines, France.

Lee, D.H.; Kim H.Y.; Gatelier, N. & Amantini, E. (2003). Numerical Study on the Estimation of the Temperature Profile and Thermo-mechanical Behaviour in Rock around the Taejon LNG Pilot Cavern, In: *International Symposium on the Fusion Tech. of Geosystem Eng., Rock Eng. And Geophys. Exploration*, pp. 233-237, Seoul, Korea.

Lee, D.H.; Lee, H.S.; Kim, H.Y & Gatelier, N. (2005). Measurements and Analysis of Rock Mass Responses around a Pilot Lined Rock Cavern for LNG Underground Storage, In: *Proceedings of EUROCK 2005*, Konecny, P.(Ed), pp.287-292, Balkema, Brno.

Monsen, K. & Barton, N. (2001). A numerical study of cryogenic storage in underground excavations with emphasis on the rock joint response, *International Journal of Rock mechanics and Mining Sciences* 38 (7), pp.1035-1045.

Park, E.S. (2006). Thermo-mechanical consideration for the cryogenic storage, In: *International workshop on the Underground Storage Facilities in conjunction with the 4th ARMS*, pp.81-98, Singapore.

Park, E.S.; Chung, S.K.; Lee, D.H. & Kim, H.Y. (2007a). Experience of operating an underground LNG storage pilot cavern, In: *11th ACUUS Conference*, pp. 157~162, Athens, Greece.

Park, E.S.; Chung, S.K.; Lee, H.S.; Lee, D.H & Kim, H.Y. (2007b). Design and Operation of a Pilot Plant for Underground LNG Storage, In: *Proceedings of 1st Canada-US Rock Mechanics Symposium*, Eberhardt, E., Stead, D. & Morrison, T. (Eds), pp.1221-1226, Taylor & Francis, Vancouver.

Park, E.S.; Chung, S.K.; Kim, H.Y. & Lee, D.H. (2011). The new way to store LNG in lined rock caverns, In: *The2011 World Congress on Advances in Structural Engineering and Mechanics(ASEM'11)*, pp.3951-3956, Seoul, Korea.

Park, E.S.; Jung, Y.B.; Song, W.K.; Lee, D.H. & Chung, S.K. (2010). Pilot study on the underground lined rock cavern for LNG storage, *Engineering Geology* 116(##), pp.44-52

SKEC; Geostock; SN Technigaz. (2004). Proceedings of International Symposium on LNG Storage in Lined Rock Caverns, Seoul, Korea.

Yi, M.J.; Kim, J.H.; Park, S.G. & Son, J.S. (2005). Investigation of ground condition change due to cryogenic conditions in an underground LNG pilot plant, *Exploration Geophysics* 36, pp. 67–72.

Chapter 9

City Natural Gas Metering

Liji Huang

Additional information is available at the end of the chapter

http://dx.doi.org/10.5772/45809

1. Introduction

Natural gas as a city energy supply is well established in the developed countries while in the developing countries the usage of natural gas in city utility industry has significantly increased in recent years. The "town gas" or the "water-coal gas" generated from the process manufacturing of coal is vastly being replaced by natural gas. According to the worldwide energy consumption outlook released in 2011 by US Energy Information Administration, the worldwide annual natural gas consumption is at an average growth rate of 1.6% and will reach to 186.7 trillion cubic feet in 2035. The amount is almost doubled compared to that in 2003. This is not only from the requirements of environmental pollution control but for a better life style as natural gas is normally directly delivered to the residence providing much cleaner energy with higher efficiency. Metering the usage of natural gas is therefore a basic requirement for gas companies so that tariff can be fairly applied to customers and in return the revenue can further support gas company operation. This is particularly important today as current energy costs have skyrocketed. Some governments of Eastern European countries that were used to subsidizing or even providing free natural gas for residents are now starting to install gas meters under their new tariff system.

The first metering system for gas companies in history could be traced back to the very beginning of the 19th century when the gas was made for street and home lighting. The charges were initially based on contract and the usage was calculated on an hourly basis regardless of the actual consumption. Later gas charges were billed by the approximate numbers of gas burners counted by gas company workers touring streets every night. The first real dry gas meter that provides quantitative measurement was invented in 1843 in UK by William Richards and improved by Thomas Glover who established the first gas meter company in 1944. The meter was constructed with sheepskin diaphragms and a sliding valve enclosed with steel. Gas that enters into and fills up the first diaphragm chamber is pushed out for delivery while gas is filling the second chamber. The meter is hence named "diaphragm meter". This great invention provides excellent measurement range and is operated by pure mechanical structure without additional power. The invention virtually saved the gas companies and made them profitable for their services and continued growth. The dramatic increase of the gas meter business however started ironically after the invention of electricity bulbs in 1879 that took away the lighting business of the gas companies who were forced to develop new business for home cooking and heating in order to survive. Ever since almost all gas meters made for city residential gas metering are based on the same volumetric principle, although several attempts have been made but none proved good for mass deployment. Gas meters installed at residential place often experience many unexpected conditions including privacy constrains that make it very difficult for repair and maintenance. In addition, the pure mechanical character of the meter makes it almost impossible for *in situ* adjustment of measurement with respect to the change of environmental factors, i.e., temperature and pressure that in fact have great impact to the volumetric capacity of gas. The mechanical gas meters also require the direct human intervention for the data collection, which is often inconvenient as most of the meters are installed inside residential private area. To develop an alternative gas meter, it must have the capability of long life time, sustained accuracy, no service requirement during its life time and operation without external power. Despite the challenges, the advancement of the electronics and infrastructure as well as the ever increasing energy cost has undisputedly led gas companies into the urgency of a better gas management system.

In addition to the diaphragm meters for residential gas metering, commercial and industrial users are the other two types that most of the gas companies have to deal with. For commercial applications, rotary meters are commonly used as the diaphragm meters cannot withstand high gas pressure and handle the required high gas flow rate. Most of the diaphragm meters currently available are designed for measurement of a flow below 100 cubic meters per hour and a pressure below 100kPa. In addition, the size of the diaphragm meters for higher flow rate becomes very bulky creating many logistical difficulties. Rotary meters are also used as positive displacement (volumetric) gas meters and are built based on the lobed impeller principle proposed by Roots Brothers in 1846. The rotary meters are sometimes also called Roots meters. The first rotary meter for gas appeared in 1920s and became popular for commercial gas metering about 60 years ago. The rotary meters have better accuracy, higher pressure rating and smaller in size. In particular they can be equipped with a flow computer for temperature and pressure compensation for better performance. For industrial users with even higher pressure and larger flow rate that rotary meters cannot handle, turbine gas meters are often used. Unlike the rotary meters, the turbine gas meters are "inferential meters" as gas flow rates (volume) are calculated from the measurement of gas flow speed. The word "turbine" was derived from Latin word meaning spinning. The first turbine meter was invented by Reinhard Woltman in 1790, but the applications in utility industry only happened in late 1950s, it is likely because that turbine gas meters cannot handle the earlier manufactured gases with high humidity and impurity, only perform well in clean and dry natural gas. Turbine gas meters can be applied to high pressure and are acknowledged as primary standard with temperature and pressure compensation for custody transfer. However, they have a small dynamic measurement range and are not suitable for commercial or residential gas metering.

Since 1980s, many efforts toward the above mentioned objectives have been made for the development of all-electronic gas meters that can provide solutions to the drawbacks of the mechanical metering scheme for the city gas applications (Kang, 1992; Otakane, 2003; Matter, 2004; Kono 2006; Huang, 2010). The development and deployment of ultrasonic gas meter starting in late 1980s has yet to overcome the cost barrier and the progress has been slow. MEMS (micro electro mechanical system) mass flow meters for city utility industry have been developed in several countries since late 1990s. With the dramatic advancement of the electronic technology in this century, the all-electronic gas meters become more and more close to reality and commercialization of such meters is already evident. In this chapter, we will review the existing technology, current market demand and focus on the current city utility gas meter technology with MEMS mass flow sensing as well as the future outlook of the technology.

2. Market demand and smart gas meters

Current utility gas markets are driven by multiple factors. The global energy shortage makes the energy cost constantly increasing; global warming calls for usage of clean energy and reduction of carbon dioxide emission that city coal energy is one of the major contributors; success in electricity smart meter system (replacing the mechanical meter by electronic meter with networking) adds the pressure for the current man power based management system in utility gas industry; privatization and ever reduction of government subsidization forcing a more accurate tariff system. As such, better natural gas management is critical for gas companies. In the past few years, particularly after the world economic crisis starting in 2008, governments worldwide have made many initiatives to promote a better city energy system including those for natural gas management.

One of the initiatives related to the utility meters is the "smart metering infrastructure" that provides instant data access and management of the actual usage of electricity, water and gas. As of today, the electricity smart metering projects have proven to be effective, and wide replacement of the mechanical electricity meters with all electronic electricity meters has been executed worldwide. The "smart gas meters" are however still the diaphragm meters with an electronic data process device that converts the mechanical index into digital value and transmits to the data centre. According to the "smart meter project map" compiled by the UK Energy Retail Association, there are approximately 311 smart metering projects or initiatives worldwide as of 2011, among which only 37 or 11.9% are smart gas metering projects or initiatives. Table 1 and 2 summarizes the data from the above mentioned resources. The data indicate that the smart meters are still in their infancy and opportunities are huge as the market is now ready for taking new technologies. From the data one can notice that most of the projects or initiatives are in North America and Europe. Data summarized from the same database further indicated that the gas and water smart meter installation are less that 12% of the total smart meters in the fields. From the annual market data reported by the ABS Energy Research in 2008, there are 396 million gas meters installed in the world while the demand for new installation and replacement in 2012 was estimated over 35 million units with an annual growth of 4.0%. The fast growth of the new installation market is in Russia and China where the natural gas usage was previously subsidized by government or limited by resources. The growth rate in China is expected to be 8.7% as the country is converting the "town gas" into a much cleaner natural gas energy. Another high growth market with approximately a 15% growth rate would be in India where the construction of the natural gas delivery system is in progress. In North America and Europe the market is dominated by the replacement demand, where the US market is expected to have a 5.5% growth rate as promoted by the demand of AMR (automatic meter reading) or the AMI (automatic metering infrastructure). Because of the high replacement cost, difficulties in maintenance and long meter life time, gas companies waiting for replacement are aggressively searching for new technologies for the best cost structure and future management considerations. This could also be part of the reasons that the current smart gas meter installations are less than 2% of the total meters in the field. With the current existing technologies, the smart metering projects are also limited by the actual values that can be added.

	Gas		Electricity		Water	
	Projects	%	Projects	%	Projects	%
North America	16	43.2	103	47.9	40	67.8
Europe	19	51.4	78	36.3	13	22.0
Others	2	5.4	34	15.8	6	10.2

Table 1. World smart metering projects (2011), data can be found from Google map search.

	Gas	Electricity	Water
Smart meters	7.4	60.2	1.2
Total meters	396	1,584	736
% of smart meters	1.9	3.8	0.2

Table 2. Current worldwide field installed smart meters as of 2011(units in million).

As discussed above, current smart gas meter is simply the addition of an electronic device to the existing diaphragm meters in most cases. There are no changes in the actual metering technology itself. By nature, it facilitates elimination of the meter reading labour cost. However, the current electronic convertor and transmitter is very costly, even much higher than that of the meter itself, which is also a serious drawback for the deployment of the smart gas meters. The overall cost would not result in any savings while the labour elimination fosters unemployment as a side effect. Unlike the electricity smart meters that can provide hourly data, the transmitted gas data are an average of a few days of usage that greatly reduced the viability for analysis of customer consumption pattern. It also cannot compensate the undesired gas volumetric changes due to environmental condition variations. Reports of high installation errors for the smart gas meters in some earlier US deployments, as well as, some erroneous meter performance even initiated local referendum against the smart gas meter installations. Some experts also expressed concerns about the security readiness for data transmission with the current smart meter system and call for the current government initiatives as "money trumps technology".

The market demand for a better utility gas meter to meet the current challenges opens the golden doors for new technologies in this very traditional industry segment.

3. All-electronic gas meters for city natural gas metering

With the advancement of the electronics in late 1980s, development of all-electronic gas meters for utility industry was initiated. For obvious competition reasons for gas meters, the new technology must excel the old ones before it can be massively deployed. The diaphragm meters have their well acknowledged performance in large measurement range, long term reliability, reasonable accuracy, applicability for various gas compositions, low cost to manufacture, as well as, operation without external power. It is nontrivial to outperform these features for any of the existing gas measurement technologies.

The first prototype of an all-electronic gas meter for residential applications was reported by Kang et al. in 1992. The thermal time-of-flight measurement technology was used to build the prototype. A thermal pulse was sent from the transmitter while the pulse carried by the gas flowing through the transmitter was received by the receiver that was placed precisely at a pre-set distance from the transmitter downstream. A venturi structure was used for enhancing flow stability and boosting sensitivity at low flow range. The measurement of the travel time of the pulse can then determine the flow speed of the gas. The time-of-flight measurement scheme is by theory a pure flow speed detection that shall be independent of the gas composition as it is one of the variables that must be considered for the influence on measurements. The reported data indeed showed that the prototype calibrated in air can be readily applied to measure argon as well as natural gases with variable compositions. This prototype was however limited by the electronic technology in late 1980s, and the complicated construction of the transmitter and receiver with very thin hotwires which made it impossible for demonstration of reliability in field and capability in mass production regardless of its cost considerations. Nonetheless, this prototype successfully demonstrated the feasibility of an electronic gas meter for natural gas applications.

In addition to the above mentioned features of the mechanical meters that need to be matched by new technology, the all-electronic gas meters are expected to provide the capability of data safety, remote data transmission and management, and elimination of environmental conditions such as temperature induced metrology variations. In recent years, the efforts are made with ultrasonic and MEMS thermal mass technologies.

3.1. Ultrasonic gas meters

Ultrasonic flow meters were first introduced in 1963 by Tokyo Keiki and were used for industrial natural gas measurement in late 1970s by Panametrics (Yoder, 2002). There are two fundamental measurement principles in the ultrasonic gas meter technology, transit time or time-of-flight and Doppler shift. Either one is classified as the inferential measurement similar to that for the turbine meters. In the time-of-flight measurement configuration, a pair of transmitter/detector was placed at a distance apart inside the flow channel and close to the channel wall in most cases. The time difference between the signal transmitted from upstream to downstream and the one from downstream to upstream is proportional to the gas flow rate. For the meters employed with the Doppler shift principle, the ultrasonic signal sent via the ultrasonic transmitter across the flow channel is deflected by the particles inside the flow stream. The measured Doppler frequency shift is proportional to the flow speed of the particles that are traveling at the same speed along with the flow stream. Most of the ultrasonic gas meters are made with the time-of-flight technology as "particles" in many gases may not even be present to reflect the ultrasound for measurement.

With more technical understandings of the ultrasonic gas meters in natural gas industry, and particularly after the development of the multipath ultrasonic sensing measurement technology that two, four or even six pairs of the ultrasonic transducers are installed inside the same pipeline and the averaged data significantly enhanced the measurement accuracy. These advancements helped substantial increase of the ultrasonic meter deployment in utility gas industry. In the second half of 1990s, both Europe and USA have started to establish the standards for the technology, and subsequently published regulations for use of the technology in natural gas custody transfer applications. Both the Technical Monograph 8 by European Association of Natural Gas (GERG) and AGA-9 by American Gas Association are the milestones for ultrasonic meters for natural gas applications. However, due to the high cost and difficulties for the transducers configured in small gas pipelines, the ultrasonic gas meters were largely limited to the usage in large pipelines replacing turbine meters or orifice flow meters for custody transfer.

The first attempt to use ultrasonic meter for residential applications started in 1991 in UK. About 200 meters manufactured by Siemens and Gill Electric R&D were installed together with newly calibrated diaphragm meters in serial connection throughout the country. These meters were about half the size of that for the diaphragm meters with same flow range. The meters were powered by a battery with the anticipated life of 10 years and equipped with a safety shut-off valve. The measurement range of these ultrasonic meters was comparable to that of diaphragm meters and the accuracy was slightly better. In addition, these meters were also readily applicable for AMR and pre-payment schedule. The new technology hence showed good improvements for their performance as compared to those of the diaphragm meters. Japanese major gas companies including Tokyo Gas, Osaka Gas and Toho Gas led the way for the development of ultrasonic gas meters for residential applications in 2001, and field tests of 100 meters similar to that in UK started in 2003 and concluded in 2005 without showing any inferior performance as compared to those diaphragm meters in series. US meter manufacturer Sensus introduced the ultrasonic meter for residential applications in the same period of time. Both UK and US models only covered G4/G6 in equivalence to diaphragm meters while Japanese models do include the very low flow range models of G1.6/G2.5. In 2007, European Committee for Standardization (CEN) published the EN14236-2007 for ultrasonic meters for residential applications that significantly boosted the usage of the technology. However, due to substantially higher cost compared to those for diaphragm meters, residential ultrasonic meters are very limited in field installation and are mostly in the developed countries (approximately 1.3M in UK; 0.03M in Japan and 1.2M in USA). For more than 15 years after the first deployment, only less than 0.7% of the total existing meters worldwide are now ultrasonic meters in the regime of residential applications.

Ultrasonic gas meters for city residential or commercial gas metering by principle measure directly gas speed. With the current electronic technology, it has the advantage of readiness for data transmission, data safety, remote access and management. In addition, the measurement is independent of gas composition. On the other hand, the ultrasonic meters also have the same disadvantages of temperature and pressure dependence and thus bear the similarity of the volumetric measurement character that are currently adopted by city gas industry. Therefore it would have less public sensitivity when even partial replacement of the existing meters in a specific area takes place. Customers would not notify any differences in principle if both the diaphragm meters and ultrasonic meters are installed in the same neighbourhood. Although the advantages of the ultrasonic meters seem very compelling and attractive to gas companies, their high cost is nonetheless a huge barrier for the market penetration. Moreover, the desired compensation for the environmental variations could only be achieved with the addition of temperature and pressure sensors which not only add to the cost that is already higher acceptance but also it will introduce additional metrology errors.

3.2. Differential pressure gas meters

Differential pressure gas meters are the oldest technology for gas flow measurement. They utilize the differential pressure sensor to measure the pressure drop across the designated gas pipeline and calculate the corresponding flow rate. It is therefore also an inferential type of flow meter. The performance of this type of meter depends on the accuracy of the differential pressure sensor incorporated, as well as the design of the pressure dropper inside the flow channel that shall serve as the source of the accuracy for the measurement. The higher pressure drop will be easier for the pressure sensor to resolve the differences. Pressure sensors with high accuracy and large measurement range are often costly and not readily available on market. Consequently, the differential pressure gas meters usually could not have a wide measurement range, in most cases with a turn-down ratio (maximum detectable flow rate over minimum measurable flow rate) smaller than 5:1. With additional temperature compensation, these meters can provide fairly accurate measurement and hence they are traditionally used in gas stations for custody transfer together with a large pump since the flow rate is relatively stable at the gas stations. This type of meter is therefore not suitable for city gas distribution purpose. The mandatory presence of the pressure dropper or orifice within the flow channel creates additional pressure loss other than the normal ones due to pipeline transportation, which is very much undesirable for the city gas distribution where the commercial pipeline gas pressures are often within a few kilopascal. High pressure loss is detrimental to the gas delivery capability for the end user applications.

In recent years, gas meters for residential applications based on the differential pressure sensing principle have been developed. One of the products currently available on market was manufactured by Betar Meters of Russia. As of today, there is only one model available that is equivalent to the model of the smallest diaphragm meter G1.0 with a maximum flow rate of 1.6m^3/hr. The meter body was made of casted metal while the flow channel was constituent of several stamped aluminium plates that make the meter very compact and very low cost. It is the smallest available gas meter on market. Because the silicon based differential pressure sensors consume very low power, the battery powered differential pressure gas meters are expected to have life of over 10 years. The silicon pressure sensors can be mass produced with excellent consistency that can be an advantage in meter manufacture cost reduction. However, because of the sensitivity of the pressure sensor, this meter has a small turn-down ratio of 40:1 that is far inferior to the 160:1 of the current diaphragm meter technology. In addition, the market share for such a small flow ranged gas meter model is small, and such model has actually already been phased out in many countries. Furthermore, compensation of environmental variations such as temperature for this type of meter adds significantly to cost as well as metrology errors.

As discussed above, meters with the differential pressure sensing technology are not the best solution for the city gas distribution applications for its small measurement range and sensitivity to the environmental conditions. Differential pressure measurement in a larger pipeline may not be useful due to the presence of large undesirable pressure loss.

3.3. MEMS gas meters

It is generally recognized that MEMS technology was born after Christmas in 1959 when Richard Feynman delivered his speech, "There is plenty of room at the bottom", at the American Physical Society Meeting, which inspired and promoted the technology worldwide. However, the terminology of "MEMS" only appeared in 1987 when a series of workshops on microdynamics was held in California. The European society is used to name the technology as "microsystems" while the Japanese scientists call it "micromachines". But today MEMS is widely accepted by the international community. MEMS utilizes the device process technologies similar to those used in integrated circuitry to build a comprehensive microsystem that can execute designated electronic, mechanical, optical, magnetic and/or thermal functionality. They can perform multitasks in a small form factor with minimal requirements of energy consumption at a nominal cost. As of today, devices made by MEMS technology have penetrated into our everyday life - TV sets, automotive, computer and peripherals, cell phones, projectors, medical devices, scientific instruments, just to name a few. These devices have significantly changed the way we live, saved thousands of lives and enhanced our understandings to the world surrounding us.

MEMS gas meters utilize the MEMS mass flow sensing technology to measure gas flow rate. With the state-of-the-art electronics for the signal process, MEMS gas meters have extended dynamic range, enhanced data safety and are easy for network and remote data transmission. The measured flow rate has automatic temperature and pressure compensation and it

	Ultrasonic	Differential pressure	MEMS Mass flow
Temperature compensation	Additional	None	Included
Pressure compensation	Additional	None	Included
History (as of 2011)	20+ years	3 years	18+ years
Models	Residential and commercial	Small residential	Commercial and industrial
Markets	UK, US, Japan	Russia	China, Japan, Italy
Calorific value	No	No	Possible
Integration	No	No	Yes
Cost	Medium to high	Low	Low

Table 3. Comparison of current all-electronic gas meter technologies.

is also possible for direct calorific value measurement. These meters also have a substantially smaller form factor for better logistics. These features are ideal for the current demand of better gas energy management. In Table 3, the current all-electronic gas meter technologies are compared with respect to their characteristics.

The actual deployment of MEMS mass flow meters in utility industry started in mid 2000, and limited shipment of the commercial models to UK has started since 2011 that aims to replace the small turbine meter currently installed. Residential models of the MEMS mass flow meters made by MEMS AG had been extensively sampled in many European countries since 2003 while actual installations are few. One model of this meter for measurement of flow rate up to $6m^3/hr$, equivalent to G4 of diaphragm meter is available from Diehl Gas Metering.

In the following sections, we will discuss in detail the all-electronic gas meters made of MEMS mass flow sensing technology.

3.3.1. MEMS mass flow sensing technology

Using the silicon planar technology to make mass flow sensing devices was first proposed by Hutton in 1971 and the first device made on a 50 μm thick silicon substrate was reported in 1974 by Putten et al. The sensing principle of this sensor was energy dissipation (anemometer) and the sensing elements were made of diffused p-type resistors close to the four edges of a 1.5x1.5 mm silicon surface. Using the Wheatstone bridge circuitry it conceptually demonstrated that the sensor can be used to measure the gas flow. This novel device would be the proximity of the MEMS mass flow sensor commercialized some 15 years later. Current commercial MEMS flow sensing products are made based on the principles of calorimetry, energy dissipation and thermal time-of-flight measurement. Others are most in the stage of research. A review of these research activities can be found in the published book chapters (Haasl 2008, Bonne 2008).

The calorimetric flow sensors (Gehman 1985, Bonne 2008) have a microheater at the middle and two temperature sensors are placed symmetrically with respect to the microheater. The temperature difference, ΔT, is a measure of mass flow rate, q_m, thermal capacitance, C_p, and thermal conductivity, ξ, of the fluid: $\Delta T \propto P(\xi)/q_m \times C_p$. The thermal conductivity can be measured from consumed power, P, of the microheater. The calorimetric flow sensors are best for low flow rate as the sensing is also limited by the boundary conditions.

The construction of the anemometric sensor (Bruun 1995; van der Wiel 1993) is relatively simple as it only needs to measure the energy dissipation of the microheater when the flow fluid passing through the microheater. Therefore, the measured ΔT between the microheater and that of the gas temperature is proportional to the dissipated energy P and the mass flow rate of the fluid: $\Delta T \propto P(\xi)/(q_m)^k$, where k is the factor that related to the meter design. The heat dissipation is usually insensitive at low flow rate and therefore it is best for applications for high flow rate measurement.

The time-of-flight (TOF) flow sensors (Ashauer, 1999; Shin 2006) have the similar configurations to those of anemometric sensors, but the operation mode is not to measure the power of the microheater. The upstream thermistor sends out continuous pulses while the downstream thermistor measures the time of the heat that the pulse carried from the upstream to downstream. As the distance can be precisely made or can be further precisely calibrated, the flow speed of the fluid can then be precisely determined. Obviously this technical is best for very small flow rate. In practice, the frequency phase shifts between the two thermistors are measured for enhanced accuracy and easy data process.

A typical structure of a commercially available MEMS flow sensor is shown in Figure 1. As shown in the figure, the sensing elements and the microheater are built on the membrane that is normally made of silicon nitride or silicon oxide and silicon nitride combination. The underneath cavity provides excellent thermal isolation that shall boost the sensitivity at a low operational power. This cavity is made by front wet chemical etch in some of the earlier products, but later with the invention of deep reactive ion etching (DRIE) technology, backside opening of the cavity was applied for better yield and efficiency. The slots on the membrane shall provide the pressure balance to minimize the deformation due to internal gas pressure. Many materials that have a high temperature coefficient of resistance (TCR) can be used to make the sensing elements and the microheater, but for long term reliability, the most commonly used materials are platinum or doped polycrystalline silicon. The thermistor placed on the silicon substrate is used for measurement of environmental temperature that can provide feedback to microheater such that a constant temperature or constant power operation mode can be maintained for the performance. The elongated design of the sensor is to place the interface connection pads away from the sensing elements and the connections via wire bonding can be subsequently sealed with package materials to ensure no shortage or damages from possible deposits of conductive materials or impact of particles carried in flow fluid. The space between the sensing element and the interface is also designed to be large enough so as to ensure that the flow profile disturbance due to the package is minimized.

The design shown in Figure 1 can be used for either calorimetric, or anemometric or time-of-flight flow sensors, but most of the current commercially available sensors are using

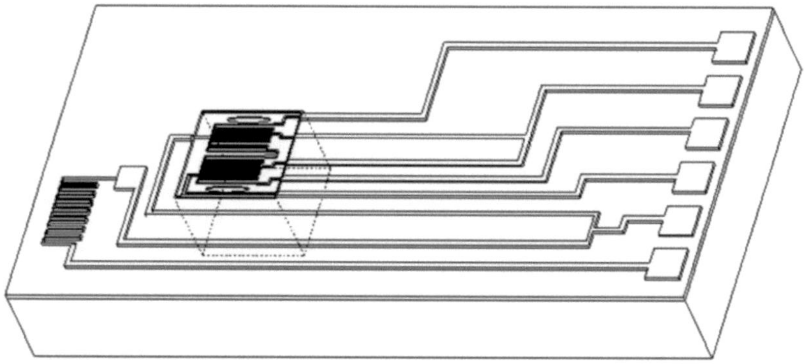

Figure 1. MEMS flow sensor designed by Siargo.

calorimetric measurement principle. Among the current commercial product suppliers, the first MEMS flow sensor was manufactured by Honeywell (Higashi 1985) in the late 1980s. Honeywell's MEMS chip utilizes the calorimetric principle having a footprint of approximately 2x2 mm. Its suspended membrane with openings was made by front wet chemical etch. The sensing elements were made by FeNi alloy or platinum. Because of the sensor size, the wire bonded at the interface is exposed to the fluid. This has limited the applications of its AWM series of flow sensors to clean and dry gases only. In mid-1990s, Bosch (Hecht 1997) released the MEMS air flow sensor for automotive applications. This sensor has also utilized the calorimetric sensing but with independent thermistor to measure the temperature of microheater. This sensor has no openings on the membrane which would however not a problem for the designated applications in automotive electronic control unit. In late 1990s, Sensirion (Mayer 2004) released a MEMS flow sensor that integrated the mass flow sensing elements with the control electronics on a single chip. This design uses doped polycrystalline silicon thermal piles as the up- and downstream thermistors featuring very low power consumption and good sensitivity. The sensor's membrane was similar to that of Bosch's made by back side DRIE process without open slots which results in a limitation for higher pressure applications. In additional, since it is undesirable to place the exposed integrated circuitry to the fluid, the sensor has to be packaged to a tiny channel that in return limits its applications in high flow rate. Yamatake (Azbil) shared the patents with Honeywell (Nishimoto 1991) and thus the sensor structure is very much the same. Omron (Fujiwara 2005) introduced its mass flow sensor products in earlier 2000s, the sensor chip has a similar footprint as that of Honeywell but its sensing elements were provided by polycrystalline silicon thermal piles. The MEMS mass flow sensors by Siargo (Wang 2009; Huang 2011) were designed for use in utility gas flow meters with a wide dynamic range to cover requirements from residential to industrial applications in city gas distribution. The sensor integrated calorimetric and anemometric sensing technology, and further enabled the self-cleaning capability. The openings on the membrane ensure the buffer to withstand medium gas pressure in industrial applications. There are a few other MEMS flow sensor manufacturers, but none of them are directly applicable to natural gas metrology.

3.3.2. Current MEMS gas meters

The first MEMS gas meter for city gas applications was developed by MEMS AG using Sensirion's MEMS flow sensor in 2000 (Matter 2004). This meter was designed to replace diaphragm meter G4 for residential applications. The compact all-electronic gas meter was a sensation for the industry and field tests were conducted extensively. However, as previously discussed, the bypass design would always be a concern for reliability although the manufacturer had data to show the otherwise. There may be additional concerns about the gas composition dependence as well as lack of industrial standards that shall be a barrier for the custody transfer applications. Further the one model product shall be difficult for the customers to manage their tariff system. In 2007, the product was transferred to Swiss Metering for sales and marketing but today it is available under its parent company, Diehl Gas Metering. This one model meter had a different appearance but at least the gas composition dependence as indicated by the user guide would have the same limitation. (www.diehl-gas-metering.com) As some critical hazardous protection ratings are still pending, this product at current remains a prototype. Yamatake released its "μF" series of MEMS gas meters for industrial applications in 2004 that was powered by external supply. A battery operated correspondence with the same performance was released in 2007. (www.azbil.com) The design of the meter utilized due chip package that enhanced the dynamic range. Due to its MEMS chip design as discussed earlier in this chapter, the meters have to have excessive protections at the inlet of the flow resulting in huge pressure loss which is very much undesirable in the city gas distribution. For a higher pressure, however, its maximum flow capability is rather disappointing. Further, missing of the hazardous protecting would add additional concerns for city gas metering. In 2011, Metersit announced its four models of MEMS gas meters matching to the equivalents of diaphragm G4, G6, G16 and G25, these models are particularly in response to Italian government's smart gas meter initiatives and regulations. They however have the same gas composition limitation as the same MEMS sensor by Sensirion was used. (www.metersit.com) Siargo since 2007 introduced its series of MEMS gas meters for both industrial (4 series, 10 models) and commercial (4 series, 7 models) gas metering applications. The products have been shipped to four countries as of 2011. (www.siargo.com) The MEMS gas meters manufactured by Siargo have the excellent dynamic measurement range by integration of the calorimetric and energy dissipation sensing elements onto a single MEMS chip. These battery powered models have an Ex ia IIC T4 rating that satisfies the city gas metering requirements. Table 4 compares the key performance parameters of the current available MEMS gas meters designed for city natural gas distribution by different manufacturers.

In this table, C means calorimetric and E is for energy dissipation. LF stands for "low flow model" while HF stands for "high flow model". Siargo's LF models include residential and commercial applications. Detailed ranges will be discussed later. Gas group H is based on European standard EN437; 12A/13A are two gas types in Japanese supply system. Siargo's meters can be used for most of the natural gases with optional automatic composition variation compensation. The pressure loss listed in the table is for low flow models. For higher flow, the pressure drops are in line with the counterparts of corresponding diaphragm meters or rotary meters. In the following sections, we will discuss the detailed design, performance and reliability of Siargo's gas meters.

	Diehl	Yamatake	Metersit	Siargo
Release date	2003	2004	2011	2006
MEMS sensor provider	Sensirion	Yamatake	Sensirion	Siargo
Technology	C	C	C	C+E
Max. flow (LF) (m^3/hr)	-	160	6	4/160
Max. flow (HF) (m^3/hr)	6	1600	40	3600
Min. pressure (mbar)	-	100	-	-
Max. pressure (mbar)	100	9800	150	3000/7000
Pressure loss (mbar)	200	>1200	200	200*
Gas Group	Group H	12A/13A	Group H	Any
Hazard rating	tbd	n/a	tbd	Ex iaIICT4

Table 4. Comparison of key performance parameters of current MEMS gas meters.

3.3.3. *MEMS gas meter design*

To accommodate the current requirements in city gas metering, the meters are designed into three series that cover the industrial, commercial and residential applications. Ideally the design shall have all current features of their mechanical counterparts while the new functions shall add significant values. In particular, the followings are among the design considerations:

- Similar or better dynamic flow range
    - To achieve this, calorimetric and energy dissipation sensing elements are integrated onto a single MEMS chip. The detection limit was then extended down to 0.008 m/sec and up to 75 m/sec. The theoretical dynamic range for this MEMS sensor shall have a turn-down ratio over 2000:1, which in principle is far better than the 160:1 turn-down for the diaphragm meters. And this single MEMS sensor chip can then be used to cover all dynamic ranges in city gas metering. The meter algorithm shall automatically determine the transitional point where the measurement scheme shall be switched from calorimetric to energy dissipation. The larger dynamic range shall require a longer manufacture (calibration) cost, therefore the actual dynamic range shall be a balance of performance and cost.
- Temperature and pressure compensation
    - This is certainly the advantages for the mass flow sensing principle as it does not require additional sensing elements to separately measure temperature and pressure. The additional measurement elements shall also incur additional metrology errors. This however requires the design for pressure balance on both MEMS sensor chip and the package.

- Leakage detection for enhanced gas safety
  - As the chip can measure very low flow rate, the algorithm can determine whether a leakage could be present if a predetermined constant low flow rate is to be measured continuously for a certain period of time, and an alarm can then be displayed or transmitted via the communication port to the data center.
- Stand-alone operation without external power
  - The beauty of mechanical metering technology is that it is operated by mechanical movement without external power. The all-electronic gas meter has to be powered by electric sources. Therefore the key would be that the power required must be low enough that a single battery can be used for sustained operation through its whole product lifetime.
- Data safety
  - With today's stat-of-the-art electronics, three individual data storage units are designed and placed on the electronic control board. Programmable data record is provided as an option for desired data safety regulations. And the data record will automatically trigger when the unexpected interrupt of the control sequence takes place such as battery failure or sensor fault.
- Build-in communication and network capability
  - The standard build-in communication protocol is Modbus (RS485) that is ready for external network and/or wireless modules such as GPRS. Optical port shall be an option, as well as other internal communication ports such as I^2C or SPI.
- Calorific value assessment and compensation
  - The integrated elements can measure the relative value of thermal conductivity and thermal capacitance from which the calorific value can be assessed. Detailed discussions can be found later in this chapter.
- Compact design for cost reduction
  - As the MEMS sensor chip is miniature, the sensor assembly including the electronic control board can be designed into a compact form that is substantially smaller than the mechanical counterpart. This will provide additional benefits for the reduction of the cost not only in manufacture but for overall gas distribution management.
- Hazard rating
  - As the applications demand, the design must meet minimal requirements for hazardous protection to ensure safety.

The series of the current products are shown in Figure 2. The flanged meter series (a) are designed for applications in the gas pipeline with medium pressure up to 1.5MPa with a pipe diameter from 25 to 150mm and maximum measurable flow rate from 125 to 3600 m^3/hr. The meters for commercial applications have a pipe diameter of 20 to 80mm covering maximum flow rate from 6 to 160 m^3/hr and maximum working pressure of 1.0MPa. The residential gas meter series have a pipe diameter of 15 to 20 mm for a maximum flow rate of 2.5 to 4 m^3/hr. The residential series have a maximum working pressure rating of 0.3MPa. The flow rates for all these meters are calibrated at the standard conditions of 101.325kPa and a customized temperature of 0, 15, or 20°C.

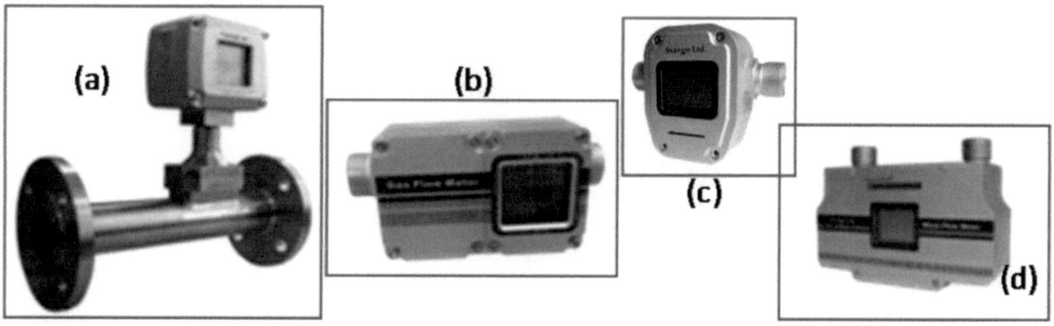

Figure 2. MEMS gas meter series manufactured by Siargo. (a) industrial applications; (b) commercial applications; (c) residential applications and (d) residential applications with pre-paid card reader option.

The meters are powered by a lithium ion battery of 19Ahr for a life time of four (industrial), six (commercial) and twelve (residential) years. For all models, both the instant mass flow rate and accumulated flow rate at the standard conditions (101.325kPa, 0/15/20°C) are displayed simultaneously. The battery indicator on the LCD will flash approximately three months before the end of the battery life. The battery pack was placed in a separate chamber that enables the integrity of the metrology during change of the battery. For data safety purpose, the meters have three separate nonvolatile memories to record the operation status of time, instant flow rate and accumulated flow rate as well as the alarm status. The clock is maintained by a crystal and can be remotely synchronized as well as adjusted if daylight-saving time should be accounted. Each memory can store up to 3000 items that are programmable by users for their specific application requirements. For remote data transmission, RS485 with Modbus protocol provides connections to the local concentrator and further transmission could be via wired or wireless network. Optical communication port is an option.

3.3.3.1. MEMS gas meter mechanical design

For all models, the sensors are inserted at the center of the flow channel that is manufactured with a venturi structure for flow stability. Figure 3 shows the assembly structure of a commercial gas meter. The lithium ion battery pack provides the power for the meter at the separate chamber for the purpose of safety while the electronics itself is designed to be intrinsic safe. The meter body is either made of stainless steel or aluminum alloy. The flow conditioner assembly is placed at the inlet of the flow which adds to a maximum pressure loss of less than 200Pa for the smallest pipe diameter at the ambient working conditions. The optional connectors usually shipped with the meter provide easy connection to the existing pipelines. For the medium pressure flanged series and the residential series, the components that formed the meters are basically the same except for a different dimensions and package.

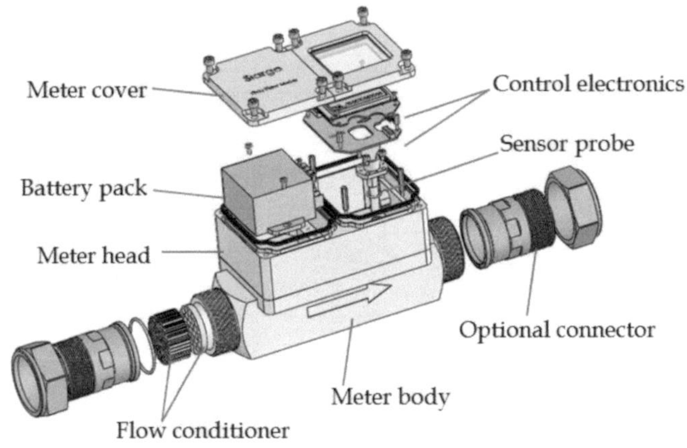

Figure 3. Component schematics of the commercial gas meters

3.3.3.2. MEMS sensor assembly

The sensor probe assembly is shown in Figure 4. The sensor probe is made of stainless steel. For large flow applications and meters with diameters more than 50mm, dual or tri-sensors are packaged on the same probe so that the mass flow value can be averaged from multipoint measurements resulting in an enhanced accuracy. The probe is shaped into a plate with a thickness about 1 mm that shall form a boundary layer in the flow channel. The sensor/plate surface is parallel to the gas flow directions such that a redistribution of the gas forces the flow into a laminar formality across the sensor probe. This laminar flow shall be helpful for maintenance of gas conversion and flow stability for the sensing signals. When the dual or tri-sensors are on the same probe, the installation of the probe will ensure that the sensor at the tip is placed at the center of the flow channel (master sensor) and the other ones (slave sensors) will be placed at one fourth (dual sensors) and one third (tri-sensors) of the flow channel diameter, respectively. As the sensor surface direction is in parallel to the flow direction, the edge of the probe is made with a sharp slope so that any particle impact onto the sensor assembly will have a good chance to be impelled away from the sensor surface reducing the head-on collision induced sensor reliability.

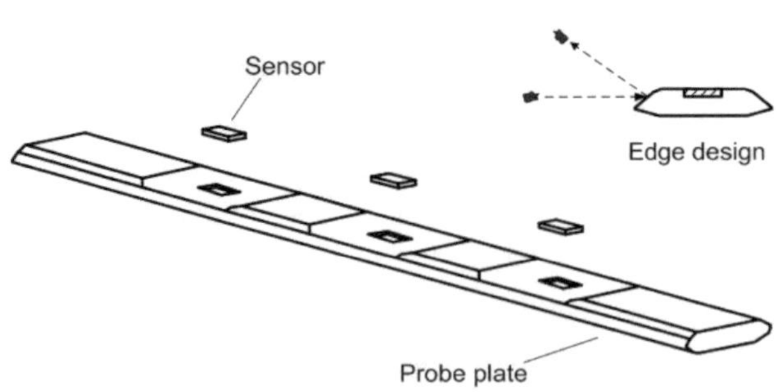

Figure 4. Schematics for sensor probe assembly.

3.3.4. MEMS gas meter performance

3.3.4.1. Permissible errors

The meters were calibrated by a sonic nozzle system that has an uncertainty of ±0.2%. The uncertainty of the sonic nozzle was custody transferred via a Bell Prover with an uncertainty of ±0.05% and traceable to a national standard. The measured permissible errors for the meters were obtained by another independent sonic nozzle system that has the same uncertainty of the one used for the meter calibration. Figure 5 shows the measured data that indicated the maximum permissible errors of the meters are well within the general requirements for city gas distribution, benchmarking to those by the traditional mechanical meters with temperature and pressure compensator.

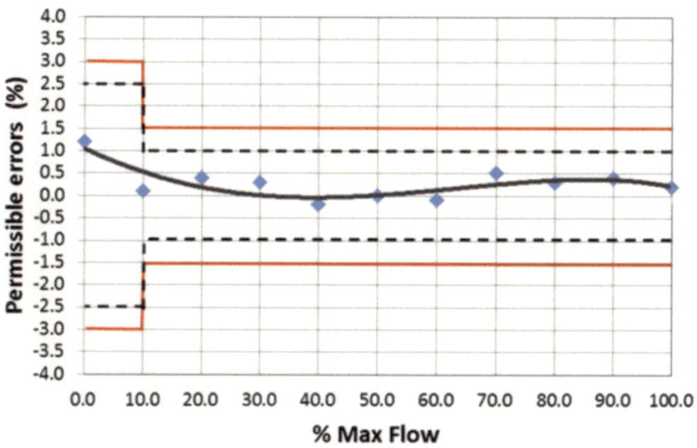

Figure 5. MEMS meter permissible error measurement.

While the MEMS meters by principle are automatically compensated for variations of temperature and pressure, the design of the sensor assembly may however introduce additional effects from the pressure changes in the pipeline. This is due to that the sensor chip has a free standing membrane with a cavity underneath. If the sensor design and its assembly cannot provide the pressure balancing configuration, pressure variations in the flow channel may lead to minor membrane deformation that would be sufficient for altering the sensor accuracy. Therefore in the sensor chip (Figure 1) and the assembly design, pressure balance via the openings on sensor chip as well as via the openings on the support and below the sensor cavity is made to eliminate the pressure change induced by sensor membrane deformation. In addition, high pressure could also introduce measurement errors as the thermal properties of the gas under high pressure may be substantially different from those at ambient calibration conditions. Another factor that would impact the meter performance is the temperature compensation of the electronic circuitry as it could produce additional errors due to the component temperature performance deviations. However, this could be removed by additional temperature calibration, and normally a temperature coefficient of 0.015%/ºC could be achieved that shall be in line with the city gas distribution requirements.

3.3.4.2. Pressure loss

Pressure loss is one of the major undesirable factors for the city gas metering as the pipeline pressure is often low when reaching to end users. A higher pressure loss will not only introduce additional energy consumption for the distribution system but may lead to customer issues, *e.g.*, the gas may not be sufficient for burner operation or even may fail to fire residential ranges.

In the current design, the sensor probe is directly inserted at the center of the flow channel instead of a bypass configuration that requires a pressure dropper between the inlet and outlet of the flow channel. Additional flow stability provided by the venturi structure of the meter body in the design also would not introduce additional pressure drop as it has been well demonstrated in literature. Figure 6 shows the measured pressure loss in air at 20ºC and 101.325kPa for the commercial models. These values are compatible with those by the diaphragm meters. For the actual usage in natural gases, the pressure loss shall be even smaller for about 40% as for the differences in the densities of air and natural gases.

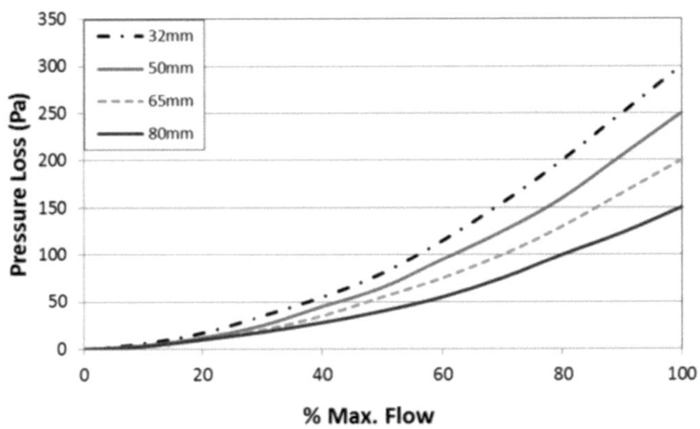

Figure 6. Pressure loss for the commercial gas meter models measured in air.

3.3.4.3. Installation conditions

For mechanical meters particularly for turbine meters, it is necessary to have enough long straight pipe lines before and after the meter installed so that the flow stability can be ensured and the measurement permissible errors can be within the requirements for city gas distribution. This limitation makes the installation costly and sometimes it simply cannot be met due to field space restrictions. In the current MEMS meters, a pair of flow conditioners (a combination of a flow straightener and a flow profiler) is designed to create a controllable flow profile regardless of the flow conditions. The plate design of the sensor assembly and its position in the flow channel result in the boundary conditions making a laminar redistribution, which again help the stability and reproducibility of the flow measurement. Therefore the straight pipeline requirement in these meters would not be a crucial factor for flow measurement uncertainties. Figure 7 shows the test and verification results by placing different bended pipes before the inlet of the flow meter. The meter was connected to a sonic nozzle system as the reference as shown in the figure. At each condition, permissible errors were taken against the original calibration, and Figure 7 is the summary of all data in this experiment. It can be observed that a straight pipeline with a length of 5 times of the pipe diameter would be sufficient for the specified permissible error (±1.5%) at all different pipe connection/conditions before the inlet of the meter. This length is much shorter than that required by turbine meters.

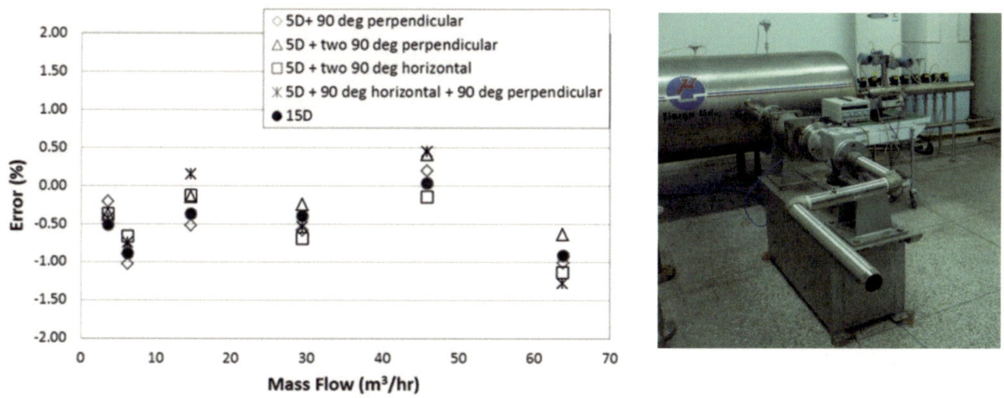

Figure 7. Measured permissible errors for different installation conditions.

3.3.4.4. *Calorific value assessment and compensation*

In USA, natural gas sold to residential customers is billed by "therms" that is based on the volumetric metering from the installed diaphragm meters and the monitoring of the gas properties at the central gas station. Each month a different factor shall be applied to gas bill to match the gas company's totalized data. In this case, it is almost impossible for customer to challenge the bill by looking at the record of the installed diaphragm meter as the monthly factor is set by the gas companies and varies. Although it is the fact that customers are consuming the thermal value of the natural gases not the volumetric value, it would be more reasonable if the installed residential meters can actually measure the thermal value of the natural gas instead of the volumetric value. In the previous report (Otakane 2003), it was found that the measurement of natural gases with different compositions by calorimetric MEMS gas meters shall have deviations but if the data were plotted against the calorific value, the permissible errors can be controlled irrespective of the gas compositions except for inclusion of high concentration of non-calorific gases such as nitrogen. However, the report neither specifies which calorific values were used to correlate output nor provide solutions to compensate the composition variation induced deviations so that the current acceptable tariff standards can be met as the volumetric value is still the base in most of the countries' tariff system. This could also be the reason for the limited applications for the existing products in Europe, for which a particular gas group has to be specified as discussed earlier in this chapter. For a wider spectrum of applications, a compensation scheme has to be provided for the purpose of fairness unless all meters with the same metrology capability can be changed overnight. The calorific value measurement at present shall only serve as an added-value for reference or for future tariff system development.

The current capability of the calorimetric MEMS meters for calorific value comes from its measurement principle. Therefore the measured flow rate shall depend on both the mass flow rate and the calorific values or the compositions of the gases. In order to decouple the calorific values from the flow rate measurement, additional sensing elements or schemes are then necessary for acquiring the relevant parameters while the mass flow rate is measured. Since the flow rate is proportional to the gas thermal conductivity and thermal capacitance, and both of them are related to the thermal values. Therefore, if the thermal conductivity and the thermal capacitance can be measured independently, it would be possible to differentiate the gas calorific value or gas composition induced mass flow rate variation. The compensation scheme can then be applied to the measured mass flow rate. Consequently, the measurement could be adjusted *in situ* to be in consistent with the current tariff system. For this purpose, additional two thermistors were integrated onto the previously discussed MEMS sensor chip, both located on the membrane for better thermal isolation. These two thermistors can measure the thermal conductivity and thermal capacitance of the gases. To demonstrate the capability of the MEMS mass flow meters incorporated with the integrated sensors, natural gases with five different compositions or thermal values were selected for tests by the meters that are pre-set with the gas calorific value compensation scheme and calibrated with air. Table 5 lists the selected gas compositions, thermal values and other data to be discussed.

	CH_4	C_2H_6	C_3H_8	N_2	GCF	HHV	R. Gravity	P/C_p
A	89.29	7.57	2.29	0.25	0.7959	44.12	0.6138	20.85
B	94.65	0.05	0.02	5.25	0.8735	37.89	0.5773	21.86
C	89.00	8.00	3.00	0.00	0.8199	39.18	0.6511	21.19
D	80.71	6.68	1.90	0.00	0.7815	39.59	0.7137	20.68
E	97.55	0.43	0.06	1.02	0.8778	36.81	0.5761	21.92

Table 5. List of the test gas properties and measured data.

In this table, only the concentrations for major constituent components of the gases are listed. Other minor constituents make up to the remaining concentrations of each gas. The HHV stands for high heating value with a unit of MJ/m^3. The HHV values were obtained using gas chromatography-mass spectrometer (GCMS) and hence are not *in situ* values. The relative gravity (R. Gravity) values were referenced to that of air and are also measured *ex situ*. GCF is the gas conversion factor that was obtained by referencing to the volumetric values in air. The meters calibrated in air were connected to a high precision standard volumetric rotary meter with the maximum permissible error of ±0.5%. The correlation between the readings of the two types of meters, if it is linear, shall establish the gas conversion factor. Figure 8 shows the measured data for gas A from MEMS gas meter with a maximum flow rate of $4m^3/hr$ (G2.5). The excellent linearity further established that the real gas calibration would not be required in this type of meter. This allows a substantially reduction in manufacture. In other words, the GCF varies with the gas composition confirms that the meters could not be simply applied to the existing tariff system without knowing the gas compositions, but if the specific GCF is preset in MEMS mass flow meters, it can be used to measure the corresponding natural gas with the mass flow value that is equivalent to the temperature and pressure compensated volumetric value. Compared to the traditional thermal mass flow technology, the MEMS gas meters have the advantage that a unique gas conversion factor can be established for each gas with respect to air. Specifically, after applying the gas conversion factor to the meter, it will retain the accuracy in the full measurement dynamic range. The data shown in Figure 8 and the verifications justified the proposed applications in the existing tariff system. It is expected that the GCF is not a universal factor but meter design dependent. Different manufacture may have slightly different value for each gas conversion factor.

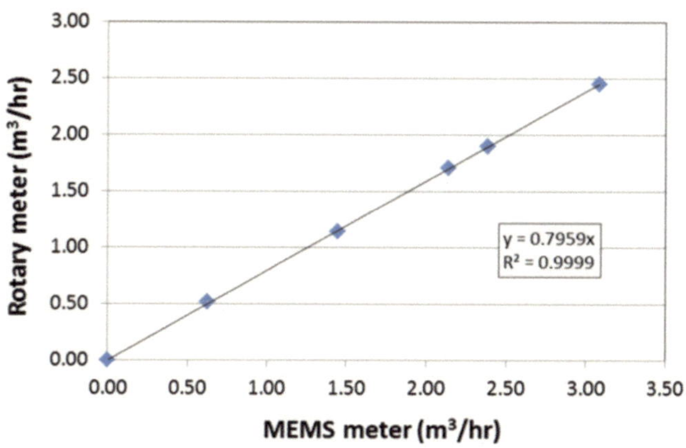

Figure 8. Gas conversion factor measurement.

The values P/C_P are obtained from the *in situ* measurements of the thermal values via the additional thermistors integrated into the calorimetric MEMS flow sensor. These values can be measured either dynamically or statically which makes it possible for an instant compensation scheme. In the previous report (Otakane 2003), the actual calorific value was not reported, and the data show strong dependence on nitrogen inclusion. From the data in Table 5, it can be observed that the GCF can be correlated to the gas thermal value (Figure 9) and the nitrogen inclusion seems less important compared to that in the previous report if the current thermal values are used for correlation since one of the gases (Gas B) has 5% nitrogen inclusion. The linear correlations of the *in situ* measured data, P/C_P, with the GCF (Figure 9) suggested that these values can be used for dynamic compensation of the gas thermal value (composition) variation. To verify this assumption, a MEMS gas meter incorporated with the compensation algorithm that is calibrated based on Gas D (which was pre-calibrated in air and applied GCF) was used for the measurement in Gas A. One can observe from Figure 10 that a large error was found without implementing the compensation scheme, while the compensation could be successfully eliminating the thermal value or composition induced errors based on current volumetric tariff system. On the contrary, one would suggest that the meter has the capability for thermal value measurement and the positive deviation shown in Figure 10 was due to that Gas A has a higher thermal value, but in order to compliance with the current tariff system such compensation scheme must be implemented.

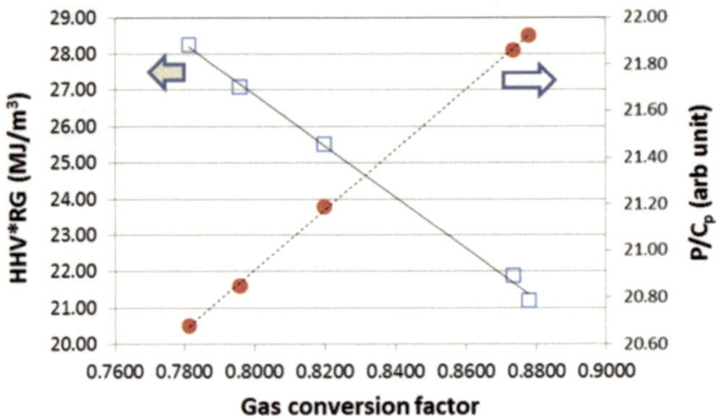

Figure 9. Correlation of the gas high heat thermal value HHV*RG (relative gravity) and *in situ* measured thermal value with respect to the gas conversion factor.

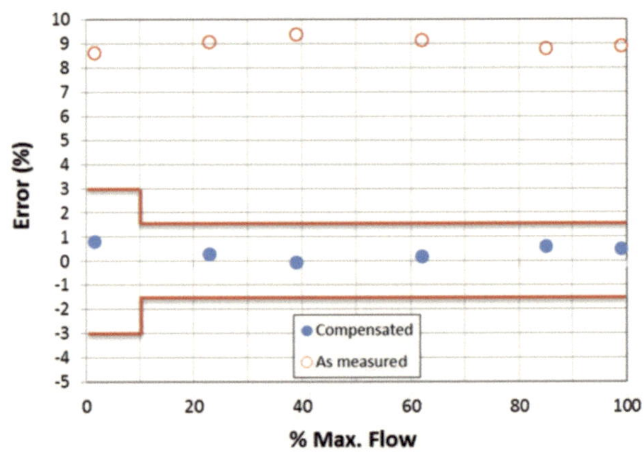

Figure 10. Comparison of as measured and compensated errors for a MEMS gas meter calibrated with Gas E but applied for measurement of Gas A.

3.3.5. MEMS gas meter field tests

3.3.5.1. Long term stability

One of the key concerns for the electronic meters is the reliability for long term operation. Figure 11 shows the comparison of the 41 day's daily accumulated flow data recorded one year apart. The meter was installed at a ceramic manufacturer where natural gas was used by the kiln for making ceramics. The process for the kiln required small fire initially for a preheat process and then heat-up with medium fire followed by large fire for shaping and then slowly reduced the heat for cooling down. This gas usage pattern was a typical application that requires a meter with large dynamic range in order to accurately record the gas consumption. With the existing mechanical meters, the metering during the preheat process can be completely missed, costing revenue for the gas suppliers. It is however an excellent application for the present MEMS meters. In this particular case, a DN50 flanged medium pressure meter with the flow range of 0~400Nm³/hr was installed. From the data shown in Figure 11, no performance degradation or accuracy deviation could be observed, since one can reasonably assume that the process of the ceramic making would consume pretty much the same amount for each run. For further validation, the meter was re-verified after the one-year installation by measuring its maximum permissible errors and the results confirmed the above observation.

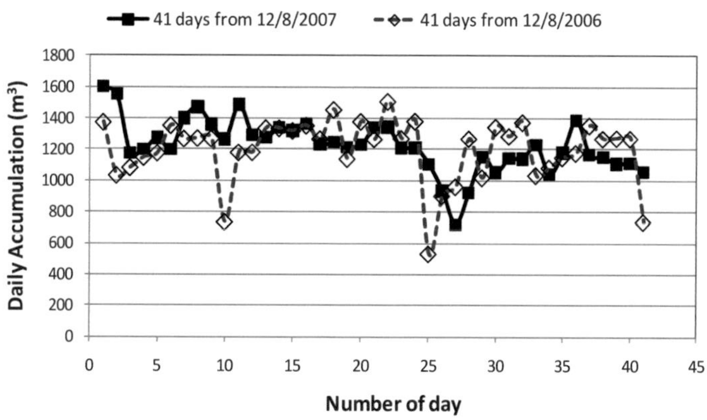

Figure 11. The daily accumulated flow apart from one year for a meter installed at a factory.

3.3.5.2. Project for distribution accuracy enhancement

To avoid inaccurate metering for small flow, a gas supplier had installed seven diaphragm meters (G25) at a public school where the gas consumption was much less in the morning as only breakfast was served. In the evening the gas consumption substantially increases due to gas burner operation for supply of hot water and other heating requirements. The seven diaphragm meters could meter a maximum flow of 280 m³/hr that was well covered by a DN50 flanged MEMS gas meter for industrial applications for current design (0~400Nm³/hr). Since the local standards required the meter set at 20°C and 101.325kPa at calibration, the data collected within one year indicated that the MEMS meter matched closely with the total gas consumption of the seven diaphragm meter cluster during summer time but when temperature dropped, the MEMS meter recorded the gas consumption more accurately since the diaphragm meter could not be compensated with environmental temperature change.

Figure 12. One MEMS meter can replace seven diaphragm meters with better accuracy.

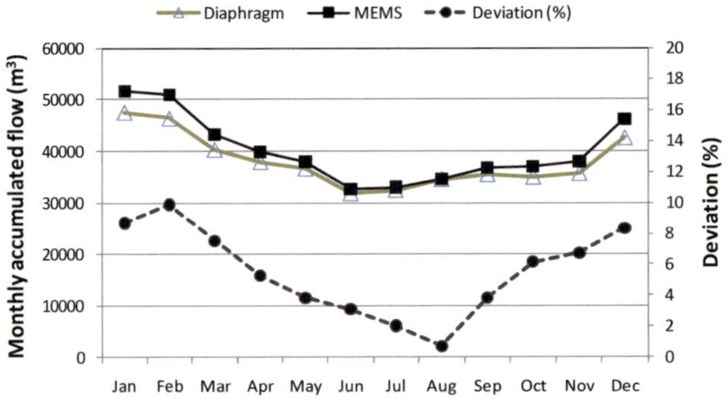

Figure 13. MEMS meter can better record the data while temperature varies.

4. Future development

Among the current all-electronic gas meter technologies for city gas metering, MEMS mass flow sensing technology would be the best candidate for future development and enhancement. It provides much better gas data acquisition and management. The MEMS mass flow meters feature a large dynamic range with automatic temperature and pressure compensation. Its capability in gas safety with leakage detection, readiness in data safety and remote data management adds significant values to current energy management. The small form factor provides significantly reduction in manufacture cost and logistic cost. The technology is also promising for direct calorific value measurement without any additional cost.

For the existing issues, one of the key improvements in near future would be further power consumption reduction. Although the current designed lifetime for residential meters was claimed for as long as 15 years with battery by Metersit, few battery manufacturers would provide such a warranty. Battery life itself shall be dependent on many environmental factors which might significantly reduce the specified lifetime. Once the power fails, the all-electronic meter will stop working although such failure will not create detrimental results for end users but for gas companies it is a concern. Remote data logging shall help to monitor each meter's status, a remedy to power failure is however still very much desired. One of the promising directions could be the current energy harvesting activities utilizing the flow energy inside the pipeline to generate electricity to power electronics. MEMS harvesters for vibrational energy have been emerging for commercialization. Similar approaches for flow measurement (Schmidt 1997; Kim 2000) would suggest the clues in this aspect. Another issue for gas metering is the unexpected gas composition variations or *in situ* gas composition information. MEMS flow sensing chip definitely has rooms for integration of gas sensors that would very likely provide such value. Additional desired functions shall include *in situ* calibration or verification, self-cleaning of possible additives on sensor surface and onsite exchange of the metrology components during maintenance.

The MEMS TOF technology can be integrated with the calorimetric MEMS flow sensors. This would help to simplify the leakage detection electronics as it has the advantage in ultralow flow sensing. The TOF technology shall also further simplify the gas composition compensation scheme since it could provide pure gas flow speed measurement without the process of decoupling gas calorimetric value from gas speed. The frequency measurement might provide better contamination resistance as signal attenuation would be easier to take place for the amplitude measurement by calorimetric sensing. Therefore it is expected that TOF sensing scheme might have better reliability in case of presence of contaminants.

The future development and massive deployment of the MEMS gas meters for city gas metering also require establishment of internationally acceptable standards. Current guideline could come from gas meter recommendation by International Organization of Legal Metrology, OIML R137-2011. European standards for residential ultrasonic meters EN14236-2007 also provide valuable references to the MEMS gas meters especially for the electronics of the meters. However, the MEMS gas meters have many features/functions that are missing in the above mentioned standards for the measurement schemes and principle, particularly for the calorific value related measurement. For a gas meter used for city gas metering or gas distribution, a well-defined standard is necessary.

MEMS gas meter for city gas metering is at its very earlier stage. The history of this industry asks for patience and persistence. With more and more manufacturers to participate in the development and commercialization process, and with today's state-of-the art electronics and advancement of the MEMS technology, a comprehensive gas energy management system based on the MEMS flow sensing technology shall eventually emerge and assist us for better natural gas resource management and conservation that ultimately benefits not only gas suppliers but all gas consumers.

Author details

Liji Huang
Siargo, Inc., USA

Acknowledgement

The author would like to thank his colleagues Chih-Chang Chen, Gaofeng Wang, Yahong Yao, Sugang Jiang, Jack Xuan, Yong Feng, Wenhong Deng, Xiaozhong Wu, Jianwei Fang, Xiang Lan, Changming Jiang, Xiangyou Yang, Jiliang Ruan, Rui Liu, Min Ao, Hong Luo, Rongzhong Yang, Min Du, Pei Zhou, Yaling Tao, Xueming Huang, and many others for their dedication to the very challenge process of MEMS gas meter commercialization. Financial supports from Entrepreneur Investment and Acorn Campus are also greatly appreciated. Special thanks to Mr. Richard Qu, David Tsang, Ga-Lane Chen and Chester Wang for their continued supports and inspiration. Mass production of the current products and future residential meters will be provided by Meter Technologies Limited.

5. References

ABS Energy Research (2008). World Gas Meter Report 2008. www.absenergyresearch.com. Data can also be found in www.meterpedia.com.

Ashauer, M.; Glosch, H.; Hedrich, F.; Hey, H.; Sanmaier, H. & Lang, W. (1999). Thermal flow sensor for liquids and gases based on combination of two principles. Sensors and Actuators, Vol. 73, pp. 7-13.

Bonne, U. (1992). Fully compensated flow microsensor for electronic gas metering, *Proceedings International Gas Research Conference*, Orlando, November 16–19, Vol. III, pp. 859 – 862.

Bonne, U. (2008). Gas Sensors, *Comprehensive Microsystems*, Vol. 2, Chapter 8, Elsevier Science, pp. 375-427. ISBN 978-0-44452-194-1.

Bruun, H.H. (1995). Hotwire anemometry: principle and signal analysis, Oxford University.

Fujiwara T. & Sasaki, Sho. (2005). Flow sensor and flow rate measuring method, US Patent 6,871,538, March 29, 2005.

Gehman, R.W.; Bohrer, P.J.; Johnson, R.G. & Higashi, R.H. (1985). Design and packaging of a highly sensitive microtransducer for airflow and differential pressure sensing applications, *Third International Conference on Solid State Sensors*, Philadelphia, June 11-14, pp. 207-211.

Haasl, S. & Stemme, G. (2008), Flow Sensors, *Comprehensive Microsystems*, Vol. 2, Chapter 7, Elsevier Science, pp. 209-272. ISBN 978-0-44452-194-1.

Hecht, H.; Tank, D. & Konzelmann, U. (1997). Method for correcting the output signal of an air mass meter, US Patent 5,668,313, September 16, 1997.

Higashi, R.E.; Johnson, R.G. & Bohrer P.J. (1985). Flow sensor, US Patent 4,501,144, February 26, 1985.

Huang, L.J.; Chen, C.; Yao, Y.; Wang, G.; Feng, Y.; Wei, K.; Deng, W.; Jiang, C.; Ruan, J. & Jiang, S. (2010). All-electronic MEMS flow meters for city gas applications. *Proceedings of 15th International flow measurement conference 2010*, Taipei, October 2010, pp. 376-383. ISBN 978-1-61782-472-2.

Huang, L.J.; Chen, C.; Yao, Y. & Wang, G. (2011). Micromachined thermal mass flow sensor with self-cleaning capability and methods of making the same, US Patent 7,878,056, February 1, 2011.

Hutton, S.P. (1971). Flow metering and future aims. *Proceedings of conference on modern development in flow measurement*, Harwell, September 21-23, pp. 401-405. ISBN 978-0-90122-330-2.

Kang, S.S. & Nutt, W.E. (1992). Development of a compact flow meter for residential natural gas metering, *Final Report*, Creare Incorporation.

Kim, D.; Kang, S.; Sim, J.; Shin, J.; Choi, P. & Lee, J. (2000). Characteristics of piezoresistive mass flow sensors fabricated by porous silicon micromachining, Jpn. J. Appl. Phys. Vol. 39, pp. 7134-7137.

Kono, K.; Nagai, N.; Yuasa, K.; Fujimoto, T.; Suzuki, M. & Fujii, Y.; Hiroyama, T. (2006). Development of intelligent domestic ultrasonic gas meters, *23rd World Gas Conference*, Amsterdam, 2006.

Matter, D.; Kramer, B.; Kleiner, T.; Sabbattini, B. & Suter, T. (2004). Microelectronic domestic gas meter with new technology, *Technisches Messen*, Vol. 71, pp. 3-22.

Mayer, F. & Haeberli, A.M. (2004). Method and device for measuring the low of a fluid, US Patent 6,763,710; July 20, 2004.

Nishimoto, I.; Kurosawa, T. & Yamamoto, T. (1991). Microbridge flow sensor, US Patent 5,050,429, September 24, 1991.

Otakane, K.; Sakai, K. & Seto, M. (2003). Development of the thermal flow meter, *SICE Annual Conference in Fukui*, August 4-6, pp. 2031-2034.

Putten, A. F.P. van & Middelhoek, S. (1974). Integrated silicon anemometer. *Electronics Letters*, Vol. 10, No. 21, pp. 425-448.

Schmidt, H.J.; Holtkamp, F. & Benecke, W. (1997). Flow measurement in micromachined orifices, J. Micromech. Microeng. Vol. 7, pp. 189-192.

Shin, W.C. & Besser, R.S. (2006). A micromachined thin-film gas flow sensor for microchemical reactors, Journal of Micromechanics and Microengineering, Vol. 16, pp. 731-741

US Energy Information Administration (2011). International Energy Outlook 2011. *US Department of Energy Report* DOE/EIA-0484, September 2011.

Wang, G.; Chen, C.; Yao, Y. & Huang, L.J. (2009). Micromachined thermal mass flow sensors and insertion type flow meters and manufacture methods, US Patent 7,536,908 May 26, 2009.

van der Wiel, A.J.; Linder, C.; de Rooij, N.F. & Bezinge, A. (1993). A liquid velocity sensor based on the hot-wire principle, Sensors and Actuators, Vol A37-38, pp. 693-697.

Yoder, J. (2002). Ultrasonic flowmeter market. www.Flowresearch.com.

Section 3

End Use, Financial Markets

Chapter 10

Natural Gas Fired Reciprocating Engines for Power Generation: Concerns and Recent Advances

Sreenath B. Gupta, Munidhar Biruduganti, Bipin Bihari and Raj Sekar

Additional information is available at the end of the chapter

http://dx.doi.org/10.5772/45992

1. Introduction

Concerns about the dwindling US energy deposits, uncertainties about the global energy supply chains, human health concerns, Green House Gas emissions and other environmental concerns have brought energy efficiency and clean technologies to the forefront. The energy supply matrix for the United States (see Figure 1) is a diverse mix - fossil fuels such as coal, natural gas, and petroleum are interspersed with other traditional and non-traditional resources. Out of these, due to its availability in abundance, easy transportation through pipelines, and clean burning nature, natural gas features prominently with total annual consumption of 22.5 Quads (1 Quad = 1 x 10^{15} BTU), i.e., 23.2% of the total US energy consumption. Approximately a quarter of this natural gas is used for electricity generation by peaking power plants and various distributed generation centers spread throughout the country. Additionally due to the fact that high efficiency (as high as ~ 80%) can be achieved via. Combined Heat and Power (CHP), the fraction of natural gas used in Distributed Generation continues to increase.

For centralized power generation, the prime movers of choice are large gas turbines as they offer very low maintenance. These turbines tend to be > 20 MW in size and are typically ~ 30% efficient. For most of the distributed power generation applications, the prime mover requirements are smaller than 20 MW and the choices remain reciprocating engines, fuel cells and microturbines (See figure 2). Fuel cells are low-polluting, and highly efficient (~ 60%), but often require very high capital costs. Microturbines, on the other hand, are low-polluting but have very low efficiencies (~30%). Reciprocating engines offer very low capital costs and further have very high efficiencies (~42%) but NO_x emissions are a concern. Also, the maintenance requirements are higher as compared to the other two prime movers.

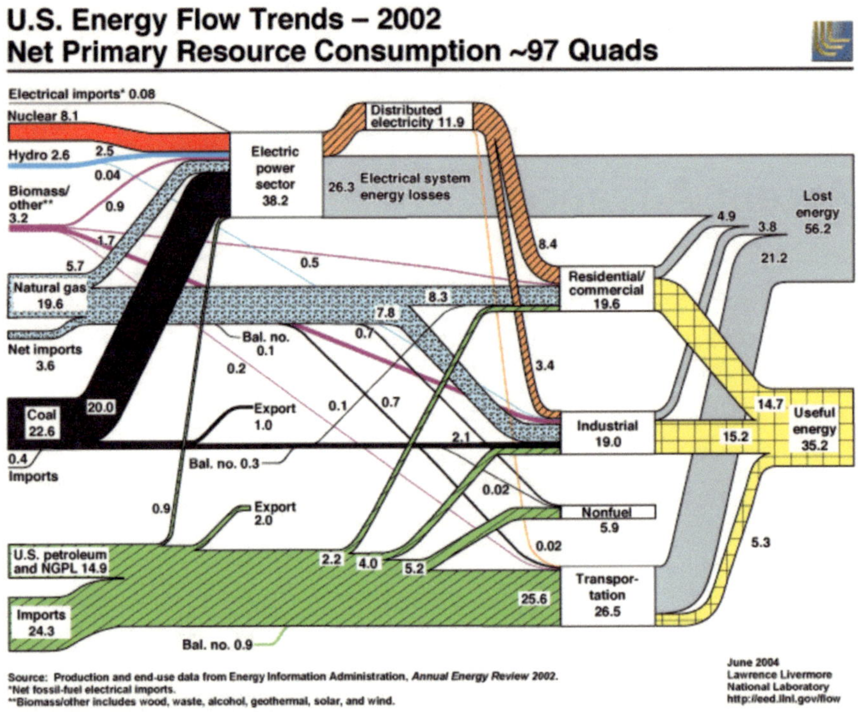

Figure 1. Sanke diagram showing the net US energy flow trends as of 2002. Courtesy: Lawrence Livermore National Lab.

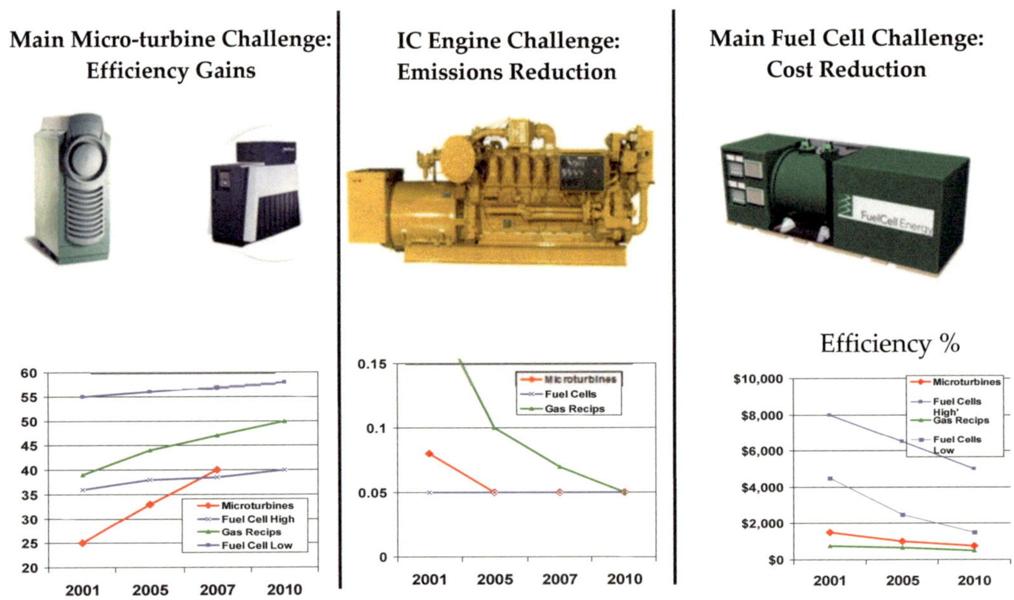

Figure 2. A performance comparison of prime movers used for distributed power generation. (Projections based on 2002 data)

Per recent DOE estimates, over 10,000 stationary reciprocating engines fueled by natural gas are already deployed in various parts of the US for distributed power generation. These are usually large bore engines, with bore sizes > 6.5 inches, and usually are within the power range of 0.5 to 20 MW. It is estimated that over 80% of such engines are of the power range 1.5 MW or lower. In order to accommodate the fact that these engines need to have high reliability, ensuring availability of >95 % throughout the year, most of these engine are of robust design with large lubrication oil reservoirs and very high life components such as bearings and spark plugs equipped with noble metal electrodes.

2. Combustion characteristics of natural gas

Before we proceed further, it behooves us to review the combustion characteristics of natural gas and the indicies commonly used in the gas engine industry to quantify them.

2.1. Natural gas combustion characteristics

The composition of natural gas as distributed in the United States varies from region to region and throughout the year. The variation of the main components of the gas as a result of 120 samples drawn at different locations is shown in Table 1 [1]. As noticed from this tabulation, methane is the primary component with ethane, propane and butane as the major components. Hydrocarbons larger than C_4H_{10} occur in trace quantities. Propane is added by some companies as a peak shaving measure and can reach concentrations as high as 23%. Nitrogen is added to aid in the pumping of natural gas through pipelines.

	Methane	Ethane	Propane	Butane	Pentane	Hexane	Nitrogen	CO_2
Mean	93	3	1	0.5	0.1	0.5	1.5	0.5
Std. dev.	5.5	2.6	1.4	1	0.3	0.1	2.9	0.5
Min.	73	0	0	0	0	0	0	0
Max.	99	13	8	7	3	1	17	2

Table 1. Summary of natural gas variation within the US [1]

With such a varying composition, the following indices are used to characterize natural gas as a fuel for reciprocating engines.

Heating value

In the US, natural gas distribution companies try to hold the heating value, HHV at 1028 BTU/cu.ft., the volume being measured under standard conditions of 60°F and 14.73 psia. This corresponds to a LHV of 910 BTU/cu.ft.

Flammability limits

For natural gas the flammability limits are widely accepted to be between 5% and 15.6% by volume in air corresponding to equivalence ratios of 0.486 and 1.707 [2]. These limits marginally widen with increased concentrations of higher hydrocarbons.

Ignition limits

The auto ignition limits (which closely correlate with knock limits) and the spark ignition limits of methane-air mixtures at 490°C as determined using a Rapid Compression Machine are given in Figure 3. As noticed, though the lean flammability limit of methane-air mixtures is $\phi = 0.5$ ($\lambda = 2.0$), mixtures leaner than $\phi \sim 0.65$ ($\lambda \sim 1.54$) are very difficult to ignite using conventional ignition systems. A similar limit for stoichiometric mixtures diluted with exhaust gas is a EGR fraction of 22%. Significant efficiency and emission benefits can accrue with the extension of such limits.

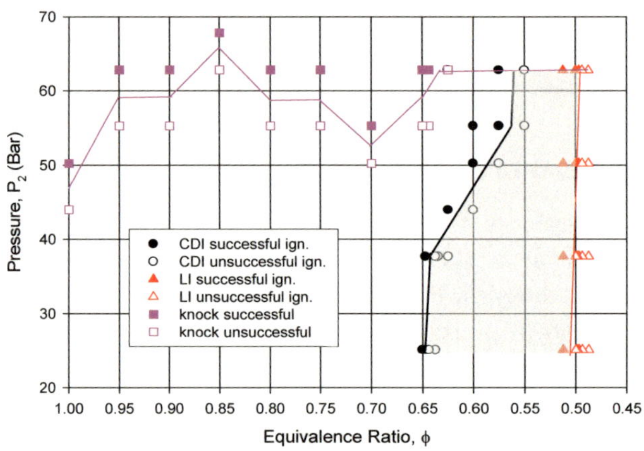

Figure 3. Ignition limits of methane-air mixtures at 490oC as determined using a RCM. [3]

Laminar Flame speed, SL

The laminar flame speed of natural gas as compared to other gaseous fuels is shown in Figure 4 below. As seen, due to the fact that the primary constituent of natural gas is methane, a very stable molecule, its flame speed closely mimics that of methane itself. Even for peak shaving conditions, wherein propane concentrations can be as high as 23%, the laminar flame speed is marginally larger. On the other hand, hydrogen laminar flame speed (not shown in the figure) is approximately 6 times that of methane. Significant engine efficiency improvements can be garnered by improving the flame speed of natural gas.

Methane Number (Knock resistance of fuels)

Knocking is a phenomenon closely associated with autoigntion of fuel air mixtures. In reciprocating engines, the end gas is compressed by the expanding flame front initiated by the spark kernel and auto ignites potentially at more than one location. This phenomenon could be damaging to the engine hardware and is manifest in pressure oscillations observed in the combustion pressure traces. As gasengines are operated close to the knock limit to extract maximum efficiency, knock resistance of the fuel is of prime concern. The most widely used index in the automitive industry to detrmine the knock potential of liquid fuels

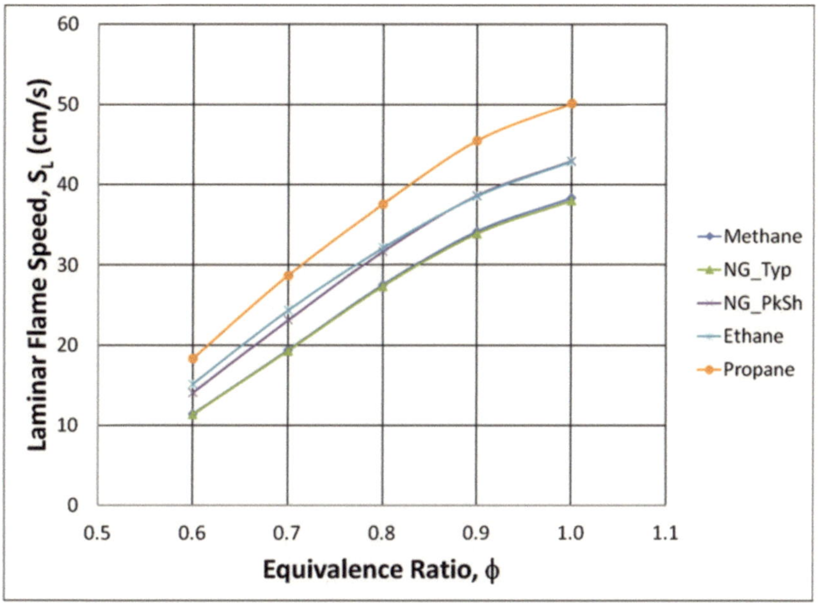

Figure 4. Laminar Flame Speed of typical natural gas and that representative of peak shaving composition as compared to other gaseous fuels. (Data points estimated using CHEMKIN software and GRIMECH 3.0 kinetic data.)

is Motor Octane Number (MON), which is determined as the knock tendency of a fuel between that of iso-Octane and n-Heptane. However, such an index is not representative of the range of values observed in gaseous fuels. As a result, the most widely used index in the industry is Methane Number (MN), which rates the knock resistance of the fuels between 100% methane and 100% Hydrogen. The other less widely used index is Butane number (BN), determined using methane and butane as primary reference fuels.

Gas	MON	MN	BN
Methane	122	100	0
Ethane	101	44	7.5
Propane	97	34	10
Butane	89	10	100

Table 2. Knock ratings of normal paraffins [4]

The knock ratings of various single component fuels are shown in Table 2. As seen, methane by itself being a stable molecule improves knock resistance, whereas long chain hydrocarbons like propane and butane inhibit it significantly. The MON equivalent of natural gas has been related to the reactive hydrogen to carbon ratio of the fuel as follows

$$MON = -406.14 + 508.04\,(H/C) - 173.55\,(H/C)^2 + 20.17\,(H/C)^3 \quad (1)$$

Further, empirical correlations relate the MON equivalent of natural gas to the methane number, as follows

$$MN = 1.624\ MON - 119.1 \quad (2)$$

Quadratic expressions are also available, relating the concentrations of individual component gases to the Methane number of the fuel [4]. Overall, the concentration of larger hydrocarbons in natural gas has a dominating effect and it significantly reduces the MN. As observed by Hack and McDonell, larger hydrocarbons also increase the heating value and the general emission propensity of natural gas [5].

As gas engines are operated close to their knock limit to maximize efficiency, the current compression ratios of these engines are limited to $CR < 12$. Only in the case of diesel pilot ignited engines, wherein diesel autoignition is encouraged, larger compression ratios are used.

Wobbe Index

This is defined as energy content per relative density. It is commonly used to determine compatibility of fuel systems with fuel type, especially in gas turbines and industrial burners. Wobbe index is less relevant to reciprocating engine combustion.

Among the combustion characteristics presented above for natural gas, poor ignitability and low flame speed are of chief concern. Significant improvements in engine performance can be garnered by addressing these issues.

3. Main operating concerns of reciprocating engines

The main concerns for natural gas fueled engines are

3.1. Emissions

Natural gas fueled engines are very low polluting as compared to other hydrocarbon fueled engines. However, as they are operated round the clock and throughout the year, the emissions emitted per year total to a very large quantity. Of chief concern are NO_x , CO and UHC emissions. NO_x emissions tend to increase with engine efficiency and have pathways of thermal, prompt, NNH and other such mechanisms. Under low-temperature conditions, NO_x formation is inhibited, whereas CO and VOC emissions increase. The final engine operation is optimized for maximum efficiency, while being compliant with local emission regulations.

In the US, the stationary engines were exempt from emission regulations till June 2006. The regulations instituted thereafter are summarized in Table 3. As noticed, the applicable regulations are based on the power of the engine, as well as, the intended application.

3.2 Efficiency

The prime operating cost of these stationary engines is the fuel cost, which can total up to six times the initial cost of the engine itself within a year of operation. As a result, the efficiency of the engine is of prime concern to the engine operator.

Stationary Spark-Ignited Generator Set Emission Regulations — (NOx+HC) / CO (g/kWm-hr) — NOx / CO / VOC (g/kWm-hr)

	kWe, 60Hz	Gross engine power kW	(HP)	2008	2009	2010	2011	2012
Standby	15–75 NG	19–97	(25–130)	—	(13.4) / 519 Part 90 V			
Standby	80–150 NG	97–179	(130–240)	—	(2.7) / 4.4 Part 1048 V[1]	or	2.0 / 4.0 / 1.0 Part 60 V	
Standby	15–75 LP	19–97	(25–130)	—	(13.4) / 519 Part 90 M			
Standby	80–150 LP	97–179	(130–240)	—	(2.7) / 4.4 Part 1048 M[1]			
Prime	15–50 NG	19–75	(25–100)	—	2.0 / 4.0 / 1.0 Part 60 V	or	(2.7) / 4.4 Part 1048 V[1]	
Prime	60–150 NG	75–179	(100–240)	—	(2.7) / 4.4 Part 1048 V[1]		1.0 / 2.0 / 0.7 Part 60 V	
Prime	15–150 LP	19–179	(25–240)	—	(2.7) / 4.4 Part 1048 M[1]			

M: Manufacture certification required
V: Manufacture voluntary certification or end user mandatory certification
1: Alternately can use emissions limit formula: (NOx+NMHC) x CO < 8.57 rounded to the nearest 0.1 g/kW-hr.
You may not select an HC+NOX emission standard hight than 2.7 g/kW-hr or a CO emission standard higher than 20.6 g/kW-hr.

Legend: Non-regulated / Interim / Final

Table 3. US emission regulations for stationary SI engines.

In order to meet the competing demands of emission regulation, fuel efficiency and system reliability, two prominent engine operational modes have evolved.

The first referred to as "Rich-burn" engines are primarily operated at equivalence ratios ≈ 1 in conjunction with a three-way catalyst to reduce CO, HC and NOx emissions. Rich-burn engines often use Exhaust Gas Recirculation to reduce in-cylinder combustion temperatures and as a result the engine-out NO_x emissions. To offset the loss in power density, intake air is boosted using a turbocharger. Efficiencies of Rich-burn engines are ~ 36% as of 2011. The primary advantage of Rich-burn engines is that they easily achieve compliance with emission regulations.

The second operational mode that has prominently evolved over the last two decades is "lean-burn." Such engines are often operated at very low equivalence ratios, $\phi < 0.65$ ($\lambda > 1.54$). The inert nitrogen gas in the excess air that is used acts as a diluent and reduces the in-cylinder combustion temperature for subsequent NO_x mitigation. Lean-burn engines are often boosted using a turbocharger to offset the loss in power density. The primary advantage of lean-burn engines is that they can achieve thermal efficiencies as high as 42% leading to significant fuel savings to a typical operator. Also, the engine-out emissions are so low that there is no need for an aftertreatment system for easy compliance with emission regulations.

3.3. Maintenance

Ignition: Of prime concern to the operation of stationary natural gas engines is the maintenance associated with spark plugs. Modern engines are optimized to meet the demands of increased efficiency, low emissions and high specific power, and are operated under conditions that are difficult to ignite. Problems with ignition can result in significant misfires leading to increased hydrocarbon emissions, and most importantly increased fuel consumption.

Friction: On account of the fact that stationary engines are not produced in large volume, technologically their designs have lagged the advancements in the automotive field. Besides leading to energy loss that can total up to ~ 7%, friction leads to reduced hardware life.

4. Recent technological advancements

Addressing the above concerns, US-DOE has embarked on a research program called Advanced Reciprocating Engine Systems (ARES) since 2002, with the research activities being conducted at various universities, National Labs and industry. The main goals of the ARES program are to achieve >50% efficiency, < 0.1 g/bhp-hr NO_x, increased fuel flexibility, and lowered capital costs by 2013. In the remainder of this chapter, some of the technologies pursued under the ARES program at Argonne will be presented, along with some important technologies pursued elsewhere.

4.1. Pretreatment technologies

4.1.1. Nitrogen Enriched Air (NEA)

In this technology, the preferential diffusion of oxygen over nitrogen through certain polymers is advantageously used [6]. Nonporous Polymeric membranes are instituted as micron thick coatings on the inner walls of capillary tubes. The coated capillaries, in turn, are packaged into bundles (see Fig. 5) with optimized geometries for low pressure drop and high yield. When air is passed through such membrane bundles, it is split it into a nitrogen-rich stream and an oxygen-rich stream. When the nitrogen-rich stream is used for engine combustion purposes, it leads to lower in-cylinder temperatures by inert gas dilution thereby reducing NO_x formation. NEA avoids harsh and corrosive exhaust emissions, such as particulates and acidic compounds, from being introduced into the engine intake manifold and thereby offers a clean alternative to the traditional Exhaust Gas Recirculation. Advantages that accrue from NEA include improved engine hardware and significant extensions to lubrication oil life.

The basic principle of membrane operation and various designs and operating characteristics are described in Poola et al. [7]. Research work done on a locomotive two stroke diesel engine by Poola and Sekar [8] corroborate variable air composition as a technique to reduce emissions.

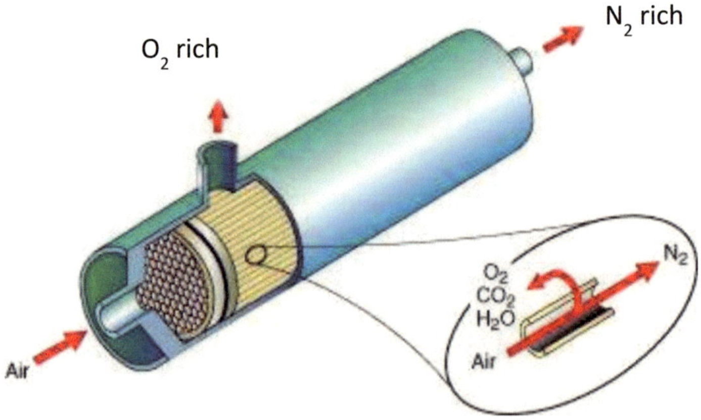

Figure 5. Schematic representation of Air Separation Membrane

Nitrogen enrichment tests with gas blends

Initial tests were conducted in a 1.87 Liter, 33 kW, 1800 rpm SI natural gas single-cylinder research engine. Nitrogen enrichment was achieved by inducting bottled nitrogen into the air stream [9]. Before proceeding further, let us define a few terms:

The amount of nitrogen enrichment (NEA) is expressed as the volume percent of nitrogen above the usual 79.05% in the intake air.

To account for the fact that the composition of air varies with nitrogen enrichment, it helps to define an oxygen based equivalence ratio,

$$\psi = \frac{\left(\dot{m}_f / \dot{m}_{O_2} \right)}{\left(\dot{m}_f / \dot{m}_{O_2} \right)}. \tag{3}$$

With nitrogen enrichment, the local combustion environment is starved of oxygen and leads to increased levels of ψ. This in turn, leads to reduced bulk gas temperatures and results in lower NOx formation. The test results for varying amounts of nitrogen enrichment for a fixed ignition timing of 20°BTDC, 1800 rpm and 12 bar BMEP are shown in Figure 6. As noticed with increasing nitrogen enrichment, NOx emissions decrease monotonically. Also, as shown in Figure 7, for a fixed ignition timing, the fuel conversion efficiency (FCE) decreases concomitantly with NOx emission. Analysis of the cylinder pressure traces shown in Figure 8 provides further insights. With increasing nitrogen enrichment the combustion event is both delayed as well as decelerated, which points to the possibility of further gains by optimizing the ignition timing.

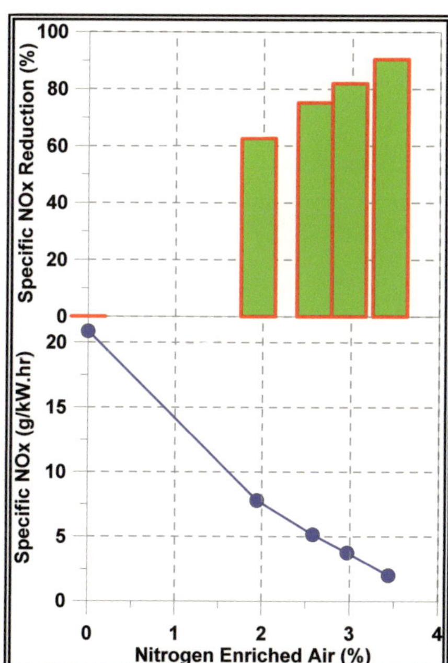

Figure 6. Specific NOx as function of Nitrogen Enrichment.

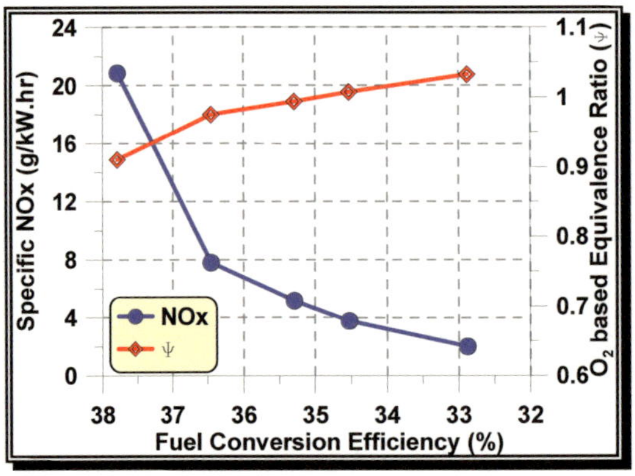

Figure 7. Spec. NOx and ψ as a function of FCE

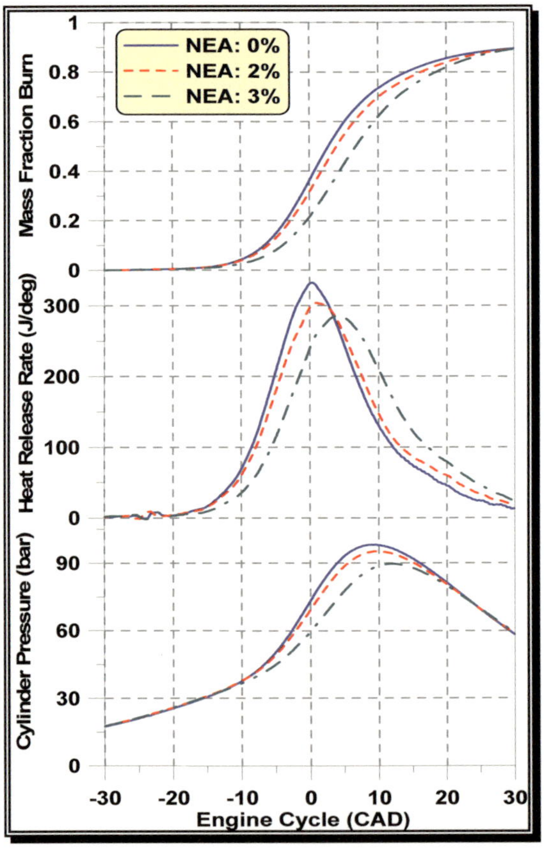

Figure 8. Combustion related data as a function of engine cycle. Ignition timing = 20°BTDC

The optimal performance results were obtained through ignition timing sweeps performed between 25° and 45° BTDC. Within these limits, based on the equivalence ratio, ignition timing could be varied between knock limit and that value limiting the COV of ignition below the industry accepted 5%. In most of the cases, engine knock was avoided with nitrogen enrichment. The optimal performance results identified through such timing sweeps are shown in Figure 9. It is interesting to note that for similar air/fuel ratio, identified by the circles on the graph, NOx decreases by approximately 50% with less than 0.5% FCE penalty. Similar NOx reduction, around $\psi = 1.08$ is an appreciable 84%, but is accompanied by ≈ 2% FCE. Operation for intermediate values of ψ is not foreseen due to unacceptably high NOx emissions.

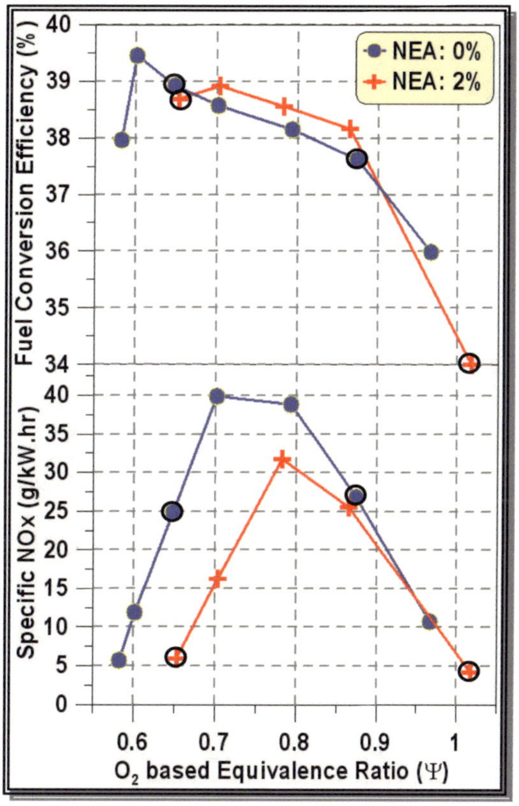

Figure 9. Specific NOx and FCE as a function of ψ

Nitrogen enrichment tests with air separation membrane

Encouraged by the above results, tests were performed by using nitrogen enriched air provided by an air separation membrane. As a compromise between the parasitic power requirements to achieve high levels of nitrogen enrichment and the amount of reduction in NOx emissions, it was decided to fix the nitrogen enrichment at 2%. For further details of the experiment, the reader is referred to reference 27. In these set of tests, both laser ignition as well as conventional CDI were used. The test results are as shown in Figure 10. For either method of ignition, low BSFC and specific NOx were obtained for lean operation. A combination of LI and NEA led to NOx reductions up to 89%, as compared to 70% reductions achievable just by using NEA [10].

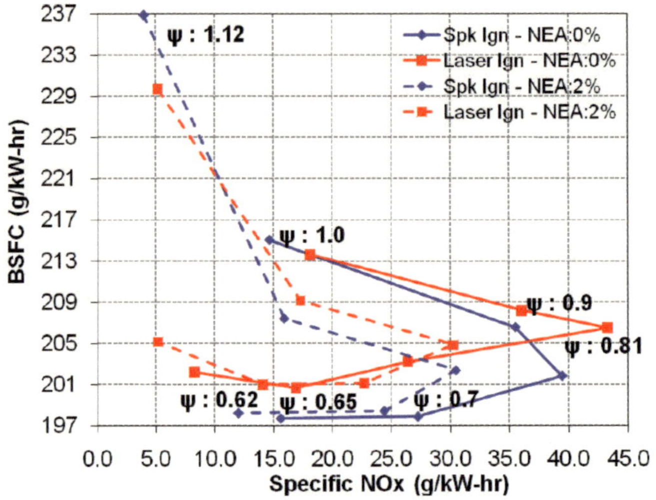

Figure 10. BSFC versus Specific NOx at different ψ.

Nitrogen enrichment vs. EGR

Tests were also conducted to compare the efficacy of NEA and EGR to reduce NOx while maintaining the combustion efficiency. To facilitate simulated EGR under turbocharged conditions in a single-cylinder engine, an EGR system using a reciprocating compressor was developed. Such a system facilitated exhaust gas recirculation up to 29%. Results from such tests are summarized in Figure 11. It is seen that though both EGR and NEA offer comparable efficacy towards NOx reduction, their ability to retain concomitant FCE could not be estimated accurately. Better models for auxiliary power requirements in either case are required to facilitate such a comparison.

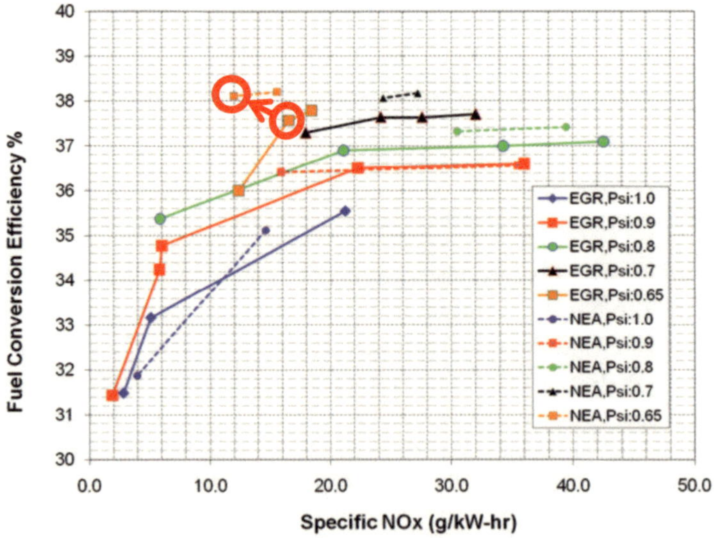

Figure 11. Performance tradeoff with EGR or NEA.

4.1.2. Hydrogen addition

Though natural gas is characterized by high octane number of 122 and clean burning nature, its combustion in IC engines is limited by low flame speed and poor ignitability. The low flame speed limits the rate at which the flame spreads from the ignition kernel to consume the complete in-cylinder charge. This is manifest in the form of low rates of heat release, and results in lower engine efficiencies. Ignitability, on the other hand, is limited by the capabilities of the ignition system used and at the outer extremes is limited by the flammability limits of the fuel. On the other hand hydrogen gas has flame speeds that are 6 times larger than those of natural gas. Also, hydrogen has a lean flammability limit that extends up to◎◎ = 0.1. Significant operational gains can be achieved, with the addition of hydrogen to the natural gas, as it improves both ignition, as well as, flame speed.

In a majority of gas engines, natural gas is inducted in to the intake plenum, and as a result occupies $\phi < 9.5\%$ volumetric portion of the intake charge. Such a large fraction contributes to the breathing losses, resulting in low efficiencies, especially at low-loads. To offset such an effect, gaseous fuel is injected directly in to the cylinder, usually optimizing the parasitic loss associated with compression of the fuel. As a result, some of the hydrogen addition tests have used in-cylinder injection.

Wang and coworkers [11] have conducted tests in a engine fueled with various fractions of hydrogen addition. They observed that with increasing amount of hydrogen addition, the phase of the heat release curve advanced, and combustion duration decreased (see Figure 12). This phenomenon was more pronounced at low engine speed suggesting efficacy of flame speed enhancement at relatively low cylinder air motion. As the flame speed increases exponentially with hydrogen content, maximum mean gas temperature and rate of pressure rise increased rapidly. As a result, HC, CO and CO_2 emissions decreased rapidly, whereas NO_x emissions increased. Optimal hydrogen concentrations were found to be around 20% as a best compromise between emissions and efficiency.

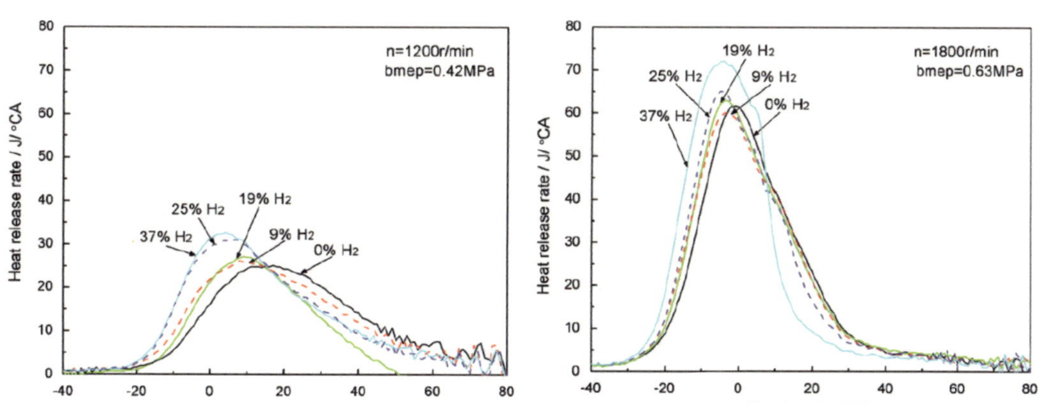

(a) (b)

Figure 12. Combustion pressure curves with the addition of Hydrogen at (a) low speed and low loads, and (b) high speed and high load [11]

To offset the increase in NO_x emissions, low-temperature combustion may be promoted using Exhaust Gas Recirculation. Recently Hu et al. [ref. Hu 2009] conducted studies in a stoichiometric engine equipped with a three-way catalyst using EGR rates up to 40% and hydrogen fractions up to 40%. They have observed that for a given hydrogen fraction, NOx concentrations decrease with increasing EGR fraction. CO and CO_2 while remaining less influenced by EGR, decreased significantly with hydrogen addition. As shown in Fig. 13, an optimization of EGR fraction, hydrogen addition, and ignition timing results in high efficiencies and low emissions. Optimal engine operation at 2000 rpm was achieved, for EGR fractions in the range 10-20% and hydrogen fractions in the range 30-40%.

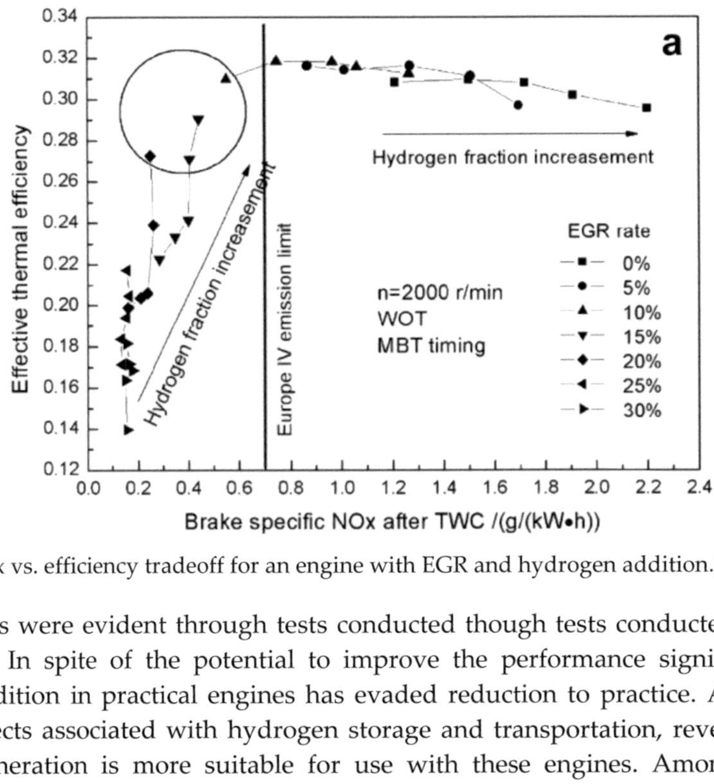

Figure 13. NOx vs. efficiency tradeoff for an engine with EGR and hydrogen addition.[12]

Similar results were evident through tests conducted though tests conducted on lean-burn engine [13] . In spite of the potential to improve the performance significantly, use of hydrogen addition in practical engines has evaded reduction to practice. A review of the practical aspects associated with hydrogen storage and transportation, reveals that on-site hydrogen generation is more suitable for use with these engines. Among the possible methods of hydrogen generation (a) Steam methane reforming, (b) partial oxidation (POX), (c) autothermal reforming, and (d) exhaust gas fuel reforming [14]. All of these methods use a catalyst and are somewhat sensitive to the temperature, carbon poisoning, and sulfur poisoning and it appears the former two being less complex are most reliable for use with stationary engines. Gas Technology Institute has evaluated the prospect of using engine exhaust heat, which for a natural gas engine can be in excess of 550°C, to reform a slipstream of natural gas fuel [15]. As shown in Figure 13, the slipstream is reformed to H_2 and CO mixture which is further enriched by a water shift reaction further downstream. Though excellent gains in emission reductions and efficiency improvements up to 3% points were possible, long term performance of the catalyst degraded due to carbon poisoning. An alternative path, though less pursued, is to use two engines in tandem: A PO_x engine that operates under rich conditions, partially oxidizes the fuel, the exhaust of which is fed to a larger engine that consumes additional fuel [16]. Research continues in identifying a potential pathway for on-site hydrogen generation.

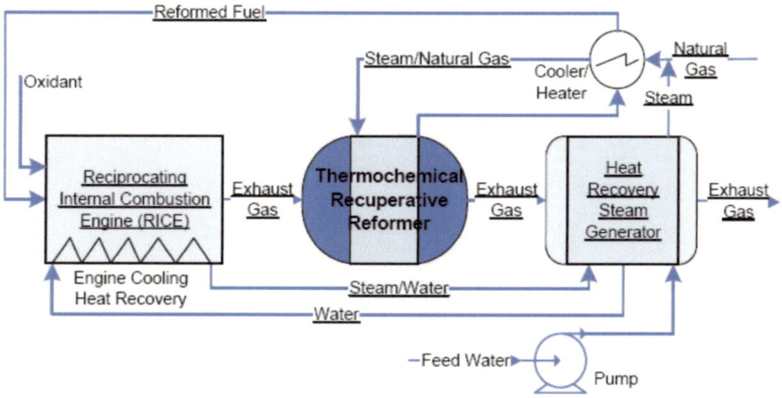

Figure 14. Schematic representation of TCR use with engine.[15]

4.2. Aftertreatment

Catalyst aftertreatment is a logical approach to achieve emission requirements (0.1g NOx/hp-hr: <7 ppm NOx) with minimal impact on engine design. In addition, catalyst technology may be able to abate other pollutants such as formaldehyde, residual VOC and unburned CH_4 from natural gas engines at the same time as it controls NO_x. Though the performance of a three-way catalyst under stoichiometric operation is proven, the associated engine efficiencies are somewhat low. Unfortunately, the presence of excess oxygen in the exhaust precludes using three-way catalyst technology for lean-burn operation.

Today, the only proven aftertreatment technology that can provide substantial NO_x reduction (90+%) in lean burn exhaust is the SCR catalyst system that requires supplemental addition of ammonia/urea as a reductant to the exhaust stream. However, there are a significant number of implementation issues associated with urea SCR that make it a less than ideal technology to control NO_x emissions. These issues include storage, corrosion, transportation, infrastructure, and concerns over release (slip) of ammonia into the air or oxidation of ammonia at high temperatures which yields further NO_x emissions.

A more attractive aftertreatment technology is NSR (NOx Storage and Reduction) catalysis (also referred to as NO_x Adsorbers), wherein NO_x is stored by reaction with an alkaline earth to form the solid nitrate. In a reduction step under rich condition, an injection of reducing agents releases and reduces the stored NO_x, converting it to N_2. Therefore, the catalyst needs to be regenerated occasionally to restore the NO_x storage capacity. The preferred reducing reagents; CO or H_2 are not present in sufficiently high amounts in natural gas engine exhaust. The challenge is to regenerate the catalyst with an available reductant (CH_4). Because NSR catalysts consist of alkali metal or alkaline earth metal oxides, sulfur oxides in the exhaust stream readily form stable sulfates with the catalyst material, causing a loss in deNOx performance. In spite of the best efforts under the ARES program, efforts to improve the NO_x reduction efficiency while improving the resistance to sulfur poisoning and reducing the fuel penalty at reasonable levels have met with partial success [17, 18].

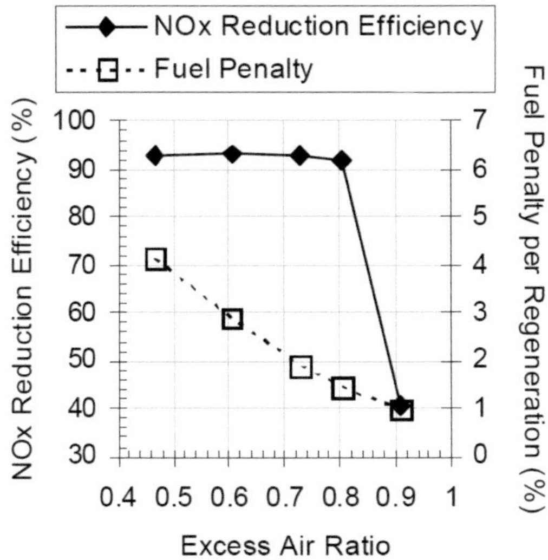

Figure 15. NOx reduction efficiency during sorption and fuel penalty per regeneration as a function of excess air ratio (◎) [17].

4.3. Friction reduction

In a typical reciprocating engine, approximately 7% of fuel energy is wasted due to frictional losses. Approximately half of these losses can be attributed to the piston Ring-Cylinder Liner (PRCL) interface where mixed mode lubrication occurs at top-dead center and bottom-dead center. It is estimated that approximately 50% of the friction at the PRCL interface can be reduced with (i) the use of improved lubricants, (ii) improved ring pack designs, and (iii) Laser Surface Texturing (LST). In the case of stationary engines that operate throughout the year, this can lead to significant fuel savings and significant GHG emissions reduction. Research in improved lubricants continues as and will not be discussed here. Ring packs were optimized through modeling and improvements up to 0.5% points in efficiency were observed through tests on a 6-cylinder engine [19].

LST on the other hand, stands out as one of the promising tribological advancements in the last 30 years. In this technology, sections of the cylinder are covered with a 2-D array of micron size dimples, approximately 100 μm dia and 5 μm deep, which in turn act as micro-hydrodynamic bearings in the case of full/mixed lubrication, micro-reservoirs for lubricant in the case of starved lubrication, and as micro-traps for absorbing wear particles. Through numerous modeling and experimental efforts the depth, size and spacing of the dimples have been optimized to result in friction reduction at the PRCL interface up to 50% [20, 21].

Figure 16. Photograph of laser textured surface.

4.4. Ignition

As mentioned earlier, to reduce in-cylinder NO_x formation, Low-Temperature Combustion conditions are preferred. This can be achieved by using excess combustion air as used under lean-burn conditions, or with the use of EGR as used under rich-burn conditions. In either case, to offset the loss in specific power of the engine, engine intake pressures are boosted using a turbocharger. Under such conditions, the in-cylinder pressures at the time of ignition tend to be high. The voltage required to strike a spark across the spark gap, which is governed by Paschen's law ($V_{discharge}$ ∝ pressure x distance between electrodes), increases to a point that it surpasses the 40 kV design limit of commercial Capacitance Discharge Ignition (CDI) Systems. Also, with the use of high voltage ignition systems the erosion of spark electrodes is accelerated (see Figure 17). Depending upon the manufacturer, typical Iridium and Platinum tipped spark plugs are currently replaced every 1000-3000 hrs of operation. To reduce the maintenance burden, engine manufacturers would like to extend this interval to 8000 hrs of engine operation. Additionally, confirming to market demands for higher specific power from a given engine frame, manufacturers would like to operate these engines under high in-cylinder pressures (see Fig. 18). Such conditions, along with the need for reliable ignition, and the concomitant need for reduced maintenance warrant the use of advanced ignition strategies.

Numerous entities around the world have proposed and evaluated numerous advanced ignition strategies to extend ignition limits, enhance the ignition reliability, improve ignition stability and reduce maintenance requirements. Most prominent ones will be discussed below in the context of their power draw, initial cost, long-term durability and ability to ignite lean/ highly diluted mixtures.

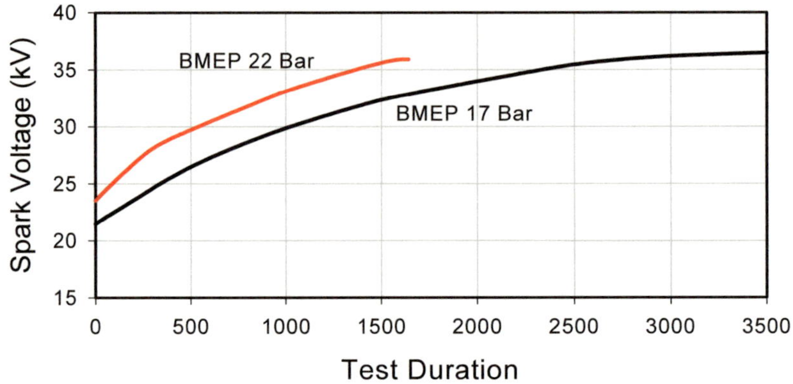

Figure 17. Ignition voltage vs. engine operational duration. [22]

ASME ICE-Vol. 35-2, 2000]

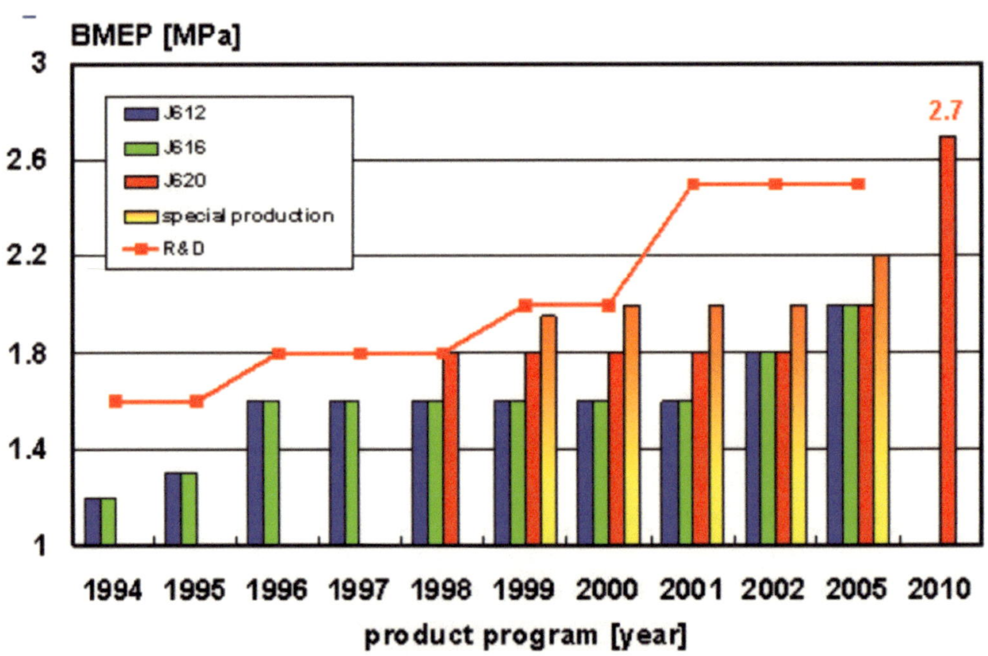

Figure 18. Engine BMEP trend at GE-Jenbacher [23]

4.4.1. Plasma jet / Rail plug / Corona discharge ignition

A number of studies have been performed evaluating the high energy plasma ignition systems [24, 25]. Most of such systems prove promising in extending the lean ignitability limits and thereby improving the performance of the engine. However, their power draw tends to be very large and unattractive. Also, high-energy plasma generation results in accelerated wear of the spark plug electrodes.

To offset the above shortcomings, several groups have tried using the corona discharge phase of gas breakdown which ionizes the gas prior to a full arc discharge [26 - 28]. Though Corona discharge offers promise, the system needs to be carefully designed specific to an engine to avoid inadvertent arcing between the electrode and the engine cylinder head.

4.4.2. *Diesel pilot ignition*

In this method of ignition, a very small quantity (< 4% of total fuel) of diesel is injected directly into the cylinder, which under goes auto-ignition and serves as a high-energy ignition kernel (up to 5000 times that achieved in SI) to ignite the lean natural gas-air charge introduced in to the cylinder. The load is varied by regulating the natural gas inducted into the intake air instead of throttling the intake charge. As a result, higher compression ratios can be used resulting in higher efficiencies. The amount of NO_x resulting from such a method of ignition though very low is limited by the NO_x formation resulting from the diesel pilot itself. Krishnan et al. [29] have tried using very advanced timing for the diesel pilot in order to reduce the NO_x formation. Others have tried to reduce the amount of the diesel pilot as a fraction of the total fuel introduced in to the cylinder with promising results [30].

4.4.3. *Laser ignition*

Among all of the ignition technologies discussed so far, laser ignition remains unparalleled in its ability to extend the ignition limits. As seen in Figure 3, laser ignition can extend the lean ignition limits of methane-air mixtures all the way to the lean flammability limit of $\phi = 0.5$.

Another advantage of laser ignition is the fact that it improves ignitability of fuel-air mixtures as pressure increases. As shown in Figure 19, for conventional ignition systems, the spark voltage which determines the success of ignition events, increases with pressure. In the case of laser ignition, laser pulse energy which becomes the governing factor decreases with increase in pressure. Such a trend proves beneficial in increasing the BMEP of the engines and thereby allows enhancing thermal efficiency.

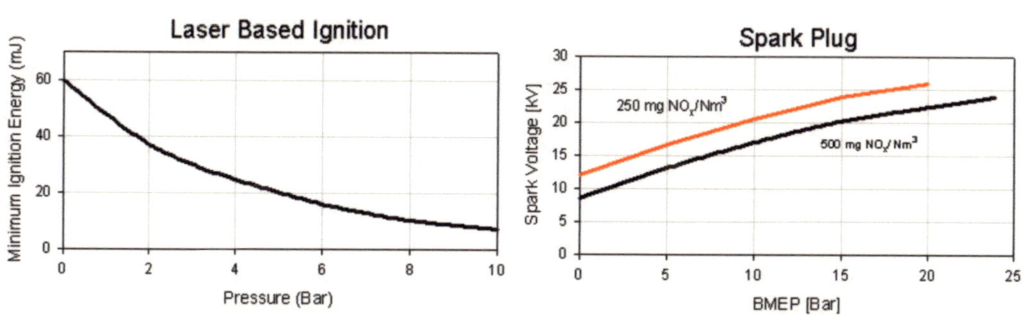

Figure 19. Pressure dependence of laser ignition (LI) and conventional spark ignition (SI). Please note that BMEP is given to be representative of in-cylinder pressure. [22]

Laser ignition tests performed in natural gas fueled single-cylinder engines have shown (i) shorter ignition delays, (ii) accelerated rates of combustion, and (iii) lean ignition limit extensions. However, as shown in Figure 20, the most significant advantage is the fact that laser ignition can result in NOx reductions up to 70% for a given engine efficiency, or alternately up to 3% point improvement in efficiency for a given NOx level [31]. The associated cost savings due to reduced fuel consumption, and avoidance of expensive after treatment systems, have provided an impetus for the development of practical laser ignition systems.

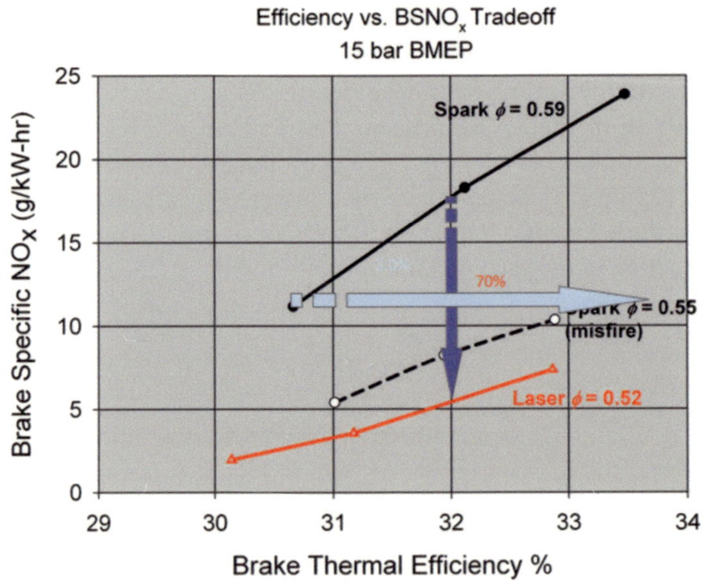

Figure 20. BSNOx vs. thermal efficiency improvement with laser ignition. [31]

The earliest known attempt at laser ignition was that performed by Dale et al. in 1978 [32], when they used a 14 ft. long CO_2 laser. At that time, laser ignition was considered impractical due to the high-cost and large size of the lasers, and interest in laser ignition remained luke warm. Since then, numerous developments in optics and lasers have resulted that make lasers efficient, of small size, low-cost and relatively insensitive to heat and vibration. As a result, interest in this field reemerged around 2000 and various organizations have made attempts to develop a viable laser ignition system. Kroupa et al. [32] and simultaneously Woodruff & Dustin McIntyre et al. [33] have tried to develop an end-pumped laser-per-cylinder Nd:YAG system. Weinrotter, Herdin et al. [34] have tried to develop a split system wherein a long laser pulse from a pump laser is transmitted through a optical fiber to end pump a laser gain medium equipped with a saturable absorber installed in the spark plug well of the engine. Bihari et al. through their attempts found it difficult to transmit high-power laser pulses through optical fibers and have embarked in developing a system that uses free-space transmission [31]. Recently, Taira et al [35, 36] have successfully demonstrated microlaser ignition systems, by using new age lasing materials called optical ceramics, and by using pulse trains of low-energy pulses to ignite instead of a single high-energy laser pulse. As of 2011, attempts to reduce laser ignition to practice continue throughout the world.

5. Conclusions

Technologies for improving the efficiency and reducing emissions from stationary natural gas engines have been presented. Notable omissions from this review are (i) use of advanced combustion cycles, such as, Miller cycle [37], (ii) use of high velocity compression, (iii) use of stratified charge [38], and (iv) use of enhanced turbulence levels [39]. As natural gas becomes widely available in abundance, development of technologies for further improvement of these engines is likely to continue well into the future. Advancements in materials and manufacturing processes for engine components, newer technologies for preprocessing natural gas as a fuel, and robust aftertreatment technologies are likely to yield efficiencies > 50%, while achieving easy compliance with emission regulations. Though not discussed in this chapter, technologies that harness energy from waste heat will further improve thermal efficiencies of these engines.

Author details

Sreenath B. Gupta, Munidhar S. Biruduganti, Bipin Bihari, Raj R. Sekar
Distributed Energy Research Center, Argonne National Laboratory, Argonne, IL, USA

Acknowledgement

The authors research and develop novel ways of improving the efficiency and reducing emissions from stationary natural gas fueled engines that are primarily used for distributed power generation. Of particular importance are their efforts to develop a novel laser based ignition system, and nitrogen enrichment of engine intake air using polymeric membranes. These researchers are currently involved in developing engine strategies for efficient energy conversion from various alternative gaseous fuels – various digester gases, as well as, synthesis gases.

6. References

[1] Griffiths J. C., Connelly, S., M., and DeRemer, R. B., "Effect of Fuel Gas Composition on Appliance Performance," GRI Contract 5011-345-0100, GRI report No. 82/00337, Dec. 1982.

[2] Liao, S. Y., Cheng, Q., Jiang, D. M., Gao, J., "experimental Study of Flammability limits of Natural Gas-air Mixtures," J. of Hazardous Materials, B119, pp. 81-84, 2005.

[3] Klett, G. M., Gupta, S. B., Bihari, B., and Sekar, R. R., "Ignition Characteristics of Methane-air Mixtures at Elevated Temperatures and Pressures," ICES2005-1064, ASME Spring Technical Conference, Chicago, IL, 2005.

[4] Ryan, T.W., Callahan, T. J., and King, S. R., "Engine knock Rating of Natural Gases – Methane Number," JEGTP, October 1993, Vol. 115, pp. 769.

[5] Hack, R., McDonell, V. G., "Impact of Ethane, Propane and Diluent Content in Natural gas on the Performance of a Commercial Microturbine Generator," JEGTP, jan. 2008, Vol. 130, pp. 011509-1

[6] Winston Ho, W.S., and Sirkar, K. K., "Membrane Handbook," Chapman & Hall, New York, 1992.

[7] Poola, R.B., Stork, K.C., Sekar, R.R., Callahan, K., Nemser, S., "Variable Air Composition with Air Separation Membrane: A New Low Emissions Tool for Combustion Engines," SAE Transactions, Journal of Engines, Section 3, Vol. 106, pp. 332-346, 1998.

[8] Poola, R., Sekar, R., "Simultaneous Reduction of NOx and Particulate Emissions by Using Oxygen-Enriched Combustion Air", ASME ICE-Vol. 37-1, 2001.

[9] Biruduganti, M., Gupta, S., Sekar, R., "Low Temperature Combustion Using Nitrogen Enrichment To Mitigate NO_x From Large Bore Natural Gas-Fueled Engines", ASME ICES2008-1616.

[10] Biruduganti, M., Gupta, S., Bihari, B., McConnell, S., and Sekar, R., "Air Separation Membranes – An Alternative to EGR in Large Bore Natural Gas Engines," Journal of Engineering for Gas Turbines and Power, Aug. 2010, Vol. 132, 082804.

[11] Wang, J., Huang, Z.,Bbing Liu, Y. F., Zeng, K., Miao, H., and Jiang, D., "Combustion behaviours of a direct-injection engine operating on various fractions of natural-gas hydrogen blends," International Journal of Hydrogen Energy, 32 (2007) pp.3555-3564.

[12] Hu, E., Huang, Z., Liu, B., Zheng, J., Gu, X., Huang, B., "Experimental investigation on performance and emissions of a spark-ignition engine fuelled with natural gas-hydrogen blends combined with EGR," International Journal of Hydrogen Energy, 34 (2009) pp.528-539.

[13] Shresta, S. O., and karim, G. A., "Hydrogen as an additive to methane for spark ignition engine applications," International Journal of Hydrogen Energy, 24.6 (1999), pp.577-586.

[14] Wyszynski, M. L., Megaritis, T., and Lehrle, R. S., "Hydrogen from exhaust gas fuel reforming: Greener, leaner and smoother engines," Future Power Systems Group: The University of Birmingham, 2000.

[15] Pratapas, J. M., et al., "Thermo chemical recuperation (TCR) for fuel savings and emissions reduction at compressor engines," Gas Machinery Conference, Dallas, TX, Oct. 1-3, 2007.

[16] McMillian, M. H., Lawson, S. A., "Experimental and Modeling study of hydrogen/syngas production and particulate emissions from a natural gas-fueled partial oxidation engine," International J. of hydrogen energy, vol. 31 (2006) pp. 847-860.

[17] Parks, J. E., Ferguson III, H. D., Williams, A. M., Storey, J. M., "Lean NO_x Trap Catalysis for NO_x Reduction in Natural Gas Engine Applications," Paper no. ICEF2004-0871, ASME 2004 Internal Combustion Engine Division Fall Technical Conference (ICEF2004), October 24–27, 2004 , Long Beach, California, USA.

[18] Holmgreen, E. M., Yung, M. M., Ozkan, U. S., "Dual-catalyst aftertreatment of lean-burn natural gas engine exhaust," Applied Catalysis B: Environmental, vol. 74, Issues 1-2, 18 June 2007, pp. 73-82.

[19] Quillen, K., Stanglemaier, R. H., Moughan, L., Takata, R., Wong, V., Reinbold, E., and Donahue, R., "Friction reduction by ring pack modifications of a lean-burn four-stroke NG engine: Experimental results," JEGTP, vol. 129, oct 2007, pp. 1088-1094.

[20] Nathan Bolander, "Piston ring lubrication and friction reduction through surface modification," Ph.D. Thesis, 2007, Purdue University.

[21] Cater, M. S., Bolander, N. W., and Sadeghi, F., "Experimental investigation of surface modifications for piston-ring/cylinder liner friction reduction, ASME ICES 2008.

[22] Kopecek, H., Wintner, E., Pischinger, H., Herdin, G., and Klausner, J., "Basics for a Future Laser Ignition System for Gas Engines," Paper No. 2000-ICE-316, ASME Fall Technical Conference, Peoria, US, ICE-35-2, 2000.

[23] Herdin, G., Klausner, J., Wintner, E., Weinrotter, M., Graf, J., Iskra, K., 2005, "Laser Ignition - a New Concept to Use and Increase the Potentials of Gas Engines," ASME ICEF2005-1352, ASME Internal Combustion Engine Division 2005 Fall Technical Conference, Ottawa, Canada.

[24] Dale, J. D., Checkel, M. D., and Smy, P. R., "Application of High Energy Ignition Systems to Engines," *Progress Energy Combustion Science*, Vol. 23, pp. 379-398, 1997.

[25] Gao, H., R.D. Matthews, M.J. Hall, and S. Hari, "From Spark Plugs to Rail Plugs - The Characteristics of a New Ignition System," SAE Journal Of Engines, Vol. 113, (2004), pp. 1-10.

[26] Theiss, N., Ronney, P., Liu, J., and Gundersen, M., 2004, "Corona Discharge Ignition for Internal Combustion Engines," ASME ICEF 2004-891.

[27] Freen, P. D., Gingrich, J., and Chiu, J., "Combustion Characteristics and Engine Performance of a new Radio Frequency Electrostatic Ignition System Igniting Lean Air-Fuel Mixtures," ASME ICEF2004-853, Fall Technical conference of the ASME-ICED division, Oct. 24-27, Long Beach, CA, 2004.

[28] Freen, Paul D., "Electrically Controlled Combustion Optimization System," as described on the website www.etatech.us.

[29] Krishnan, S.R., Srinivasan, K.K., Singh, S., Bell, S.R., Midkiff, K.C., Gong, W., Fiveland, S.B., and Willi, M., 2004, "Strategies for Reduced NO_x Emissions in Pilot-Ignited Natural Gas Engines," ASME Journal of Engineering for Gas Turbines and Power, Vol. 126, pp. 665-671.

[30] Goto, S., Takahashi, S., Yamada, T., and Yamada, T., "Development of High-Density Gas Engine 28G," Engineering Review, Vol. 40, No. 1, Feb. 2007.

[31] Bihari, B., Gupta, S. B., Sekar, R. R., Gingrich, J. and Smith, J., "Development of Advanced Laser Ignition System for Stationary Natural Gas Reciprocating Engines," ICEF2005-1325, ASME-ICE 2005 Fall Technical Conference, Ottawa, Canada, 2005.

[32] Kroupa, G., Franz, G., and Winkelhofer, E., "Novel miniaturized high-energy Nd-YAG laser for spark ignition in internal combustion engines," Optical Engineering, Vol. 48, pp. 014202 (Jan 26, 2009).

[33] McIntyre, D. L., Woodruff, S. D., and Ontko, J. S., "Lean-Burn Stationary Natural Gas Engine Operation With a Prototype Laser Spark Plug," J. Eng. Gas Turbines Power, Vol. 132, Issue 7, pp. 072804-810, July 2010.

[34] Weinrotter, M., Kopecek, H., Graf, J., Klausner, J., Herdin, G., and Wintner, E., "Laser Ignition in Internal Combustion Engines – A Novel Approach Based on Advanced Lasers," Advanced Solid-State Photonics Conference, Vienna, Austria, Feb. 6, 2005.

[35] Tsunekane, M., Inohara, T., Kanehara, K., and Taira, T., "Micro-solid-state laser for ignition of automobile engines," chapter in book advances in Solid-State Lasers: Development and Applications, eds. Mikhail Grishin, pp. 195-212, Feb. 2010, INTECH web publisher.

[36] Pavel., N., Tsunekane, M., and Taira, T., "Composite, all-ceramics, high-peak-power Nd:YAG/Cr^{4+}:YAG monolithe micro-laser with multiple-beam output for engine ignition," Opt. express 19(10), pp. 9378-9384, 2011.

[37] Tanaka, K., Shimoda, H., Noguchi, T., Goda, Y., Matsushita, Y., Nagamoto, T., "Development of the lean burn Miller cycle gas engine," Paper#199, CIMAC/ ASME STC 2004, Kyoto, Japan.

[38] Davey, M., Evans, R. L., and Mezo, A., "The Ultra Lean Burn Partially Stratified Charge Natural Gas Engine," SAE 2009-24-0115.

[39] Kastanis, E., and Evans, R.L., "The Squish-Jet Combustion Chamber for Ultra-Lean Burn Natural Gas Engines," SAE 2011-24-0112.

Chapter 11

LNG – Based Cogeneration Systems: Evaluation Using Exergy-Based Analyses

Tatiana Morosuk and George Tsatsaronis

Additional information is available at the end of the chapter

http://dx.doi.org/10.5772/51477

1. Introduction

The liquefied natural gas (LNG) market was established in the 1960s. During its over 50 years history, this market has grown significantly and the growth is expected to continue in the future. New technologies that lead to an increased efficiency in each step of the LNG chain are permanently in the focus of consideration. The information given in this section is based on many References (for example, EIA, 2003 and 2004; NTL, 2005; IIF-IIR, 2006; PFC Energy, 2011; GIIGNL, 2011; PLATTS, 2011; EUROGAS, 2011; Foss, 2003; Patel, 2005; Dorigoni & Portatadino, 2008; Maxwell & Zhu, 2011; Kumar et al., 2011).

The LNG chain (Figure 1) is divided into two main blocks: (a) Export terminal with associated technology for a liquefaction process, and (b) import terminal using a regasification process. Export and import terminals are connected by the LNG transport.

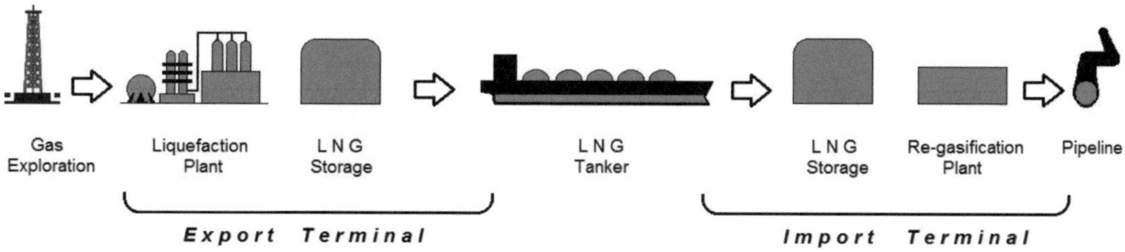

Figure 1. LNG chain

At the end of the year 2010, the World LNG market contained:

- 25 LNG liquefaction plants in 18 countries. Some of the first LNG liquefaction plants are still in operation (for example, in Algeria since 1964, in USA since 1969, and in Libya since 1970). New liquefaction plants that started operation during the year 2010 include two in Qatar, one in Yemen and one in Peru.

- 360 vessels (the oldest is in operation since the year 1969), and
- 83 LNG regasification plants (including 10 floating structures) with a total storage capacity of 38.5 million m^3 of LNG in 363 tanks. The oldest regasification plants are in operation since 1969 in Spain and Italy, and since 1972 in France and Japan. Many of the old regasification plants have been reconstructed during the last decade. The newest regasification plants include those completed in 2009 (in China, UK and Canada), and 2010 (in USA, Japan and Chile).

The average life cycle of a liquefaction and regasification plant is around 20 to 25 years. In the year 2010, the LNG trade rose with a growth rate of 21.2% compared with 2009, whereas the pipeline trade of natural gas increased only by 7%.

The following economic data can be found in the literature for the period 1992-2002. The price of LNG depended on the region and time (all values are given in constant US$ 2010 per million BTU):

- For the USA – between a minimum of 4.45 (1993) and a maximum of 5.84 (2001);
- For Europe – between a minimum of 3.06 (1999) and a maximum of 5.14 (2001); and
- For Japan – between a minimum of 4.73 (1998) and a maximum of 5.00 (2000).

In the year 2010 the LNG price was 3.75, 6.20, and 7.70 for USA, Europe, and Japan, respectively. Note that all the above cost numbers represent average annual values, while the average monthly price can vary by more than ± 1.5 US$/per million BTU. The price of LNG, obviously, depends on the price of natural gas. However, the price of LNG is also affected by the cost of the liquefaction, shipping and regasification processes.

The total capital investment associated with the LNG chain can be divided as follows: Exploration – 15% to 20%, liquefaction and storage – 30% to 45%, shipping – 10% to 30%, and storage and regasification – 15% to 25%. The capital investment for LNG storage in an import terminal is 30% to 50% of the capital investment of the import terminal, i.e. 5% to 10% of the overall capital investment of the LNG chain.

The specific values of the capital investment for the export terminals are: For the second generation (1980s) – US$ 600/ton_{LNG} and for the last generation (2000s) – US$ 200/ton_{LNG}. Significant differences are associated with the energy consumption for the liquefaction of one ton LNG, which depends on the used liquefaction process.

The range of deviation in the total capital investment cost for the import terminals is larger than that for the export terminals. For example, the total capital investment of a small import terminal constructed in the middle of 1980s is approximately US$ 100 million, while the new generation of middle-capacity import terminals (4.0 to 8.0 million ton LNG/year) have a total capital investment of US$ 200 to 300 million. Modern technologies used for the regasification process lead to an increase in the capital investment of the import terminals up to US$ 400 million (USA, 2009-2010) and they even can be higher than US$ 2 billion for a state-of-the art Japanese import terminal.

In the middle of 1990s the cost of the liquefaction process was US$ 0.8 to 1.20/million Btu, the cost of shipping was US$ 0.4 to 1.0/million Btu, and that of regasification and storage

US$ 0.3 to 0.5/million Btu. The result of various improvements in all LNG-chain processes is that the overall cost of LNG delivery has been reduced by almost 30 % in the last 20 years.

During the last decade and through significant technological advancements, the cost of the liquefaction processes has been decreased to around US$ 0.5/million Btu. Improvements in the LNG tankers, the facilities, and the import terminals led to a decrease in the cost of LNG to only approximately US$ 0.1/million Btu for each mentioned process. Note that the cost per unit of energy is higher for the import terminals than for the export terminals. Therefore, improving the regasification process contributes significantly to a reduction in the final cost of natural gas at the end of the LNG chain.

At present, the following regasification processes are widely used in import terminals:

- Open rack vaporizers using sea water. The energy consumption is approximately 0.008 $kWh/kg_{(LNG)}$ for driving the sea water circulating pumps.
- Submerged combustion vaporizers that are water baths heated by burning fuel gas. Between 1.5 and 2% of the imported gas is used as fuel gas for LNG vaporization.
- Intermediate fluid vaporizers (for example, warm/hot water/glycol case)
- Seawater heating added to the water/glycol heaters with shell and tube vaporizers, and
- Other heating sources, such as using wastewater effluent from a treatment plant, or recovering waste heat from an existing industrial facility.

Using supplemental heat from seawater, or from a wastewater treatment plant, a power plant or another industrial facility (if available), might be economically attractive as the operating costs are significantly reduced. However, the capital cost is substantially increased. Both the hot water/glycol system and the submerged combustion vaporizers have relatively low capital costs, but high operating costs. The submerged combustion vaporizers are used quite often.

Since the beginning of the LNG history, the fifth type of the regasification processes has been discussed in the literature. By using heating sources that originate from an existing industrial facility, the dissipative process of LNG regasification is converted to a productive cogeneration process. Some import terminals are already using such a technology. New related concepts are still under development. The interest in the investigation of effective regasification technologies remains unchangeably high.

2. State of art in the regasification technologies

The studies in the field of the regasification of LNG follow studies on regasification of cryogenic liquids. The main difference between these technologies lies in the mass flow rates of the working fluids: relative small and temporary for the cryogenic liquids and large and continuous for the LNG. The widely used so-called atmospheric evaporators, with direct heat transfer between environmental air and cryogenic liquids, have never been used for LNG regasification.

Direct and indirect heat transfer processes (including submerged combustion vaporizers) between water (sea water, glycol case, etc.) and LNG are known as dissipative systems because no efficiency can be defined for these processes.

The vaporization of LNG becomes part of a productive system only when the low-temperature energy (and exergy) of LNG is used for cooling purposes in another system that may or may not need to reject thermal energy to the environment. One of the first ideas to combine the regasification of LNG with the storage of food (including deep frozen products) or with air conditioning systems cannot be realized, because of the distance between an import terminal and a place where the cooling is required.

A modification of this idea is to create an industrial complex consisting of an LNG import terminal and different chemical plants. These industrial complexes, in which the LNG low-temperature energy is used, are characterized by very strong interdependencies among processes as well as high capital investment costs and operating and maintenance expenses. However, the primary energy consumption of the industrial complex is low. This makes such systems attractive for practical applications.

For example, in a project developed by Osaka Gas in Japan (Otsuka, 2006), the LNG import terminal is combined with refinery and petrochemical plants. The regasification of LNG takes place during four stages: (a) Separation of light hydrocarbons produced as a byproduct in the oil refining process (temperature level is around −100ºC; an energy source to separate olefin used as a raw material of polymer products at the petrochemical plant); (b) liquefaction of carbon dioxide produced as a byproduct in the manufacture of hydrogen (temperature level is around −55ºC), (c) low-temperature storage of butane (−8 ºC), and (d) cooling of water used to cool the intake air for gas turbines (10 ºC).

A more interesting idea is to create an LNG-based cogeneration system. The thermodynamic analysis of a power system operating as a closed cycle shows that decreasing the temperature level of heat rejection to the environment leads to an increase in efficiency. Using the low-temperature heat during the LNG regasification enables engineers to decrease significantly the lower temperature level of a power system. One of the first publications, where this idea has been described, was a paper published by Angelino, 1978.

Since the end of the 1970s several approaches for cogeneration systems for electricity generation and LNG vaporization have been developed. For example, the gas-turbine-based concepts that can be found in literature include the following:

- German scientists developed a relative simple and efficient system for vaporization of LNG using closed-cycle gas turbines (Griepentrog & Sackarendt, 1976; Krey, 1979). The factors discussed in these publications include the effects of working fluid for the closed-cycle gas turbines, cycle pressure level, compressor and turbine inlet temperature, and efficiency values of the components. The reported value of the overall energetic efficiency is around 80%. This idea had a practical background, i.e. the development of a power system in Oberhausen (Germany) based on a closed-cycle gas turbine with helium as a working fluid (Griepentrog & Sackarendt, 1976; Frutschi, 2005).

- A similar idea but with hydrogen as the working fluid has been described by Najjar, 1991.
- Japanese scientists proposed to use the low-temperature heat coming from LNG for pre-cooling and interstage cooling of air for the so-called mirror gas-turbine system (Kaneko et al., 2004). The reported energetic and exergetic efficiencies amount to 56% and 60%, respectively.
- The idea of a cogeneration system with cascade arrangement of an open-cycle gas-turbine system as a topping cycle and a closed-cycle gas-turbine system as a bottoming cycle for electricity generation and LNG vaporization has been proposed and proved in Italy at the end of 1970s by Snamprogetti ENI Group (SNAMPROGETTI, 1978). A very brief description of this system can be found in a recent publication of Italian scientists from the University of Palermo (Dispenza et al., 2009 a, 2009 b). They improved the old idea and focused the research on the use of different working fluids (helium, nitrogen and carbon dioxide) in the closed-cycle gas-turbine system. However, open track vaporizers are also involved in these schematics. The electric efficiency of the cogeneration system is between 43% and 45% and the overall efficiency is 69%. The specific electricity generation is 0.38 kW_{el}/kg_{LNG} and 0.29 kW_{el}/kg_{LNG} for helium and nitrogen, respectively. For the economic analysis no clear data can be found; the authors tried to estimate the payback time under the assumption that the cost of the generated electricity is lower than the price in the Italian market of 74.75 Euro/MWh for the year 2006. The environmental analysis is based only on CO_2 emissions: 137.22 g/MJ of electricity generated and 55.28 g/MJ of regasified LNG. The avoided emissions amount to 26.71 kt for helium and 6.49 kt for nitrogen.
- The concepts based on combined-cycles for cogeneration systems are discussed only in recent publications, for example: In a typical combined-cycle power plant, the low-temperature energy of LNG is used for pre-cooling and interstage cooling of air for the gas-turbine system and for cooling in the steam condensation process (Shi et al., 2010). Typical thermodynamic data for this cogeneration system include (a) an energetic efficiency of the conventional plant of 56.5% and of the combination with LNG of 59.3%, and (b) net generated power of $\dot{W}_{net}$ =212 MW and $\dot{W}_{net}$ =288 MW for the conventional and the LNG-combined plant, respectively.
- The discussions in the field of combined-cycle power plants with Organic Rankine Cycle (ORC) mainly focus on the selection of a working fluid for the ORC:
- The case of propane is discussed by Najjar & Zaamout, 1993. An evaluation of a 200 MW gas-turbine power plant shows that addition of a cryogenic circuit may save about 62.595 tons/yr of LNG and 39 MW of power. This idea has been applied to one import terminal (Otsuka, 2006).
- Results using ammonia have been reported by Querol et al., 2011. The cogeneration system consists on a gas-turbine system, an ORC with ammonia as working fluid and an open track vaporizer (sea water first comes to the open track vaporizer, and after that enters the condenser of the ORC). The main data include: $\dot{W}_{net}$ =8 to 13 MW, energetic efficiency 44.5 to 46.6%, and cost of the generated electricity 76.7 to 54.8 Euro/MWh.

- A mixture of CF_4 and C_3H_8 was considered as an alternative to the water-ammonia mixture (i.e., a Kalina process, but with a mixture of CF_4 and C_3H_8 as working fluid) by Liu & Guo, 2011. The energetic efficiency of this cogeneration system is equal to 66%, and the exergetic efficiency is 23.5%.
- The MATIAN cycle (supercritical Rankine cycle with CO_2 as working fluid and integrated oxy-fuel combustion) has been combined with LNG vaporization (where the heat of the condensation process is used for the LNG) by Deng et al., 2004. The reported energetic efficiency is approximately 80%, and the exergetic efficiency is 49 to 50%.
- CO_2 is used as working fluid for the cogeneration system where the so-called "CO_2 Rankine-like cycle" (term used in these publications) is combined with a closed-cycle gas turbine system. Oxy-fuel combustion and CO_2 storage are integrated in the overall system. Several schematics of such a cogeneration system have been discussed by Zhang & Lior 2006a and 2006b; Liu et al., 2009. The by-products of this cogeneration system are water (subtracted from combustion gases), nitrogen and argon (after the air separation unit). The mass flow rate of LNG is 54.7 ks/s. The energetic efficiency of the cogeneration system without LNG integration is approximately 50% and with LNG 64% to 67% (for different schematics). The corresponding exergetic efficiency is equal to 51% (without LNG integration) and 53.6 to 67.5% (with LNG integration). The specific capital investment cost is around US$ 1100/kW_{el} and the cost of the generated electricity is around US$ 0.04/kWh (The economic analysis has been conducted in CHY - the currency of the People's Republic of China, and the obtained data were converted in US$).

A different idea for the systems generating electricity and gasifying LNG has been published recently: Heat from the environment (air or sea water) is used as a high-temperature source for a power system, whereas the LNG represents the low-temperature reservoir:

- The Polish scientists (Szargut and Szczygieł, 2009) proposed such a system with a cascade (binary) of organic Rankine cycles using acetylene and ethane as working fluids. Depends on many parameters, the maximal exergetic efficiency of such a cogeneration system varies between 45% and 47%. The conclusion from the economic analysis is that the pay-back time of the investment expenditures is sufficiently short.
- Italian scientists from the University of Palermo (Dispenza et al., 2009c; La Rocca, 2010) have reported a similar but simpler (regular) organic Rankine cycle with ethane or ethylene. In these publications the value of the Coefficient of Performance (*COP*) is used as a variable for the efficiency. According to the definition of *COP* given by the authors, this value is around 1.6 for the analyzed systems.

Almost in all mentioned publications the concept of exergy is used. Not always the definition of energetic and exergetic efficiencies are given; therefore it is difficult to compare these cogeneration systems. Note that in the publications where the exergetic (second-law, availability) efficiency has been defined, the authors used the concept of "exergy output over exergy input", which is inferior to the product-over-fuel concept.

Only few of the above-mentioned publications deal with an economic or environmental analysis. The environmental analyses focus only on the CO_2 emissions through the fuel consumption.

There are some publications and technical reports (for example, IHV-2025, 2003; Malacic et al., 2008; EPA, 2010; PETRONAS, 2010; EIA, 2010) that discuss environmental issues associated with LNG terminals. These publications consider factors such as the CH_4 emissions through the leaking of LNG and NG, the CO_2 emissions through fuel consumption, the effect of decreasing the temperature of sea water after using open rack vaporizers, and the increase in the concentration of chlorine (to control bio-fouling in the open rack vaporizers). The information is always given for the entire terminal (existing or planned) without discussing the re-gasification process.

Since 2007 a group at the Institute for Energy Engineering at Technische Universität Berlin has been studying cogeneration systems for the simultaneous generation of electricity and re-gasification of LNG (Griepentrog et al., 2008a and 2008b; Tsatsaronis & Morosuk, 2009; Tsatsaronis et al., 2009; Tsatsaronis & Morosuk, 2010; Morosuk & Tsatsaronis, 2011; Morosuk et al., 2012). The research aims at developing thermodynamically efficient processes, and conducting economic and environmental-impact evaluations of these processes by using the so-called exergy-based methods. In the following sections we demonstrate how to evaluate an LNG-based cogeneration system using these methods.

3. The authors' concept

The system shown in Figure 2 represents a novel cogeneration system for generating electricity and vaporizing LNG. This cogeneration system is an advanced case of the gas-turbine-based approach.

The overall system consists of three sub-systems:

- *LNG sub-system* (process 1-2-3-4) – LNG from the storage system is (a) compressed by an LNG pump (PM), (b) vaporized in heat exchanger II (HE II) using the thermal energy rejected from the closed-cycle power sub-system, and (c) expanded in an LNG expander (EX III).
- *Closed-cycle gas-turbine power sub-system* (cycle 11-12-13-14) – After removing the thermal energy in HE II, the working fluid (a working fluid that can be used in a closed-cycle gas-turbine system) is compressed in compressor III (CM III), heated in heat exchanger I (HE I, using the thermal energy rejected from the open-cycle gas-turbine power system), and expanded in expander II (EX II).
- *Open-cycle gas-turbine power sub-system* (process 21-22-23-24+25-26-27-28) – Air after compression in compressor I (CM I) is cooled in a cooler (CL) by removing thermal energy to the environment, and is compressed in compressor II (CM II). After the combustion process in the combustion chamber (CC), the combustion gases are expanded in expander I (EX I) and rejected to the atmosphere after being cooled in HE I.

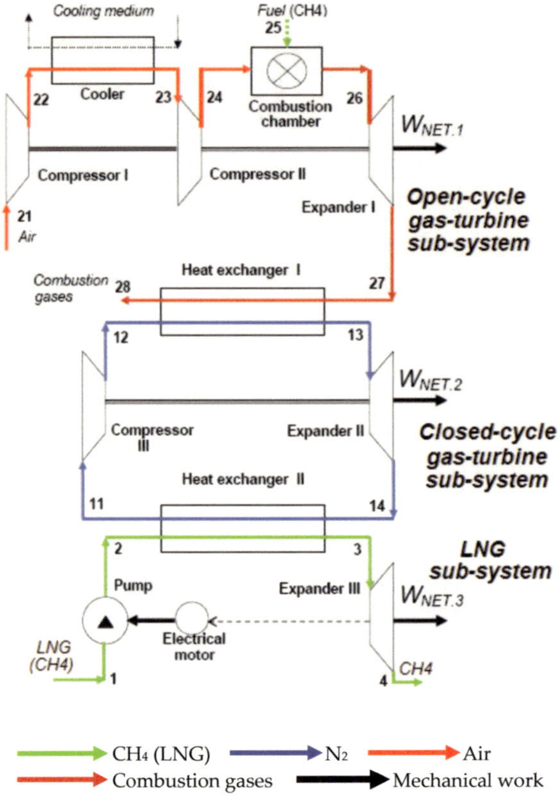

Figure 2. Schematic of the LNG-based cogeneration system

The following advantages are associated with the cogeneration system proposed by the authors:

- Complete independency of seawater. This means that for an import terminal the technical, economical and environmental aspects associated with the treatment, transport and use of seawater are irrelevant.
- The area of the import terminal that is necessary for the regasification process is relatively small (compared with other cogeneration systems using a Rankine cycle).
- The LNG storage within the import terminal is not important. For the *open-cycle gas-turbine power sub-system* the authors considered an efficient gas-turbine system (LMS-100^{TM} (Reale, 2004)) that can reach 100% capacity within approximately 10 minutes.

3.1. Simulation

For the simulation of the cogeneration system, we assumed the following:

- The *open-cycle gas-turbine power sub-system* is based on a LMS 100 gas turbine, with T_{21}=15°C; p_{21}=0.1013 MPa; T_{26}=1290°C; pressure ratio 42:1 (Reale, 2004), $\eta_{CM,I}$ =90%, $\eta_{CM,II}$=90%, $\eta_{EX,I}$ =94%. The heat loss in the combustion chamber amounts to 2% of the lower heating value of the fuel.

- *For the closed-cycle gas-turbine power sub-system* we assumed $\eta_{CM,III}$=85%, $\eta_{EX,II}$=88% (Frutschi, 2005). Five working fluids have been considered for the analysis: Nitrogen, dry air, argon, helium and carbon dioxide.
- In the *LNG sub-system*, LNG is leaving the storage space as a subcooled liquid at T_1 = –160°C and p_1 = 1.0 MPa. After vaporization, superheated vapor at p_4 = 8.0 MPa and $T_4 \approx$ 2°C is obtained. The isentropic efficiencies assumed for the LNG pump and the expander III are η_P=66.5%, and $\eta_{EX,III}$=85%, respectively. For the analysis, the LNG pump is considered together with the required electrical motor.

The generators required for producing the electric power were not considered in the analysis.

For the simulation and the energetic and exergetic analyses, the following software were used: GateCycle (GE, 1989), Gatex (GATEX, 2002), and EES (EES, 2012).

The operation conditions for the closed-cycle gas-turbine power sub-system are fixed by the temperatures at states 11 (415ºC) and 14 (–129ºC). Carbon dioxide cannot be used as a working fluid because of its thermodynamic properties: The temperature of –129ºC cannot be achieved at the state of superheated vapor at $p_{11}(p_{14})$>0.1 MPa.

State	Working fluid	T [°C]	p [MPa]	v [m³/kg]	$\dot{E}$ [MW]	$\dot{m}$ [kg/s]
1	LNG	-160	1.0	–	124.3	64.96*) for air;
2		-144	27.2	–	126.8	65.02*) for nitrogen;
3		86	27.0	–	98.9	67.36*) for argon;
4		2	8.0	–	77.7	67.56*) for helium
11	Air	-129	0.28	0.142	31.7	222.50
12		71	4.27	–	69.7	
13		415	4.06	0.049	102.9	
14		103	0.30	–	22.4	
11	Nitrogen	-129	0.28	0.147	60.6	217.20
12		70	4.27	–	133.4	
13		415	4.06	0.051	196.1	
14		101	0.3	–	42.7	
11	Argon	-129	0.28	0.103	40.9	445.30
12		64	1.94	–	79.4	
13		415	1.84	0.078	112.1	
14		102	0.3	–	31.4	
11	Helium	-129	0.28	1.052	41.2	45.32
12		66	1.94	–	81.2	
13		415	1.84	0.779	114.0	
14		102	0.3	–	32.0	
21	Air	15	0.101	–	0.62	209.00
22		242	0.67	–	86.7	

State	Working fluid	T [°C]	p [MPa]	v [m³/kg]	$\dot{E}$ [MW]	$\dot{m}$ [kg/s]
23		117	0.65	–	67.8	
24		416	4.35	–	186.1	
25	Methane	15	4.5	–	507.0	5.14
26		1290	4.19	–	523.6	
27	Comb. gases	435	0.11	–	80.9	214.14
28		90	0.102	–	10.9	

*) Depends on the working fluid used for the closed-cycle gas-turbine sub-system

Table 1. Thermodynamic data for the material streams

3.2. Energetic analysis

Selected energetic variables of the LNG-based cogeneration system with different working fluids (that can be used in the closed-cycle gas-turbine sub-system) are given in Figure 3. Note that the variables of the open-cycle gas-turbine sub-system remain unchanged. The calculated values using dry air, nitrogen, argon and helium demonstrate that all these working fluids do not affect significantly the energetic efficiency of the overall system. The most significant differences refer to (Table 1):

- The mass flow rate of the re-gasified LNG air amounts to 64.96 kg/s, for nitrogen to 65.02 kg/s, for argon to 67.36 kg/s und for helium to 67.56 kg/s,
- The mass flow rate of the working fluid for the closed-cycle gas-turbine sub-system, is for air, nitrogen, argon, and helium 222.50 kg/s, 217.20 kg/s, 445.30 kg/s, und 45.32, respectively.
- The pressure ratio within the closed-cycle gas-turbine sub-system is p_{12}/p_{11} =6.8 for argon and helium and p_{12}/p_{11}=15 for nitrogen and air.

Based on these data for the closed-cycle gas-turbine sub-system we excluded argon as a potential candidate, and we looked closer to helium.

A new definition of the energetic efficiency for the LNG-based cogeneration system has been proposed (Tsatsaronis & Morosuk, 2010) as

$$\eta_{tot} = \frac{\dot{W}_{NET,tot} + \left(\dot{H}_4 - \dot{H}_1\right)}{\dot{m}_{25} HHV} = \frac{\dot{W}_{NET,tot}}{\dot{m}_{25} HHV} + \frac{\left(\dot{H}_4 - \dot{H}_1\right)}{\dot{m}_{25} HHV} = \eta_{el} + \eta_{LNG} \quad (1)$$

where the lower heating value (HHV) of methane is 55.5 MJ/kg and $\dot{W}_{NET,tot} = \dot{W}_{NET,1} + \dot{W}_{NET,2} + \dot{W}_{NET,3}$. The values of $\dot{W}_{NET,1}$, $\dot{W}_{NET,2}$ and $\dot{W}_{NET,3}$ represent the net mechanical power generated in the open-cycle gas-turbine sub-system, the closed-cycle gas-turbine sub-system, and the LNG sub-system, respectively. $\dot{H}_4$ and $\dot{H}_1$ represent the enthalpy flow rates at the thermodynamic states 4 and 1, respectively. The values of η_{el} and η_{LNG} are given in Figure 4.

An interesting conclusion can be obtained if the values of the specific volume at states 11 (intake to the CM III) and 13 (intake to the EX II) are analyzed. For the three working fluids (air, nitrogen and argon) these values are quite similar. Much higher values are observed for helium. This significantly affects the size of compressor III and expander II and, therefore the costs of these components. The price of the working fluid is also an important factor: The prices for nitrogen and dry air are similar; argon is approximately 25 times more expensive and helium is even more. Taking into account these economic aspects, both argon and helium cannot be recommended as working fluids for these systems.

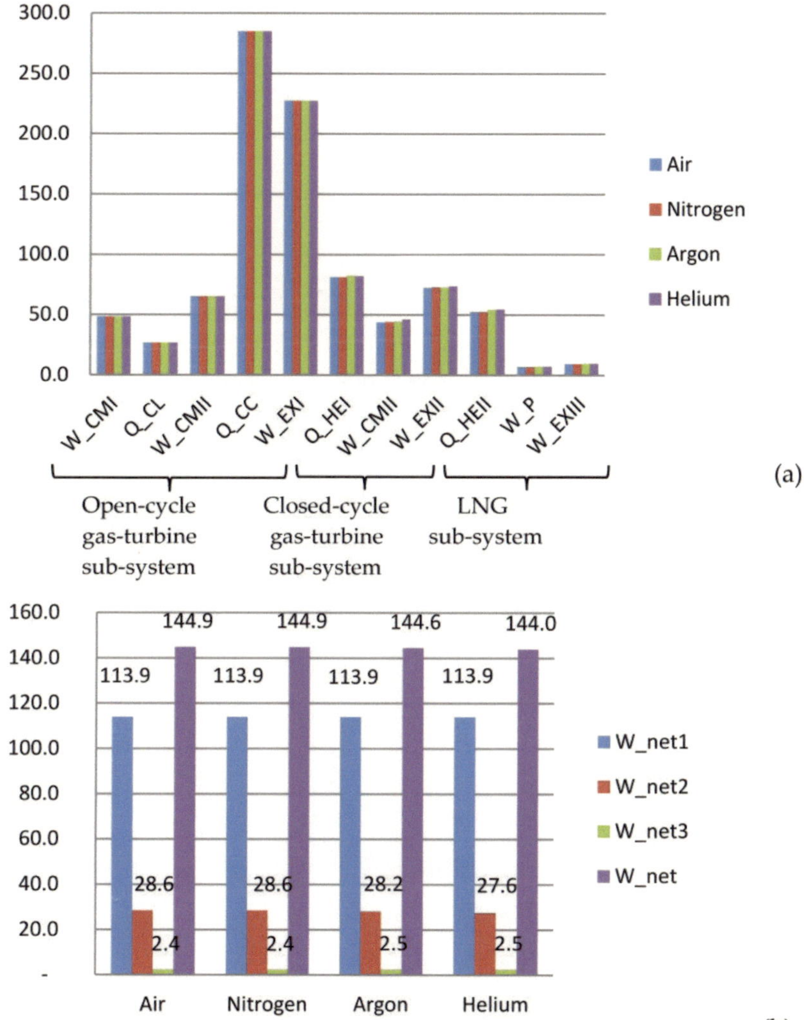

Figure 3. Results of the energetic analysis (all values are given in MW): (a) Component characteristics; (b) sub-system characteristics.

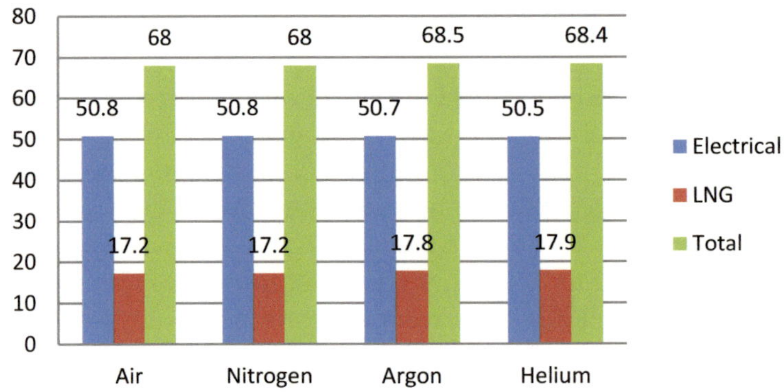

Figure 4. Electrical, LNG and overall energetic efficiency (%) of the LNG-based cogeneration systems.

4. Exergy-based analyses

4.1. Exergetic analysis

Exergy is defined as the maximum theoretical useful work (shaft work or electrical work) obtainable from an energy conversion system as this is brought into thermodynamic equilibrium with the thermodynamic environment while interacting only with this environment (Tsatsaronis, 2008).

An *exergy-based* thermodynamic *analysis* identifies the location, the magnitude, and the causes of thermodynamic inefficiencies, which are the *exergy destruction* (due to irreversibilities within the system), and the *exergy loss* (exergy transfer to the environment). In an exergy analysis we calculate the exergy associated with each energy carrier (stream) in the overall system, the exergy destruction within each system component and process, and the *exergetic efficiency* (for each process, component, or system).

The exergy balance for the overall cogeneration system is

$$\dot{E}_{F,tot} = \dot{E}_{P,tot} + \sum_{k} \dot{E}_{D,k} + \dot{E}_{L,tot} \tag{2}$$

and for the *k*th component

$$\dot{E}_{F,k} = \dot{E}_{P,k} + \dot{E}_{D,k} \tag{3}$$

The variables used for the conventional exergetic evaluation of the *k*th component in a system include the following:

- Exergy destruction rate that is calculated from the exergy balance
- Exergetic efficiency

$$\varepsilon_k = \frac{\dot{E}_{P,k}}{\dot{E}_{F,k}} = 1 - \frac{\dot{E}_{D,k}}{\dot{E}_{F,k}} \tag{4}$$

- Exergy destruction ratio

$$y_k = \frac{\dot{E}_{D,k}}{\dot{E}_{F,tot}} \tag{5}$$

The definition of the exergetic efficiency for the LNG-based cogeneration system is

$$\varepsilon_{tot} = \frac{\dot{E}_{P,tot}}{\dot{E}_{F,tot}} = \frac{\dot{W}_{NET,tot} + \left(\dot{E}_4^M - \dot{E}_1^M\right)}{\dot{E}_{25} + \dot{E}_{21} + \left(\dot{E}_1^T - \dot{E}_4^T\right)} \tag{6}$$

The modern approaches for the exergetic analysis use the general concepts of *fuel* and *product* introduced over 25 years ago (Tsatsaronis, 1984): The *exergy of product* is the desired result (expressed in exergy terms) achieved by the system (e.g., the *k*th component) being considered, and the *exergy of fuel* represents the exergetic resources expended to generate the exergy of the product. These concepts are used in a consistent way for the exergy-based analyses (Lazzaretto & Tsatsaronis, 2006; Meyer et al., 2009).

The following definitions of the exergy of fuel ($\dot{E}_{F,k}$) and the exergy of product ($\dot{E}_{P,k}$) are used for the overall system and for the components (Bejan et al., 1996; Lazzaretto & Tsatsaronis, 2006):

- Overall system – $\dot{E}_{F,tot} = \dot{E}_{25} + \dot{E}_{21} + \left(\dot{E}_1^T - \dot{E}_4^T\right)$, $\dot{E}_{P,tot} = \dot{W}_{NET,tot} + \left(\dot{E}_4^M - \dot{E}_1^M\right)$ and $\dot{E}_{L,tot} = \dot{E}_{28}$
- Compressor I – $\dot{E}_{F,CM\,I} = \dot{W}_{CM\,I}$ and $\dot{E}_{P,CM\,I} = \dot{E}_{22} - \dot{E}_{21}$
- The cooler is a dissipative component, therefore we calculate $\dot{E}_{D,CL} = \dot{E}_{22} - \dot{E}_{23}$ without defining for this component an exergetic efficiency or the exergies of fuel and product.
- Compressor II – $\dot{E}_{F,CM\,II} = \dot{W}_{CM\,II}$ and $\dot{E}_{P,CM\,II} = \dot{E}_{24} - \dot{E}_{23}$
- Combustion chamber – $\dot{E}_{F,CC} = \dot{E}_{25}$ and $\dot{E}_{P,CC} = \dot{E}_{26} - \dot{E}_{24}$
- Expander I – $\dot{E}_{F,EX\,I} = \dot{E}_{26} - \dot{E}_{27}$ and $\dot{E}_{P,EX\,I} = \dot{W}_{EX\,I}$
- Heat exchanger I – $\dot{E}_{F,HE\,I} = \dot{E}_{27} - \dot{E}_{28}$ and $\dot{E}_{P,HE\,I} = \dot{E}_{13} - \dot{E}_{12}$
- Expander II – $\dot{E}_{F,EX\,II} = \dot{E}_{13} - \dot{E}_{14}$ and $\dot{E}_{F,EX\,II} = \dot{W}_{EX\,II}$
- The pump operates below the reference temperature: $e_2^{PH} > e_1^{PH}$, but $e_2^T < e_1^T$, and $e_2^M > e_1^M$. Therefore, $\dot{E}_{F,P} = \dot{W}_P + \left(\dot{E}_1^T - \dot{E}_2^T\right)$ and $\dot{E}_{P,P} = \dot{E}_2^M - \dot{E}_1^M$
- In heat exchanger II the reference temperature is crossed: $T_2 < T_0$, $T_3 > T_0$ and $T_{14} > T_0$, $T_{11} < T_0$. Because of the pressure drop on both sides of heat exchanger II the mechanical exergy at the inlet of each stream is larger than the mechanical exergy of the same stream at the outlet: $e_2^M > e_3^M$ and $e_{11}^M < e_{14}^M$. In this way, $\dot{E}_{F,HE\,II} = \left(\dot{E}_{14}^T + \dot{E}_2^T\right) + \left(\dot{E}_{14}^M - \dot{E}_{11}^M\right) + \left(\dot{E}_2^M - \dot{E}_3^M\right)$ and $\dot{E}_{P,HE\,II} = \left(\dot{E}_{11}^T + \dot{E}_3^T\right)$
- In Compressor III the reference temperature is crossed: $\dot{E}_{F,CM\,III} = \dot{W}_{CM\,III} + E_{11}^T$ and $\dot{E}_{P,CM\,III} = \dot{E}_{12}^M - \dot{E}_{11}^M + \dot{E}_{12}^T$
- In expander III the reference temperature is also crossed. Thus, $\dot{E}_{F,EX\,III} = \dot{E}_3^M - \dot{E}_4^M + \dot{E}_3^T$ and $\dot{E}_{P,EX\,III} = \dot{W}_{EX\,III} + \dot{E}_4^T$

The chemical exergies of LNG, and NG and of the working fluid of the closed-cycle gas-turbine sub-system do not need to be considered in the exergetic analysis of this system because only the physical exergy of the working fluid is used in the corresponding sub-systems. Only the chemical exergies in the open-cycle gas-turbine sub-system, where combustion takes place, need to be considered. The chemical exergy of air is not zero because of the humidity (60%) considered in it. The physical exergies of LNG, NG and N_2 are split into their thermal and mechanical parts according to the approach presented in (Morosuk & Tsatsaronis, 2008).

For the exergetic analysis we assumed: $T_0 = T_{21}$ =15ºC (288.15 K) and p_0=1.013 bar. The results from the exergetic analysis are given in Tables 1 and 3. Figure 4 shows the exergy destruction ratio within components of the overall cogeneration system and Figure 5 shows the exergetic efficiency of the overall system. Note that the exergetic efficiency cannot be split is analogy to the energetic efficiency.

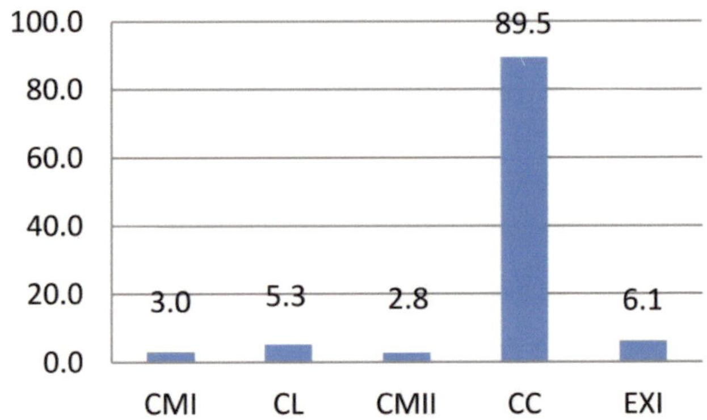

(a)

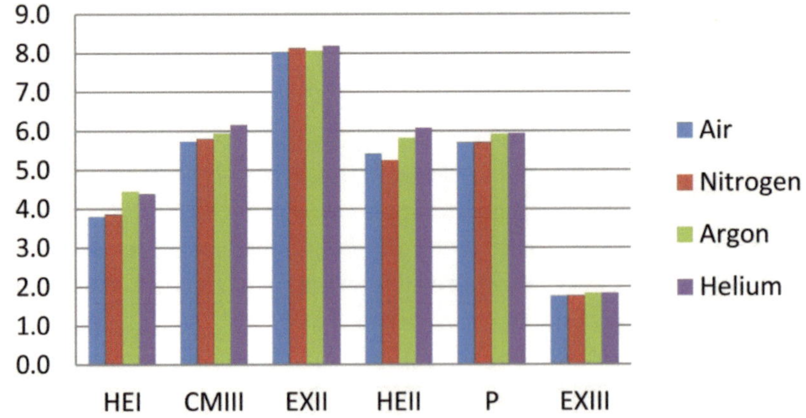

(b)

Figure 5. Exergy destruction rate (all values are given in MW) (a) within the components of the open-cycle gas-turbine sub-system, and (b) the remaining components.

The calculated overall efficiency (η_{tot} = 68% (HHV-based) or η_{tot} = 75.5% (LHV- based)) and exergetic efficiency (ε_{tot} = 52.6%) demonstrate the thermodynamic advantages of this novel cogeneration concept that combines LNG regasification with the generation of electricity, particularly if air or nitrogen is used for the open-cycle gas-turbine sub-system.

The results obtained from the conventional exergetic analysis are based on the value of $\dot{E}_{D,k}$ (Figure 5). Assuming that no changes can be conducted in the open-cycle gas-turbine sub-system (Figure 5a), the priorities for improving the components of the LNG and the closed-cycle gas-turbine sub-systems (Figure 5b) are the following:

- Expander II should be considered first because it exhibits the highest exergy destruction among all components of the considered sub-systems.
- Compressor III, heat exchanger II, and pump have comparable (values of $\dot{E}_{D,k}$), therefore they have the same relative importance for the improvement procedure. Note that the cogeneration system with nitrogen has the lowest exergy destruction within heat exchanger II.
- The exergy destructions within heat exchanger I (note the significant difference between air and nitrogen on one side and argon and helium on the other side) and expander III are relatively low, therefore these components, especially expander III, cannot significantly affect the efficiency of the overall cogeneration system.

The detailed results from the exergetic analysis and the sensitivity exergetic analysis for this LNG-based cogeneration system with nitrogen as working fluid for the closed-cycle gas-turbine sub-system have been reported by (Tsatsaronis and Morosuk, 2010; Morosuk and Tsatsaronis 2011). In this paper only the values of the exergetic efficiency of the overall cogeneration system with different working fluids for the closed-cycle gas-turbine sub-system is given (Figure 6). The exergetic analysis assists us to conclude that nitrogen should be used as the working fluid of the closed-cycle gas-turbine sub-system. In the further analyses only the case with nitrogen is considered.

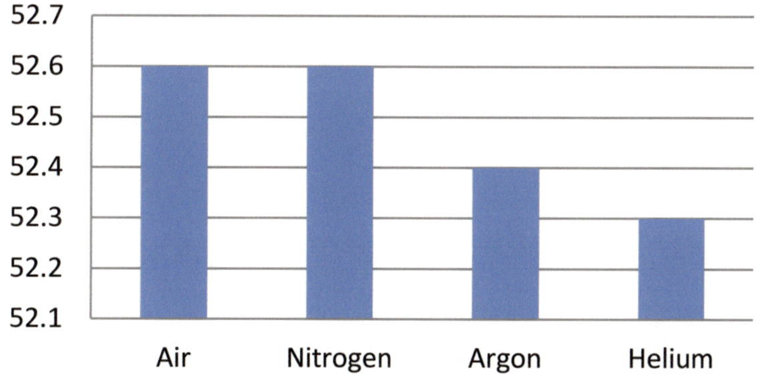

Figure 6. Exergetic efficiency of the overall cogeneration system with different working fluids for the closed-cycle gas-turbine sub-system.

4.2. Exergoeconomic analysis

Exergoeconomics (for example, Tsatsaronis, 1984; Bejan et al., 1996; Tsatsaronis 2008) is a unique combination of exergy analysis and cost analysis conducted at the component level, to provide the designer or operator of an energy conversion system with information crucial to the design or operation of a cost-effective system. A complete exergoeconomic analysis consists of (a) an exergetic analysis, (b) an economic analysis, and (c) an exergoeconomic evaluation.

The exergoeconomic model for an energy conversion system consists of *cost balances* written for the *k*th component and auxiliary equations based on the P and the F-rules. The cost balances can be written as

$$\dot{C}_{P,k} = \dot{C}_{F,k} + \dot{Z}_k \tag{7}$$

or

$$c_{P,k}\dot{E}_{P,k} = c_{F,k}\dot{E}_{F,k} + \dot{Z}_k \tag{8}$$

where

$$\dot{Z}_k = \dot{Z}_k^{CI} + \dot{Z}_k^{OM} \tag{9}$$

To simplify the discussion, we assume, that the contribution of $\dot{Z}_k^{OM}$ remains constant when the design changes, and, therefore, the changes in the value of $\dot{Z}_k$ are associated only with changes in the capital investment cost $\dot{Z}_k^{CI}$.

The real cost sources in an energy conversion system are (a) the capital investment (and the operating & maintenance expenses) for each component, (b) the cost of exergy destruction within each component, and (c) the cost of exergy loss from the overall system. The last two terms can be revealed only through an exergoeconomic analysis:

- The cost rate associated with exergy destruction within the *k*th component is

$$\dot{C}_{D,k} = c_{F,k} \cdot \dot{E}_{D,k} \tag{10}$$

- The cost rate associated with exergy loss from the overall system is

$$\dot{C}_{L,tot} = \dot{C}_{28} \tag{11}$$

The exergoeconomic model for the Base Case of the LNG-based cogeneration system is:

- Compressor I:

$$\dot{C}_{W1,CMI} + \dot{Z}_{CMI} = \dot{C}_{22} - \dot{C}_{21} \tag{12}$$

with $c_{21} = 0$.

- The cooler (dissipative component) was considered together with a cooling tower (which is not shown in Figure 1):

$$\dot{C}_{22} + \left(\dot{Z}_{CL} + \dot{Z}_{CT}\right) = \dot{C}_{23} \tag{13}$$

- Compressor II:

$$\dot{C}_{W1,CM\,II} + \dot{Z}_{CM\,II} = \dot{C}_{24} - \dot{C}_{23} \tag{14}$$

- Combustion chamber:

$$\dot{C}_{25} + \dot{Z}_{CC} = \dot{C}_{26} - \dot{C}_{24} \tag{15}$$

where $c_{25} = c_4$

- Expander I:

$$\dot{C}_{26} - \dot{C}_{27} + \dot{Z}_{EX\,I} = \dot{C}_{W1,EX\,I} \tag{16}$$

the auxiliary equation for EX I according to the F-rule is $c_{26} = c_{27}$. Note that c_{W1} is the specific cost of electricity generated within open-cycle gas-turbine sub-system.

- Heat exchanger I:

$$\dot{C}_{27} - \dot{C}_{28} + \dot{Z}_{HE\,I} = \dot{C}_{13} - \dot{C}_{12} \tag{17}$$

$c_{26} = c_{27}$ (F-rule)

- Expander II:

$$\dot{C}_{13} - \dot{C}_{14} + \dot{Z}_{EX\,II} = \dot{C}_{W2,EX\,II} \tag{18}$$

$c_{27} = c_{28}$ (F-rule), and $c_{14} = c_{14}^{T} = c_{14}^{M}$. The value of c_{W2} represents the specific cost of electricity generated within closed-cycle gas-turbine sub-system.

- Compressor III:

$$\dot{W}_{W2,CM\,III} + \dot{C}_{11}^{T} + \dot{Z}_{CM\,III} = \dot{C}_{12}^{M} - \dot{C}_{11}^{M} + \dot{C}_{12}^{T} \tag{19}$$

with $\dfrac{\dot{C}_{12}^{T}}{\dot{E}_{12}^{T}} = \dfrac{\dot{C}_{12}^{M} - \dot{C}_{11}^{M}}{\dot{E}_{12}^{M} - \dot{E}_{11}^{M}}$ (P-rule).

- Heat exchanger II:

$$\left(\dot{C}_{14}^{T} + \dot{C}_{2}^{T}\right) + \left(\dot{C}_{14}^{M} - \dot{E}_{11}^{M}\right) + \left(\dot{C}_{2}^{M} - \dot{C}_{3}^{M}\right) + \dot{Z}_{HE\,II} = \dot{C}_{11}^{T} + \dot{C}_{3}^{T} \tag{20}$$

where $c_{14}^{M} = c_{11}^{M}$ and $c_{2}^{M} = c_{3}^{M}$ according to the F-rule, and $c_{3}^{T} = c_{11}^{T}$ according to the P-rule.

- Expander III:

$$\dot{C}_3^M - \dot{C}_4^M + \dot{C}_3^T + \dot{Z}_{EX\,III} = \dot{C}_{W3,EX\,III} + \dot{C}_4^T \tag{21}$$

with $c_3^M = c_4^M$ (F-rule) and $c_4^T = c_{W3}$ where c_{W3} is the specific cost of electricity generated within LNG sub-system.

- Pump I is considered together with the required electrical motor

$$\dot{C}_W + \left(\dot{C}_1^T - \dot{C}_2^T\right) + \left(\dot{Z}_{PI} + \dot{Z}_{EM}\right) = \dot{C}_2^M - \dot{C}_1^M \tag{22}$$

with $c_1^T = c_2^T$ (F-rule) and $c_1 = c_1^{CH} = c_1^{PH} = c_1^T = c_1^M$. The chemical exergy of LNG does not change within the processes of sub-system 3, therefore $c_1^{CH} = c_2^{CH} = c_3^{CH} = c_4^{CH}$.

The value of c_W is the average specific cost of electricity generated by the overall cogeneration system

$$c_W = \frac{\dot{C}_{W1} + \dot{C}_{W2} + \dot{C}_{W3}}{\dot{W}_{NET,1} + \dot{W}_{NET,2} + \dot{W}_{NET,3}} \tag{23}$$

Figure 6 shows selected data obtained from the conventional exergoeconomic analysis. Here the cooler is considered together with the cooling tower ($\dot{Z}_{CT}$=3.25 $/h) and the pump is considered together with the electrical motor ($\dot{Z}_{EM}$=1.33 $/h).

For the economic analysis, the methodology presented by Bejan et al., 1996 is applied using the following assumptions and sources (Bejan et al., 1996 and GE, 2008):

- The cost of LNG is assumed to be 12 US$/GJ
- The average cost of money is i_{eff}= 10 %
- The plant economic life is n=15 years with 7300 hours/year
- The average general inflation rate is r_n=2.5 %.

The data are given in US$ for the year 2009.

The total purchased equipment cost of the LNG-based cogeneration system is around US$ 40.5 million. For the comparison with other systems for regasification of LNG, specific values of the purchased equipment cost should be used: (a) per kg of regasified LNG is equal to US$ 0.62 million, and (b) per MW of generated electricity is equal to US$ 0.28 million. The data obtained from the exergoeconomic analysis of the LNG-based cogeneration system have been discussed in detail by (Tsatsaronis et al., 2009).

The following conclusions regarding priorities of improvement for the LNG-based cogeneration system can be drawn from the

- economic analysis (based on the values of $\dot{Z}_k$, Figure 7a) – the capital investment cost of the open-cycle gas-turbine system is dominating, and the values of $\dot{Z}_k$ for all components of the closed-cycle gas-turbine system are comparable,

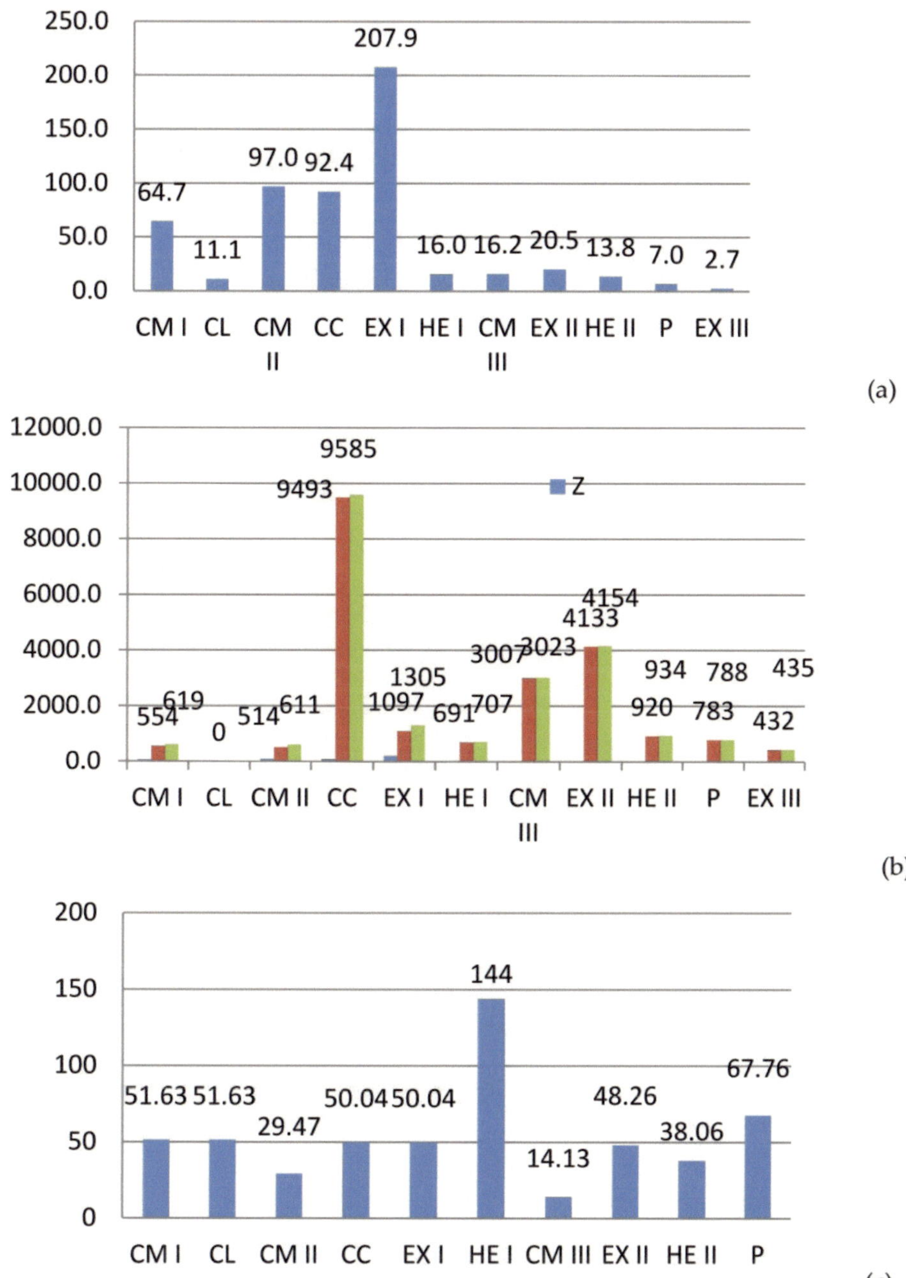

Figure 7. Selected exergoeconomic variables: (a) capital investment cost for each component ($\dot{Z}_k$, US\$2009/h), (b) total cost associated with each component ($\dot{Z}_k + \dot{C}_{D,k}$, US\$2009/h), and (c) specific cost of the exergy of fuel for each component ($c_{F,k}$, US\$2009/GJ).

- exergoeconomic analysis (based on the values of ($\dot{Z}_k + \dot{C}_{D,k}$), Figure 7b) – the total cost associated with each component mainly depends on the cost of the exergy destruction ($\dot{C}_{D,k}$) while the capital investment cost is negligible. For reducing the overall cost associated with both products of the cogenerations system, we should focus on reducing the exergy destruction within the turbomachinery of the closed-cycle gas-turbine sub-system, i.e. within EX II and CM III.

The "net" cost of the generated electricity is:

- for the open-cycle gas-turbine sub-system 0.18 US$/kWh,
- for the closed-cycle gas-turbine sub-system 0.28 US$/kWh,
- for the LNG sub-system 0.16 US$/kWh, and
- the average for the overall system is 0.20 US$/kWh (Eq. 23).

Note that the cost of the exergy losses (stream 28) should always be charged to the product, i.e. the value of $\dot{C}_{28}$ should be added to the nominator of Eq. (15).

We have two possibilities to estimate the cost of the generated electricity and regasification of LNG:

- *Assumption 1* – cost of the LNG is equal to the cost of the natural gas (i.e. all costs associated with the regasification process are charged to the cost of the generated electricity)

$$c_W = \frac{\dot{C}_{W1} + \dot{C}_{W2} + \dot{C}_{W3} + \dot{C}_{28}}{\dot{W}_{NET,1} + \dot{W}_{NET,2} + \dot{W}_{NET,3}} \tag{24}$$

  Using this assumption the average specific cost of the electricity is 0.23 US$/kWh.

- *Assumption 2* – the average cost (per unit of exergy) of the total product of the cogeneration system is calculated from

$$c_{p,tot} = \frac{\dot{C}_{W1} + \dot{C}_{W2} + \dot{C}_{W3} + \dot{C}_{28}}{\dot{W}_{NET,1} + \dot{W}_{NET,2} + \dot{W}_{NET,3} + \left(\dot{E}_4^M - \dot{E}_1^M\right)} \tag{25}$$

  In this way the average specific cost of the electricity is 0.20 US$/kWh and the difference between the cost of LNG and natural gas (after regasification) is 0.06 US$/GJ.

4.3. Exergoenvironmental analysis

Exergy analysis provides a powerful tool for assessing the quality of a resource as well as the location, magnitude, and causes of thermodynamic inefficiencies. In addition, LCA supplies the environmental impacts associated with a component or an overall system during its entire useful life. In the exergoenvironmental analysis (Tsatsaronis, 2008; Meyer et al., 2009), the environmental impacts obtained by LCA are apportioned to the exergy streams pointing out the main system components with the highest environmental impact and possible improvements associated with these components. Finally, exergoenvironmental variables are calculated, and an exergoenvironmental evaluation is carried out.

Life cycle assessment is a technique for assessing the environmental aspects associated with a product over its life cycle. The LCA process consists of goal definition and scoping (defining the system under consideration), inventory analysis (identifying and quantifying the consumption and release of materials), and interpretation (evaluation of the results).

In general, any of recently introduced indicators can be used for LCA. For this exergoenvironmental analysis, an impact analysis method called Eco-indicator 99 (Goedkoop & Spriensma, 2000) has been selected because it considers many environmental aspects and uses average European data.

In order to identify the raw materials inlet flows, it is first necessary to perform a sizing of the plant components and to collect information about the weights, main materials, production processes and scrap outputs of all relevant pieces of equipment needed to build the plant. This information is usually not very widely published (compared with the corresponding cost information), neither is given what materials are used for each equipment item. In this way, only rough calculations of the employed main materials can be conducted.

For the LCA of the system being analyzed (a detailed discussion is presented in (Morosuk et al., 2012)), we assumed in analogy with the economic analysis a life time of 15 years and 7300 working hours per year at full capacity.

The exergoenvironmental model for an energy conversion system consists of *environmental impact balances* written for the *k*th component and auxiliary equations based on the P and F-rules. The environmental impact balances can be written as

$$\dot{B}_{P,k} = \dot{B}_{F,k} + \left(\dot{Y}_k + \dot{B}_k^{PF}\right) \tag{26}$$

or

$$b_{P,k}\dot{E}_{P,k} = b_{F,k}\dot{E}_{F,k} + \left(\dot{Y}_k + \dot{B}_k^{PF}\right) \tag{27}$$

where $\dot{Y}_k$ is the environmental impact that occurs during the three life-cycle phases: Construction $\dot{Y}_k^{CO}$, operation & maintenance, $\dot{Y}_k^{OM}$, and disposal, $\dot{Y}_k^{DI}$ constitute the component-related environmental impact associated with the *k*th component $\dot{Y}_k$:

$$\dot{Y}_k = \dot{Y}_k^{CO} + \dot{Y}_k^{OM} + \dot{Y}_k^{DI} \tag{28}$$

To simplify the discussion, we assume here that the value of $\dot{Y}_k$ is mainly associated with $\dot{Y}_k^{CO}$.

To account for *pollutant formation* within the kth component, a new variable was recently introduced $\dot{B}_k^{PF}$ (Boyano et al., 2011). This term $\dot{B}_k^{PF}$ is zero if no pollutants are formed within a process, i.e. for processes without a chemical reaction (compression, expansion, heat transfer, etc.). For components, where chemical reactions occur (for example, combustion), the value of $\dot{B}_k^{PF}$ is

$$\dot{B}_k^{PF} = \sum_i b_i^{PF} \left(\dot{m}_{i,out} - \dot{m}_{i,in} \right) \tag{29}$$

where only pollutant streams which finally will be emitted to the environment are taken into account: CO, CO_2, CH_4, N_2O, NO_x and SO_x (Meyer et al., 2009).

The environmental impact of exergy destruction $\dot{B}_{D,k}$ identifies the environmental impact due to the exergy destruction within the kth component

$$\dot{B}_{D,k} = b_{F,k} \dot{E}_{D,k} \tag{30}$$

To identify the most important components from the viewpoint of formation of environmental impacts, the sum of environmental impacts $\left(\dot{Y}_k + \dot{B}_k^{PF} + \dot{B}_{D,k} \right)$ is used.

In this paper some data obtained from the exergoenvironmental analysis are given in Figure 7. Here the cooler is considered together with the cooling tower ($\dot{Y}_{CT}$ = 0.065 Pts/h) and the pump is considered together with the electrical motor, $\dot{Y}_{EM}$ = 0.363 Pts/h.

The results obtained from the LCA (value $\dot{Y}_k$ in Figure 8a) demonstrate that the component-related environmental impact associated with HE I and HE II are the highest among all components of the overall system. The exergoenvironmental analysis demonstrates that the value of the environmental impact associated with the exergy destruction dominates for all components ($\dot{B}_{D,k} >> \dot{Y}_k$). Only for the combustion chamber the value of the environmental impact of the pollutants formation ($\dot{B}_k^{PF}$) is comparable with the value of the environmental impact associated with exergy destruction ($\dot{B}_{D,k}$) (Figure 8b). Based on the sum ($\dot{Y}_k + \dot{B}_{D,k} + \dot{B}_k^{PF}$) the most important components are again EX II and CM III (assuming that the open-cycle gas-turbine sub-system cannot be improved). The environmental impact associated with the exergy destruction within EX II and CM III can be improved by decreasing the exergy destruction within these components.

The environmental impact associated with the generation of electricity and regasification of LNG can be estimated in analogy with cost (Eqs. (23)-(25)):

- The average "netto" environmental impact of the generated electricity is 31 mPts/kWh (by open-cycle gas-turbine sub-system 30.4 mPts/kWh, by closed-cycle gas-turbine sub-system 30.4 mPts/kWh, and by LNG sub-system 1.88 mPts/kWh).
- After charging the environmental impact associated with the exergy losses, we have the following data:
- If the environmental impact of the LNG is equal to the environmental impact of the natural gas (i.e. the environmental impact associated with the regasification process is charged to the environmental impact of the generated electricity), then b_W = 32.83 mPts/kWh, or
- If the average environmental impact (per unit of exergy) of the total product of the cogeneration system is calculated, then the average specific environmental impact of the electricity is 29.05 mPts/kWh and the difference between the environmental impact of LNG and natural gas (after regasification) is 0.013 mPts/kWh.

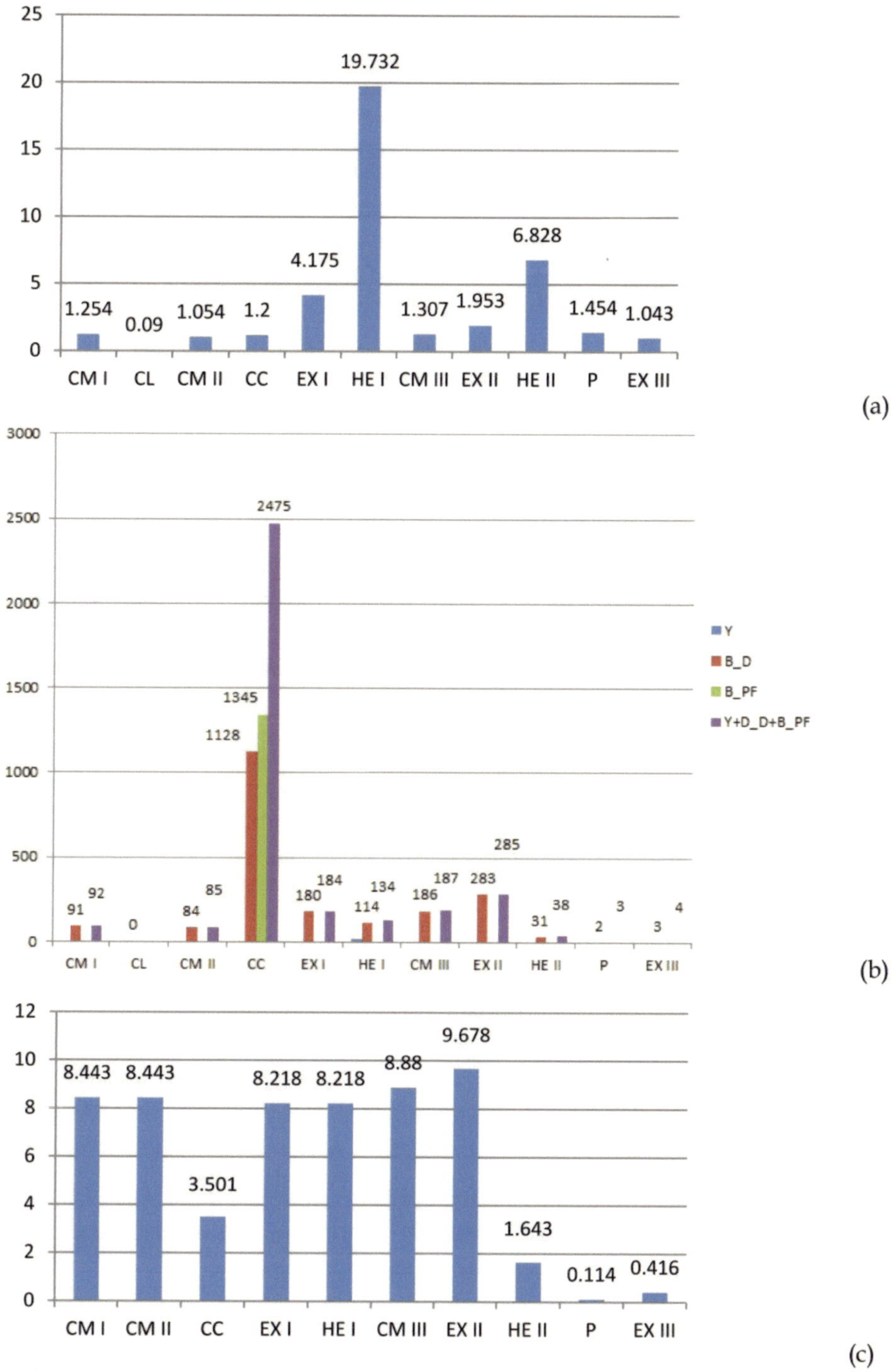

Figure 8. Selected exergoenvironmental: (a) component-related environmental impact ($\dot{Y}_k$, Pts ECO-99/h), (b) total environmental impact associated with the component ($\dot{Y}_k + \dot{B}_{D,k} + \dot{B}_k^{PF}$, Pts ECO-99/h), and (c) specific environmental impact of the exergy of fuel for the component ($b_{F,k}$, Pts ECO-99/GJ).

Note that the average valuc of the environmental impact associated with the electricity generation in Europe is 23 mPts/kWh (and, in general is varied between 10 and 62 mPts/kWh) as reported by Goedkoop & Spriensma, 2000.

None of the mentioned publications in the field of LNG-based cogeneration systems discussed neither LCA nor exergoenvironmental analysis. In order to compare the analyzed system with others, let us simplify the environmental analysis by considering only CO_2 generation within the open-cycle gas-turbine sub-system. Specific values of the CO2 generation include: (a) per kg of regasified LNG is equal to 90 g/kg, and (b) per MW of generated electricity is equal to 40 g/MW.

5. Advanced exergy-based analyses

In order to improve this energy conversion system, we should answer the following questions:

- What is the potential for decreasing (a) the exergy destruction within each component, (b) the cost and/or the component-related environmental impact of each component as well as the cost of exergy destruction and/or environmental impact associated with the exergy destruction?
- How an improvement in one component affects (positively or negatively and by how much) the remaining components?

Advanced exergy-based methods have been developed especially as a tool to find correct answers for these questions. The interactions among different components of the same system can be estimated and the quality of the conclusions obtained from an exergetic, an exergoeconomic, or an exergoenvironmental evaluation can be improved significantly, when the

- exergy destruction in each (important) system component,
- investment cost associated with such component, and
- component-related environmental impact associated with such component,

as well as the

- cost of exergy destruction within each (important) system component, and the
- environmental impact associated with exergy destruction within such component

are split into endogenous/exogenous and avoidable/unavoidable parts (all publications up to date in the field of the advanced exergy-based methods are summarized and generalized in (Tsatsaronis 2008; Tsatsaronis & Morosuk, 2008a and 2009b).

We call the analyses based on these procedures *advanced* (*exergetic, exergoeconomic,* or *exergoenvironmental*) *analyses*.

Endogenous exergy destruction is the part of exergy destruction within a component obtained when all other components operate ideally and the component being considered operates with the same efficiency as in the real system. The *exogenous* part of the variable is the difference between the value of the variable within the component in the real system and the endogenous part:

$$\dot{E}_{D,k} = \dot{E}_{D,k}^{EN} + \dot{E}_{D,k}^{EX} \tag{31}$$

Such a splitting shows the interconnections between the components.

The *unavoidable exergy destruction* ($\dot{E}_{D,k}^{UN}$) cannot be further reduced due to technological limitations such as availability and cost of materials and manufacturing methods. The difference between total and unavoidable exergy destruction for a component is the *avoidable* exergy destruction ($\dot{E}_{D,k}^{AV}$) that should be considered during the improvement procedure

$$\dot{E}_{D,k} = \dot{E}_{D,k}^{UN} + \dot{E}_{D,k}^{AV} \tag{32}$$

Combining the two splitting options gives analysts an opportunity to calculate

- the *avoidable endogenous exergy destruction* ($\dot{E}_{D,k}^{AV,EN}$), which can be reduced by improving the *k*th component from the exergetic point of view, and
- the *avoidable exogenous exergy destruction* ($\dot{E}_{D,k}^{AV,EX}$) that can be reduced by a structural improvement of the overall system, or by improving the efficiency of the remaining components.

This methodology can also be applied to the splitting of the values of capital investment cost and to cost of exergy destruction as well as to component-related environmental impact and to environmental impact associated with the exergy destruction.

The conventional exergy-based analyses suggest to initially decrease the exergy destruction within turbomachinery of the closed-cycle gas-turbine sub-system. This decrease of exergy destruction will not only increase the overall efficiency, but will simultaneously reduce both costs and environmental impact associated with the overall system.

The detailed advanced exergy and exergoeconomic analyses of the LNG-based cogeneration system have been reported by Tsatsaronis et al., 2009. Here only selected data for the compressor III and the expander II are given (Table 2).

Splitting the exergy destruction into its endogenous and exogenous parts (first two columns of Table 2) shows, that $\dot{E}_{D,k}^{EN} > \dot{E}_{D,k}^{EX}$ for CM III and EX II, therefore the interconnections among these (and the remaining) components are not very strong.

The results from splitting the exergy destruction into its avoidable and unavoidable parts show that the avoidable exergy destruction that occurs in expander II is larger than the unavoidable one, while for compressor III we have a different arrangement.

The real potential for improving the components can be obtained using the values of $\dot{E}_{D,k}^{AV,EN}$ and $\dot{E}_{D,k}^{AV,EX}$. For both analyzed components the value of $\dot{E}_{D,k}^{AV,EX}$ is much smaller than the value of $\dot{E}_{D,k}^{AV,EN}$. This means that the irreversibilities within such components can be reduced by improving the components themselves, and not other components.

Component	$\dot{E}_{D,k}^{EN}$ (MW)	$\dot{E}_{D,k}^{EX}$ (MW)	$\dot{E}_{D,k}^{UN}$ (MW)	$\dot{E}_{D,k}^{AV}$ (MW)	$\dot{E}_{D,k}^{AV}$ (MW)	
					$\dot{E}_{D,k}^{AV,EN}$ (MW)	$\dot{E}_{D,k}^{AV,EX}$ (MW)
CM III	4.461	1.338	3.037	2.762	2.273	0.489
EX II	7.693	0.434	3.722	4.405	4.182	0.223

Table 2. Selected data obtained from the advanced exergetic analysis

6. Exergy-based improvement

By summarizing all information obtained from the exergy-based methods, we know that the LNG-based cogeneration system can be improved (from the thermodynamic, economic and environmental points of view) by improving the efficiency of the turbomachinery within the closed-cycle gas-turbine sub-system. As before we assumed that the open-cycle gas-turbine sub-system is a standard unit and, therefore, cannot be improved.

Let us assume that the isentropic efficiency within turbomachinery will increase up by 0.5 percentage points (realistic assumption), i.e. with $\eta_{CM,III}$=85.5%, $\eta_{EX,II}$=88.5%. As we already found out, the values of the capital investment cost and the component-related environmental impact are very small compared with the value of the total cost associated with the component and total environmental impact associated with the component, respectively. Therefore, for the improvement we assume that $\dot{Z}_k$ and $\dot{Y}_k$ for both components, CM III and EX II, remain unchanged.

The selected data for the comparison analysis are given in Table 3. We would like to emphasize that only by using the exergy-based methods including the advanced ones, the LNG-based cogeneration system could be easily improved – through the thermodynamic improvement of the turbomachinery of the closed-cycle gas-turbine sub-system. The cost of the generated electricity and the environmental impact associated with the electricity generation decrease significantly by 10% and 12%, respectively.

Variables, units	Case 1	Case 2
$\eta_{CM,III}$ [%]	85	85.5
$\eta_{EX,II}$ [%]	88	88.5
$\dot{W}_{net,2}$ [MW]	28.6	29.5
$\dot{W}_{net,tot}$ [MW]	144.9	146.7
η_{tot} (HHV-based) [%]	68.0	68.7
ε_{tot} [%]	52.6	52.8
$\dot{E}_{D,CMIII}$ [MW]	5.8	5.6
$\dot{E}_{D,EXII}$ [MW]	8.1	7.9

$\dot{m}_{Nitrogen}$ [kg/s]	222.50	218.9
$\dot{m}_{LNG}$ [kg/s]	65.02	65.03
$\dot{W}_{net,tot}$ / $\dot{m}_{LNG}$ [kW/kg$_{LNG}$]	2.23	2.26
c_W [1)] [US$/kWh]	0.23	0.20
b_W [1)] [mPts/kWh]	32.83	28.89

[1)] All costs are charged to the electricity (Assumption 1)

Table 3. Selected data for the LNG-based cogeneration system to demonstrate the results of the improvement.

7. Conclusions

LNG will have in future a significantly larger contribution to the energy supply in the world than it had in the past. Thus, applying thermodynamically efficient, cost effective, and environmentally benign plants for the regasification of LNG is of particular importance for the use of LNG.

The concepts of the LNG-based cogeneration system developed by the authors (a) use gas turbines to keep the investment costs low and the efficiency high, (b) minimize the irreversibilities within the plant components and the losses to the environment, and (c) maximize the capability of LNG to generate electrical energy.

The application of advanced exergy-based methods to this concept allowed us to better understand their operation, i.e. how the thermodynamic inefficiencies are formed and how they affect the cost and the environmental impact of both products – electricity and regasified LNG. In this chapter, exergy-based methods for the evaluation and the improvement of an LNG-based cogeneration system were successfully applied.

8. Nomenclature

$\dot{B}$	environmental impact rate associated with an exergy stream (Points/s)
b	environmental impact per unit of exergy (Points/J) or per unit of mass (Points/kg)
$\dot{C}$	cost rate associated with an exergy stream (US$/h)
c	cost per unit of exergy (US$/J) or per unit of mass (US$/kg)
$\dot{E}$	exergy rate (W)
e	specific exergy (J/kg)
f	exergoeconomic factor [-]
$\dot{H}$	entropy rate (W)
h	specific entropy (J/kg)
HHV	higher heating value (J/kg)
LHV	lower heating value (J/kg)
k	kth component
$\dot{m}$	mass flow rate (kg/s)

p	pressure (bar)
$\dot{Q}$	heat rate (W)
r	relative difference (%)
T	temperature (ºC)
v	specific volume (m³/kg)
$\dot{W}$	power (W)
$\dot{Y}$	construction-of-component-related environmental impact rate (Points/s)
y	exergy destruction ratio (%)
$\dot{Z}$	cost rate associated with investment expenditures (US$/h)

Greek symbols

Δ	difference
ε	exergetic efficiency (%)
η	energetic efficiency for the overall system (%), or isentropic efficiency of a pump, compressor or expander (%)

Superscripts

AV avoidable; *EN* endogenous; *EX* exogenous; *M* mechanical; *PH* physical; *T* thermal; *UN* unavoidable

Subscripts

b refers to environmental impact; *D* exergy destruction; *F* exergy of fuel; *k* *k*th component; *L* exergy loss; *P* exergy of product; *tot* overall system; 0 thermodynamic environment (reference state)

Abbreviations

CL cooler; *CC* combustion chamber; *CM* compressor; *EM* electrical motor; *EX* expander; *HE* heat exchanger; *P* pump

Author details

Tatiana Morosuk and George Tsatsaronis
Technische Universität Berlin, Institute for Energy Engineering, Germany

Acknowledgement

The authors would like to thank Professor Hartmut Griepentrog (Greif-Foundation, Gelsenkirchen, Germany) for developing the initial system and for many helpful discussions regarding gas-turbine power systems and LNG vaporization.

9. References

Angelino, G. (1978). The use of liquid natural gas as a heat sink for power cycles. *ASME Journal of Engineering and Power*, Vol.100, pp. 160-77, ISSN 0742-4795

Bejan, A.; Tsatsaronis, G. & Moran, M. (1996). *Thermal Design and Optimization*, Wiley, ISBN 978-047-1584-67-4, New York, USA

Boyano, A.; Blanco-Marigorta, A.M.; Morosuk, T. & Tsatsaronis, G. (2011). Exergoenvironmental analysis of a steam methane reforming process for hydrogen production. *Energy – The International Journal*, Vol.36, No. 4, pp. 2202-2214, ISNN 0360-5442

Deng, S.; Jin, H.; Cai, R. & Lin, R. (2004). Novel cogeneration power system with liquefied natural gas (LNG) cryogenic exergy utilization, *Energy – The International Journal*, Vol. 29, pp. 497–512, ISNN 0360-5442

Dispenza, C.; Dispenza, G.; La Rocca, V. & Panno, G. (2009). Exergy recovery during LNG regasification: Electric energy production - Part one. *Applied Thermal Engineering*, Vol. 29, pp. 380–387, ISNN 1359-4311

Dispenza, C.; Dispenza, G.; La Rocca, V. & Panno, G. (2009). Exergy recovery during LNG regasification: Electric energy production - Part two. *Applied Thermal Engineering*, Vol. 29, pp. 388–399, ISNN 1359-4311

Dispenza, C.; Dispenza, G.; La Rocca, V. & Panno, G. (2009). Exergy recovery in regasification facilities – Cold utilization: A modular unit. *Applied Thermal Engineering*, Vol. 29, pp. 3595–3608, ISNN 1359-4311

Dorigoni, S. & Portatadino, S. (2008). LNG development across Europe: Infrastructural and regulatory analysis, *Energy Policy*, Vol. 36, pp. 3366–3373, ISNN 0301-4215

Engineering Equation Solver (2012), 1992–2012, V7.847,#92: McGraw-Hill.

Energy Information Administration (2003). *The Global Liquefied Natural Gas Market: Status & Outlook*, U.S. Department of Energy, Washington, USA, Available from http://www. eia.doe.gov.

Energy Information Administration (2004). *International Energy Outlook 2004*, U.S. Department of Energy, Available from http://www.eia.doe.gov/oiaf/ieo/ elctricity.html.

Energy Information Administration (2010). *Summary of the Environmental Impact Statement*. Department of the Environment and Local Government, Canada, Available from http://www. gnb.ca

Environmental Protection Agency (2010). *Greenhouse gas emissions reporting from the petroleum and natural gas industry*. U.S. Environmental Protection Agency. Climate Change Division, 24.01.2012, Available from http://www.epa.gov/ climatechange

EUROGAS (2011). *Natural Gas Demand and Supply. Long term Outlook to 2030*. Available from http://www.eurogas.com

Foss, M.M. (2003). *Introduction to LNG: An overview on liquefied natural gas (LNG), its properties, the LNG industry, safety considerations*. Centre for Energy Economics, The University of Texas, USA, Available from http://www.beg.utexas.edu/ energyecon/lng

Frutschi, H.U. (2005). *Closed-cycle gas turbines. Operating experience and future potential*, ASME Press New, ISBN 0-7918-0226-4, New York, USA

Gatex (2002) (Technische Universität Berlin, Institut für Energietechnik), 2002. Ref: Eisermann, W; Hasberg, W & Tsatsaronis, G. (1984). THESIS - Ein Rechen-programm zur Simulation und Entwicklung von Energieumwandlungsanlagen, *Brennst.-Wärme-Kraft*, Vol. 36, Nos. 1-2, pp. 45-51

General Electric (1989). *GateCycle for Windows,* Version 5.52.0.r., The General Electric Company, 1989-2004, Available from http://www.gepower.com/enter

General Electric (2008). *Gas turbine catalog,* Available from http://www.ge.com

GIIGNL (2011). *The LNG industry 2010,* International Group of Liquefied Natural Gas Importers, Available from http://www.giignl.org

Goedkoop, M. & Spriensma, R. (2000) . *The Eco-indicator 99: A damage oriented method for Life Cycle Impact Assessment.* Methodology Report. Amersfoort, Netherlands, Available from http:// www.pre.nl

Griepentrog, H. & Sackarendt, P. (1976). Vaporization of LNG with Closed-Cycle Gas Turbines. In: *Proceedings of the Gas Turbine and Fluid Engineering Conference,* March 21-25, 1976, New Orleans, USA, paper 76-GT-38

Griepentrog, H.; Tsatsaronis, G. & Morosuk, T. (2008). LNG vaporization using a novel co-generation system. In: *Proceedings of the ASME international mechanical engineering congress and exposition,* November 2–6, 2008. Boston, MA, USA, file IMECE2008-67208, ISBN 978-0-7918-3840-2

Griepentrog, H.; Tsatsaronis, G. & Morosuk, T. (2008). A novel concept for generating electricity and vaporizing LNG. In: Ziebik A, Kolenda Z., Stanek W, editors. *Proceedings of the 21th International Conference on Efficiency, Cost, Optimization, Simulation and Environmental Impact of Energy Systems,* Vol. II, June 23-27, 2008, Cracow-Gliwice: Poland, pp. 559-565, ISBN 978-83-922381-4-0

India Hydrocarbon Vision–2025 (2003). *Summary environmental impact assessment: liquefied natural gas terminal project in India.* Available from http://www.pptfun.com/Indiaoilgas/Infrastructure/Summary_Environmental_Impact_Assessment.pdf

International Institute of Refrigeration (2006). Liquefied Natural Gas: Current Expansion and Perspectives. *19-th Informatory Note on Refrigerating Technologies.* International Institute of Refrigeration, Paris, France, Available from http://www.iifiir.org

Kaneko, K.; Ohtani, K.; Tsujikawa, Y. & Fujii, S. (2004). Utilization of the cryogenic exergy of LNG by a mirror gas-turbine. *Applied Energy,* Vol.79, pp. 355–369, ISNN 0306-2619

Krey, G. (19790. Utilization of the Cold by LNG Vaporization with Closed-Cycle Gas Turbine. In: *Proceedings of the Gas Turbine Conference and Exhibit and Solar Energy Conference,* March 12-15, 1979, San Diego, CA, USA, paper 79-GT-84

Kumar, S.; Kwon, H.-T.; Choi, K.-H.; Lim, W.; Cho, J.H. & Tak, K. (2011). LNG: An eco-friendly cryogenic fuel for sustainable development, *Applied Energy,* Vol. 88, pp. 4264–4273, ISNN 0306-2619

La Rocca, V. (2010). Cold recovery during regasification of LNG part one: Cold utilization far from the regasification facility. *Energy – The International Journal,* Vol. 35, pp. 2049–2058, ISNN 0360-5442

Lazzaretto, A. & Tsatsaronis, G. (2006). SPECO: a systematic and general methodology for calculating efficiencies and costs in thermal systems, *Energy – The International Journal,* Vol. 31, Nos. 8-9, pp. 1257–1289, ISNN 0360-5442

Liu, M.; Lior, N.; Zhang, N. & Han, W. (20090. Thermoeconomic analysis of a novel zero-CO2-emission high-efficiency power cycle using LNG coldness. *Energy Conversion and Management,* Vol. 50, pp. 2768–2781, ISNN 0196-8904

Liu, Y. & Guo, K. (2011). A novel cryogenic power cycle for LNG cold energy recovery, *Energy – The International Journal*, Vol. 36, pp. 2828-2833, ISNN 0360-5442

Malacic, V.; Faganelli J. & Malej, A. (2008). Environmantal impact of LNG terminals in the Gulf of Trieste (Northern Adriatic). In: *Integration of information for Environmental Security*, H.Gonsa Coskun et al., eds., Springer, pp. 275-395

Maxwell, D. & Zhu, Z. (2011). Natural gas prices, LNG transport costs, and the dynamics of LNG imports. *Energy Economics*, Vol. 33, pp. 217–226, ISNN 0140-9883

Meyer, L.; Tsatsaronis, G.; Buchgeister, J. & Schebek, L. (2009). Exergoenvironmental Analysis for Evaluation of the Environmental Impact of Energy Conversion Systems, *Energy – The International Journal*, Vol. 34, pp. 75-89, ISNN 0360-5442

Morosuk, T. & Tsatsaronis, G. (2011). Comparative evaluation of LNG-based cogeneration systems using advanced exergetic analysis. *Energy – The International Journal*, Vol.36, No. 3, pp. 3771-3778, ISNN 0360-5442

Morosuk, T. & Tsatsaronis, G. (2008). New approach to the exergy analysis of absorption refrigeration machines, *Energy – The International Journal*, Vol. 33, pp. 890-907, ISNN 0360-5442

Morosuk, T.; Tsatsaronis, G.; Boyano, A. & Gantiva, C. (2012). Advanced exergy-based analyses applied to a system including LNG regasification and electricity generation. *International Journal of Energy and Environmental Engineering* (A Springer Open Journal - http://www.journal-ijeee.com/content/3/1/1), 3:1, ISNN 2251-6832

Najjar, Y.S.H. & Zaamout, M.S. (1993). Cryogenic power conversion with regasification of LNG in a gas turbine plant. *Energy Conversion and Management*, Vol. 34, No. 4, pp. 273–280, ISNN 0196-8904

Najjar, Y.S.H. (1991). A cryogenic gas turbine engine using hydrogen for waste heat recovery and regasification of LNG, *International Journal of Hydrogen Energy*, Vol. 16, Is. 2, pp. 129-134, ISNN 0360-3199

National Energy Technology Laboratory (2005). *Office of Fossil Energy*, U.S. Department of Energy. Available from http://www.fossil.energy.gov

Otsuka, T. (2006). Evolution of an LNG terminal: Senboku terminal of Osaka GAS. In: *Proceedings of the 23rd world gas conference*, pp. 1-14.

Patel, B. (2005). Gas Monetisation: *A techno-Economic Comparison of gas-to-liquid and LNG*, Foster Wheeler. Available from http://www.fosterwheeler.com

PETRONAS (2010), International Gas Union, news, views and knowledge on gas – worldwide. *World LNG Report*, Available from http://www.petronas.com (18.03.2012)

PFC Energy (2011). *World LNG Report 2010*, News, views and knowledge on gas – worldwide. International Gas Union (IGU), Available from http://www.petronas.com

PLATTS (2011). *LNG Price Assessments*. Platts Daily Spot, LNG FAQ, Available from http://www.platts.com

Querol, E.; Gonzalez-Regueral, B.; García-Torrent, J. & Ramos, A. (2011). Available power generation cycles to be coupled with the liquid natural gas (LNG) vaporization process in a Spanish LNG terminal. *Applied Energy*, Vol. 88, pp. 2382–2390, ISNN 0306-2619

Reale, M.J. (2004). *New High Efficiency Simple Cycle Gas Turbine – GE's LMS100™*, General Electric Company, USA, Available from http://www.ge.com

Shi, X.; Agnew, B.; Che, D. & Gao, J. (2010). Performance enhancement of conventional combined cycle power plant by inlet air cooling, inter-cooling and LNG cold energy utilization. *Applied Thermal Engineering*, Vol. 30, pp. 2003-2010, ISNN 1359-4311

SNAMPROGETTI (1978). *More Energy from LNG*. Electric Energy from LNG Regasification SP/BBC Process 2, Snamprogetti ENI Group, AMSEL, Linate, Italy, 1978

Szargut, J. & Szczygieł, I. (2009). Utilization of the cryogenic exergy of liquid natural gas (LNG) for the production of electricity. *Energy – The International Journal*, Vol. 34, pp. 827-837, ISNN 0360-5442

Tsatsaronis, G. & Morosuk, T. (2009). LNG-Based Cogeneration Systems. Part 1. Comparison of Gas-Turbine-Based Concepts, In: Proceedings of the International Mechanical Engineering Congress & Exposition, November 13-19, 2009, Lake Buena Vista, Florida, USA, Paper IMECE2009-10459, ISBN 978-0-7918-3863-1

Tsatsaronis, G. & Morosuk, T. (2010). Advanced exergetic analysis of a novel system for generating electricity and vaporizing liquefied natural gas, *Energy – The international Journal*, Vol. 35, pp. 820-829, ISNN 0360-5442

Tsatsaronis, G.; Morosuk, T. & Cziesla, F. (2009). LNG-Based Cogeneration Systems. Part 2. Advanced Exergy-Based Analyses of a Concept, In: Proceedings of the International Mechanical Engineering Congress & Exposition, November 13-19, 2009, Lake Buena Vista, Florida, USA, Paper IMECE2009-10460, ISBN 978-0-7918-3863-1

Tsatsaronis, G. (1984). Combination of Exergetic and Economic Analysis in Energy-Conversion Processes. In: *Energy Economics and Management in Industry*, Proceedings of the European Congress, Algarve, Portugal, April 2-5, 1984, England, Oxford: Pergamon Press, Vol. 1, pp. 151-157

Tsatsaronis, G. (2008). Recent developments in exergy analysis and exergoeconomics', *International Journal of Exergy*, Vol. 5, Nos. 5/6, pp.489–499, ISNN 1742-8297

Tsatsaronis, G. & Morosuk, T. (2008). A general exergy-based method for combining a cost analysis with an environmental impact analysis. Part I—theoretical development. In: *Proceedings of the ASME international mechanical engineering congress and exposition*, Boston, MA: USA, file IMECE2008-67218, ISBN 978-0-7918-3840-2

Tsatsaronis, G. & Morosuk, T. (2008). A general exergy-based method for combining a cost analysis with an environmental impact analysis. Part II—application to a cogeneration system. In: *Proceedings of the ASME international mechanical engineering congress and exposition*, Boston, MA: USA, file IMECE2008-67219, ISBN 978-0-7918-3840-2

Zhang, N. & Lior, N. (2006). A novel near-zero CO2 emission thermal cycle with LNG cryogenic exergy utilization, *Energy – The International Journal*, Vol. 31, pp. 1666–1679, ISNN 0360-5442

Zhang, N. & Lior, N. (2006). Proposal and analysis of a novel zero CO_2 emission cycle with liquid natural gas cryogenic exergy utilization. *ASME Journal of Engineering and Power*, Vol. 128, No. 1, pp. 81-91, ISSN 0742-4795

Chapter 12

Natural Gas Catalytic Partial Oxidation: A Way to Syngas and Bulk Chemicals Production

Gaetano Iaquaniello, Elena Antonetti, Barbara Cucchiella, Emma Palo, Annarita Salladini, Alessandra Guarinoni, Andrea Lainati and Luca Basini

Additional information is available at the end of the chapter

http://dx.doi.org/10.5772/48708

1. Introduction

The term Syngas is used for describing a mixture containing H_2 and CO, together with minor amounts of CO_2 and CH_4. Syngas can be produced from Natural Gas (NG), refinery off-gases, naphtha, heavy hydrocarbons and also from coal. The choice of a particular raw material depends on cost and availability of the feedstock, and on downstream use of syngas.

Main utilisations of syngas (Figure 1) are: i) the synthesis of Ammonia/Urea, ii) the oil refining operations, iii) the synthesis of Methanol and its derivatives, iv) the synthesis of liquid hydrocarbons via Fischer-Tropsch processes and v) other "minor" applications such as Iron ores reduction, fine chemistry productions, electronic, glass industry activities (Song and Guo, 2006).

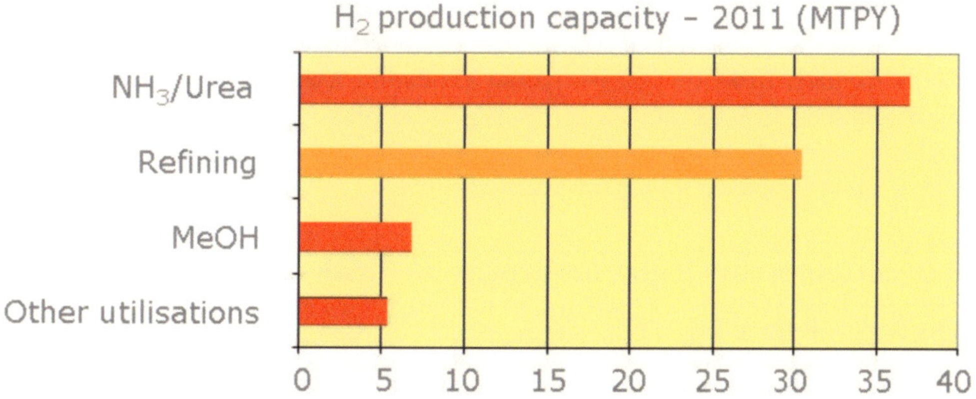

Figure 1. Worldwide H_2 production capacity (2001) devoted to main H_2 uses

The H_2/Syngas production capacity has always been increasing and it is still rising, particularly in the growing economies, in Middle East Countries and in almost all refineries spread worldwide.

Since syngas is an intermediate for feeding very different processes its composition and particularly the H_2/CO ratios are also very different (Moulijn et al., 2001). For example, in the synthesis of methanol, CO_2 and CO are both reactants and its composition is determined by a module $M = (H_2\text{-}CO_2)/(CO+CO_2)$ that should be close to 2.0 (mole/mole). Instead for Fischer Tropsch synthesis Gas to Liquid (GTL) applications, in which CO_2 is not a reactant molecule, the required synthesis gas compositions have a H_2/CO ratio of about 2.0 (mole/mole). The optimal H_2/CO ratio is lowered to 1.0 mole/mole for aldehydes productions via olefins hydroformylation. Sometimes, pure CO is required for the carbonylation process. However, the two main synthesis gas applications, the synthesis of Ammonia/Urea and the Oil Refining treatments, require the maximisation of H_2 production.

The most utilised synthesis gas production processes are:

- Steam reforming (SR)
- Partial oxidation (POx)
- Autothermal reforming (ATR)

Steam reforming is a catalytic and energy efficient technology for producing a H_2 rich syngas from light hydrocarbons like NG, refinery off-gases, LPG or Naphta. Partial oxidation is a non-catalytic technology with a unique possibility of utilising heavy hydrocarbon feedstock and produces a CO rich syngas at temperatures between 1100-1400°C. Its energy efficiency is lower than SR. ATR combines gaseous phase combustion reactions and catalytic steam/CO_2 reforming reactions; it is much less applied than SR and POx but it is the optimal choice for integration with large scale MeOH production plants and GTL processes (Holladay et al., 2009).

Despite the long term R&D and industrialisation effort that has lead to the optimisation of these technologies a relevant effort is still ongoing for defining new radical improvements allowing a reduction of the capital and energy requirements of the syngas production step. The Short Contact Time – Catalytic partial Oxidation technology, among the various proposed solutions, has reached enough reliability to promote its industrialisation and in the following chapters this promising new solution will be compared with the most consolidated technologies.

2. Main current technologies for syngas production

2.1. The Steam Reforming (SR)

SR is the most utilised technology for producing synthesis gas. This technology reacts light desulphurised hydrocarbons (S content ca. 50 ppb) with steam; for instance SR of Methane is represented with eq. (1):

$$CH_4 + H_2O = CO + 3H_2 \qquad \Delta H^0_{298K} = +206\ kJ/mol \qquad (1)$$

When utilised for H_2 production the SR step is followed by a Water Gas Shift (WGS) step for CO conversion (2):

$$CO + H_2O = CO_2 + H_2 \qquad \Delta H^0_{298K} = -41\ kJ/mol \qquad (2)$$

Subsequently H_2 is purified with a Pressure Swing Adsorption (PSA) step (Song and Guo, 2006).

SR units include two sections, namely a radiant and a convective section. Reforming reactions take place inside the radiant section. In the convective section, heat is recovered from the hot product gases for preheating the reactants feeds and for generating superheated steam.

The endothermic SR reactions are catalysed by Ni based materials in which the Ni species are deposited onto ceramic supports composed by alumina and/or alumina and magnesium spinels (Rostrup-Nielsen, 1993). Some discussions on the possible utilisation of noble metal based catalyst also including alkaline earth species are reported in literature (Faur Ghenciu, 2002; Navarro et al., 2007). The catalysts are included inside tubes inserted into the radiant furnace. The catalytic tubes have a diameter between 3 and 5 inches and a length ranging between 6 and 13 m. A radiant furnace may contain 500-600 reaction tubes. Typically gas temperatures at the exit of the tubes are higher than 800°C and pressures are comprised between 15-30 bar_g. One criticality of SR is represented by the thermo-mechanical resistance of the tubes, whose skin temperature remains at values 100-150°C higher than those of the reaction environment for allowing high heat transfer rates. For this reason the tubes are casted with alloys having a high Cr and Ni content (25-35%) and their positioning inside the radiant furnace is determined both by the necessity of increasing heat fluxes towards the reaction zone and by the requirements on the avoidance of impingement between tubes and flames produced by the burners. This impingement would lead to rapid collapse of the tubes (Ahmed and Krumpelt, 2001).

The hydrocarbons feedstock is fed into the reforming tubes after being mixed with steam at steam/carbon ratios higher than 2.3 v/v, more often higher than 2.7 v/v; this excess Steam is required both for completing the hydrocarbon reactions and for avoiding the occurrence of carbon formation reactions (3-5).

$$C_nH_{2n+2} = nC + (n+1)H_2 \qquad (3)$$

$$2CO = C + CO_2 \qquad (4)$$

$$CO + H_2 = C + H_2O \qquad (5)$$

Carbon and soot formation reactions would lead to pressure drop increase, catalyst deactivation and reaction rates reduction, causing serious heat transfer problems and tube damages (Peña et al., 1996).

Higher hydrocarbons are much more reactive towards reaction (3) than methane. For this reason, sometimes, these C2+ molecules are converted inside an adiabatic pre-reformer unit (Rostrup-Nielsen et al., 1998). This unit can be designed and operated at relatively low temperatures (ca. 550°C) leading to several advantages such as size reduction of the reformer furnace and/or increase of the production capacity (Joensen and Rostrup-Nielsen, 2002; Christensen, 1996).

2.2. The non-catalytic Partial Oxidation (POx)

The chemistry of the POx technology is based on the partial combustion of fuels that in case of CH_4 is represented with equation (6)

$$CH_4 + ½\,O_2 = CO + 2H_2 \qquad (6)$$

However this process is mainly utilised for producing syngas from heavy hydrocarbons, including deasphalter pitch and petroleum coke. These are pre-heated and then mixed with Oxygen within a burner; after ignition, reactions occur inside a high temperature combustion chamber producing an effluent that contains various amounts of soot, depending on feedstock composition. Reactor exit gas temperatures are typically comprised between 1200-1400°C. The obtained syngas has to be cooled and cleaned within a "washing" section for removing the impurities. The high temperature (1400-1100°C) heat recovery in POx is not very efficient and indeed the POx advantage over SR is in the possibility of utilising a "low value" feedstock, even containing sulphur and other compounds that would poison the SR catalysts. Currently the main utilisations of POx are: (i) in H_2 production for refinery applications, (ii) synthesis gas production from coal and (iii) in electric energy production from petroleum coke and deasphalter bottoms, through large Integrated Gas Turbine Combined Cycles (IGCC).

2.3. The Autothermal reforming (ATR)

ATR combines non-catalytic partial oxidation and catalytic steam and CO_2 reforming of light and highly de-sulphurated NG in a single reactor. The process was developed in the late 1950s by Haldor Topsøe A/S, mainly for producing syngas for methanol and ammonia plants and also for the Fischer-Tropsch synthesis (Christensen and Primdhal, 1994; Aasberg-Petersen et al., 2001). The NG is mixed at high temperature with a mixture of Oxygen and Steam and ignited in a combustion chamber originating a sub-stoichiometric flame that can be represented with eq. (7).

$$CH_4 + 3/2\,O_2 = CO + 2H_2O \qquad \Delta H^0_{298K} = -519\ kJ/mol \qquad (7)$$

Subsequently Steam and CO_2 reforming reactions (8) occur inside a catalytic bed positioned below the combustion chamber.

$$CH_4 + CO_2 = 2\,CO + 2H_2 \qquad \Delta H^0_{298K} = 247.0\ kJ/mol \qquad (8)$$

By proper adjustment of oxygen to carbon and steam to carbon ratios the partial combustion in the thermal zone (7) supplies the heat for completing the subsequent endothermic steam and CO_2 reforming reactions (Joensen and Rostrup-Nielsen, 2002). The product gas composition at the exit of the reactor results very close to the thermodynamic equilibrium of an adiabatic reactor, especially in large scale processes (Rostrup-Nielsen, 2000).

ATR is also utilised as a "secondary reformer" (for lowering the CH_4 residue) and it is placed after a primary SR in syngas plants integrated with Ammonia synthesis reactors. In this case the "secondary" ATR is fed with the syngas produced from SR and Air.

3. The Short Contact Time – Catalytic Partial Oxidation (SCT-CPO) technology

Initial observation on the occurrence of short contact time hydrocarbon oxidation processes were reported in the years 1992–1993 (Choudary et al., 1992; Hickman et al. 1992). These processes have been deeply studied since then, and the number of scientific articles published every year on this topic, is still high. They are produced by colliding for few milliseconds, gaseous premixed reactant flows with extremely hot catalytic surfaces. The fast and selective chemistry that is originated is confined inside a thin (<1 mm) solid–gas inter-phase zone surrounding the catalyst particles. Here, the molecules spend 10^{-6} s at temperatures variable between 600 – 1200° C. A key issue for the technological exploitation is in the possibility of avoiding the propagation of reactions into the gas phase, that has to remain at a "relatively low" temperature. This condition favours the formation of primary reaction products (namely CO and H_2) inhibiting chain reactions. Indeed some experimental studies whose results have been partially described in literature (Schwiedernoch et al., 2003; Hickman et al., 1993; Basini et al., 2000; Grunwaldt et al., 2001; Grunwaldt et al., 2002; Bizzi et al., 2002; Bizzi et al., 2003; York et al., 2003) indicate that partial and total oxidation products are directly produced through parallel and competing surface reactions and that the formation of partial oxidation products is favoured under SCT conditions due to the very high surface temperatures. By proper choice of the operating conditions, surface temperatures are locally much higher than those predicted by thermodynamic equilibrium calculations assuming adiabatic reactors. The occurrence of the reactions in these local environments determines in some cases conversion and selectivity values higher than those predicted by the thermodynamic equilibrium at the reactor exit temperatures (Basini, 2001). Moreover, the very high surface temperatures inhibit catalyst deactivation phenomena related to chemical poison effects (Basini, 2005 and Basini, 2006). For these and other related reasons, this chemical process is carried out in very small reactors having a very high flexibility towards reactant flow variations. It has also been found that several hydrocarbon feedstocks, even containing sulphur and aromatic compounds can be fed to a SCT-CPO reactor for producing synthesis gas.

Now a long term R&D effort is approaching the industrialization phase of a technology whose main advantages concern:

i. Small dimensions, technical and operational simplicity
ii. Possibility of modular construction of pre-fabricated and skid mounted units
iii. Flexibility towards feedstock composition & production capacity
iv. Reduction of investment costs and energy consumption
v. Reduction of CO_2 production and possibility of an almost complete CO_2 capture in case of H_2 production plants

3.1. SCT-CPO for H_2 production and CO_2 sequestration

Selectivity and conversion of SCT-CPO are determined by the high temperatures of catalytic surfaces, much higher than gas phase temperatures resulting in an adiabatic equilibrium reactor. This allows mild pre-heating of the reactants, thus reducing CO_2 emissions coming from pre-heating furnaces. Pre-heating is required only for the hydro-desulphurisation (HDS) of the feedstock, occurring in a temperature range typically comprised between 340–390°C depending on the hydrocarbon feedstock.

Figure 2 includes a schematic description of the SCT-CPO characteristics indicating that several feedstocks can be utilised for H_2 production and showing that in the absence of pre-heating furnaces great part of the produced CO_2 can be removed after the WGS section.

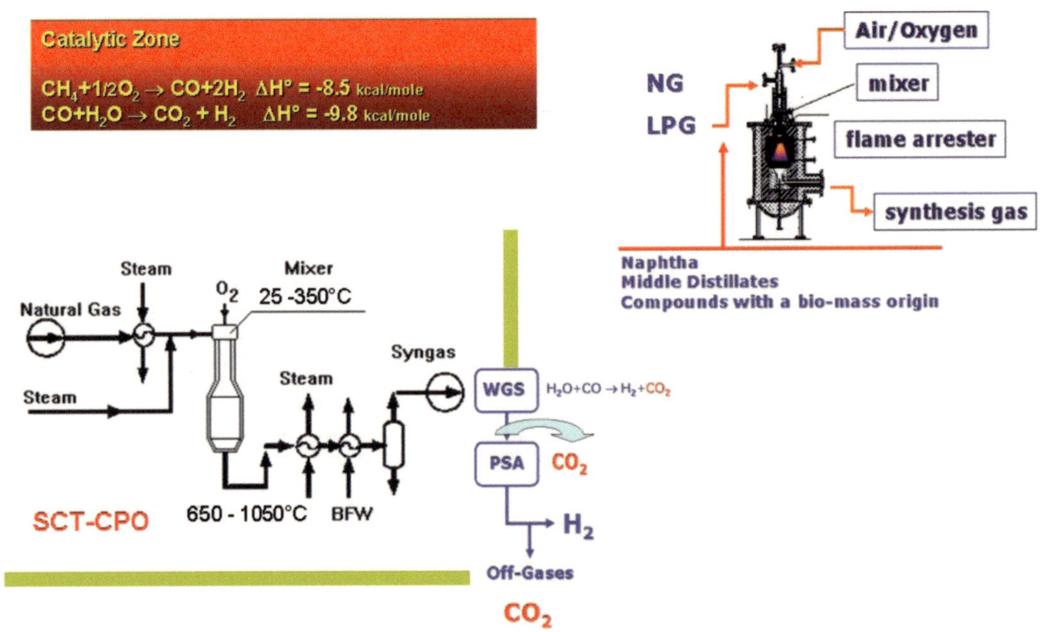

Figure 2. Main characteristics of SCT-CPO technology application devoted to H_2 production and CO_2 removal

Figure 3 includes some information for comparing the dimensions of SR and SCT-CPO reactors for producing 55,000 Nm^3/h of pure H_2. The figure shows that reactor dimensions are reduced of more than 2 orders of magnitude and that the complexity of the technology is greatly reduced too.

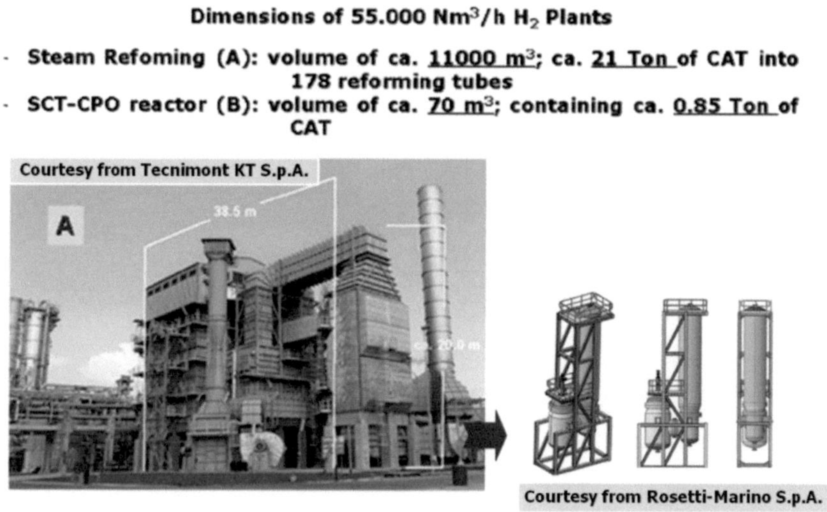

Figure 3. Comparisons between the dimensions and main characteristics of Steam Reforming and SCT-CPO technologies for H_2 production

Now it is noted that CO_2 emissions from SR result from two sources:

1. Steam Reforming and Water Gas Shift reactions producing ca. 60% of total CO_2
2. Total combustion inside the reformer furnace producing the other 40% of total CO_2

However, while the removal of CO_2 from source (1) with an absorbent medium (i.e. amine or carbonate solutions) is feasible at a reasonable cost, the removal from source (2), a Flue Gas at low pressure including a large volume of Air, would be really expensive and complex. Indeed H_2 production via SR is a very important Green House Gas (GHG) producer in refining operations (20-40% of the overall CO_2 emissions).

Nowadays, in a context wherein CO_2 taxation is increasing and refining operation are required to reduce GHG, the utilisation of SCT-CPO technology, particularly if integrated with CO_2 removal, would be really advantageous. This point is further discussed by commenting the Process Flow Diagram of Figures 4 and 5.

In the PFD shown in Figure 4 the feed is mixed with hydrogen and pre-heated for the next HDS step accounting for 2% wt of the total produced CO_2. The preheated stream is then mixed with steam and oxygen into the SCT-CPO reactor to produce synthesis gas. The hot synthesis gas is cooled in a Waste Heat Boiler (WHB) generating high pressure steam, partially utilized for a two-step WGS section, namely a high temperature shift (HTS) and a low temperature shift (LTS). Then a CO_2 removal section is included before the H_2 purification step, performed with a Pressure Swing Adsorption unit (PSA).

An alternative process scheme solution (see Figure 5) could also be utilized in situations in which a low H_2 purity (ca. 98% v/v) is required.

An effort for assessing the potential advantages of these process scheme solutions has been performed by comparing the economics of SCT-CPO and SR in several different scenarios and for different plant capacities (10,000 - 50,000 and 100,000 Nm³/h of H_2), as summarized in Figure 6 (Basini and Iaquaniello, 2011).

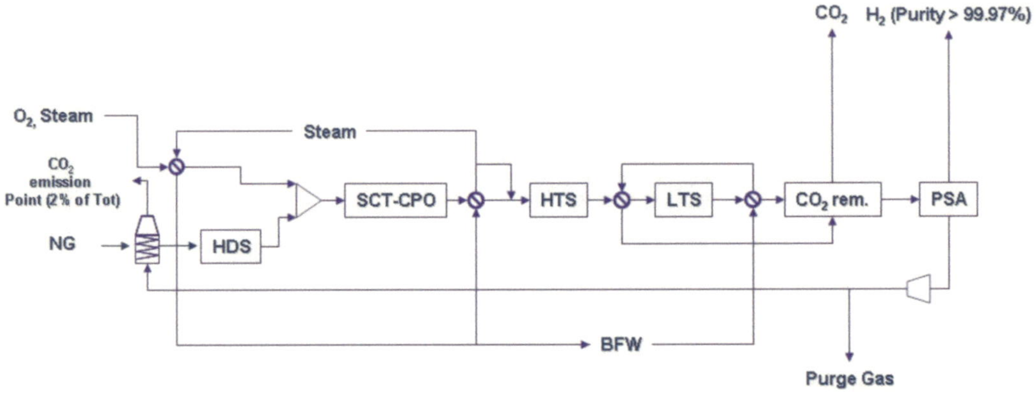

Figure 4. PFD including an SCT-CPO reactor for hydrogen production and CO_2 sequestration

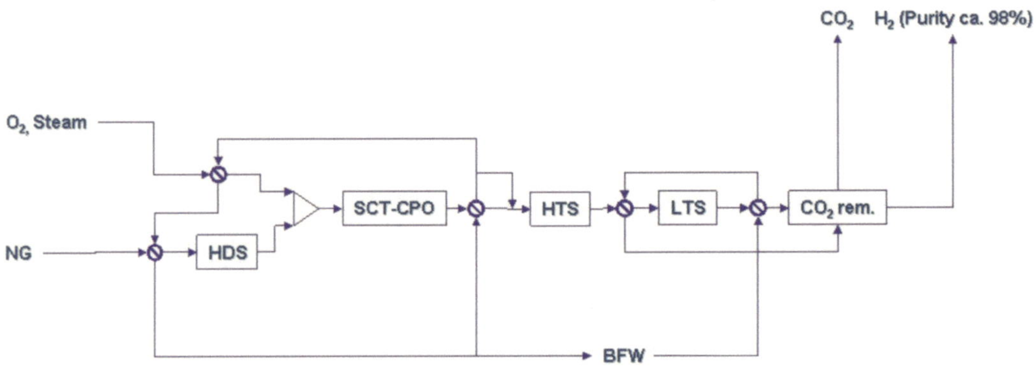

Figure 5. PFD including an SCT-CPO reactor for hydrogen production and CO_2 sequestration avoiding the PSA step

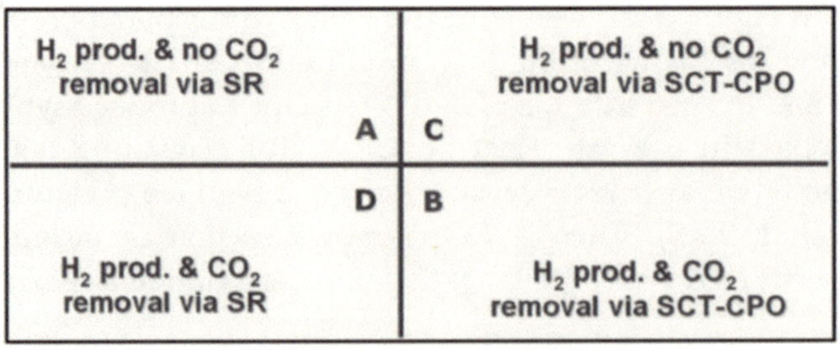

Figure 6. Possible comparison conditions between SR and SCT-CPO processes for H_2 production possibly including CO_2 sequestration

It appears evident that in any scenario case (D), namely the H_2 production with SR including a CO_2 removal section would be the less advantageous solution. Instead in the great part of the examined conditions case (B), namely the H_2 production with SCT-CPO including a CO_2 removal section is the most advantageous solution.

3.2. SCT-CPO to remove the bottleneck for H_2 production inside a refinery

Small size, technological simplicity and easy operability of SCT-CPO technology allows the pre-fabrication of the syngas production unit and the transportation of the reactor package to the utilisation site where it would be installed, requiring few interconnections minimising the interference risks with the ongoing industrial operations. Accordingly, one of the application of the SCT-CPO technology currently pursued, concerns removing the bottleneck of an existing SR (Figure 7). In this case the SCT-CPO reactor placed in parallel to an existing SR unit will utilise the same feedstock and will release the produced synthesis gas to an existing WGS and PSA "train".

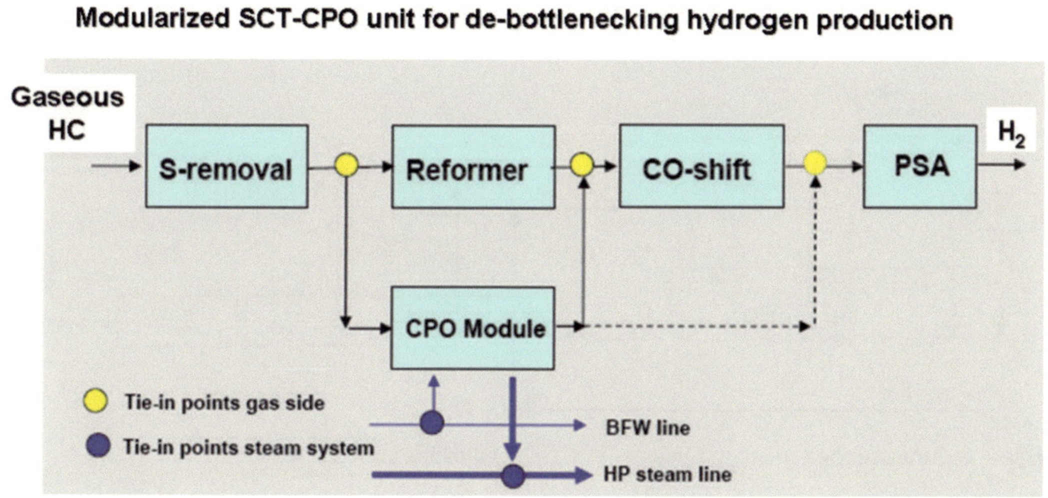

Figure 7. Scheme of the process solution in which SCT-CPO and SR are operated in parallel for increasing the H_2 production capacity inside a refinery context.

A similar concept could be used to revamp ammonia/urea plant to enhance the urea production. In this case the CPO reactor installed in parallel to the exiting train provides the extra CO_2 required increasing the urea production.

3.3. Notes on SCT-CPO integration in Ammonia/Urea production processes

Ammonia/Urea production is currently achieved with the following main steps:

1. hydro-desulphurisation
2. pre-reforming (optional)

3. primary SR section with H_2O/C ratio of about 3 v/v
4. secondary air blown ATR reformer
5. high temperature and low temperature shift conversion
6. CO_2 removal
7. CO methanation
8. syngas compression

Two different schemes can be proposed for producing ammonia and **Urea** via SCT-CPO. These schemes are reported in Figures 8 and 9 and both include the following steps:

1. hydro-desulphurisation
2. mixing of desulfurized feed with enriched air and super-heated steam
3. SCT-CPO reforming for producing the syngas
4. cooling of the syngas in a process gas boiler (PGB) and Steam addition
5. high temperature and low temperature shift conversion
6. CO_2 removal unit
7. CO methanation
8. syngas compression

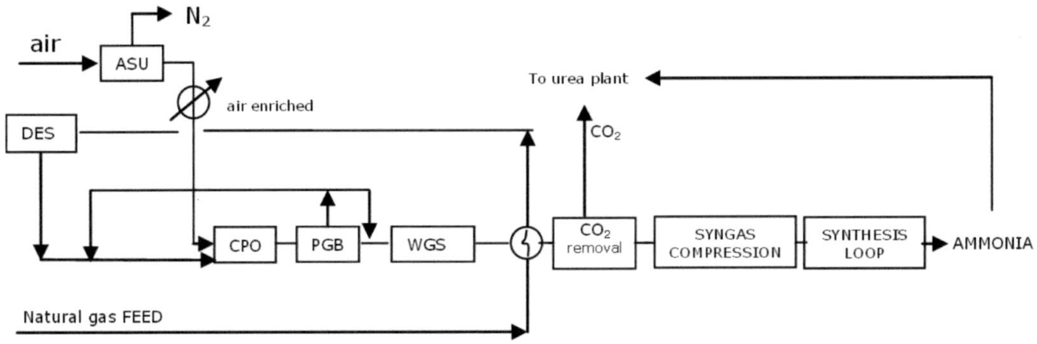

Figure 8. Ammonia and urea production via SCT-CPO utilising enriched air

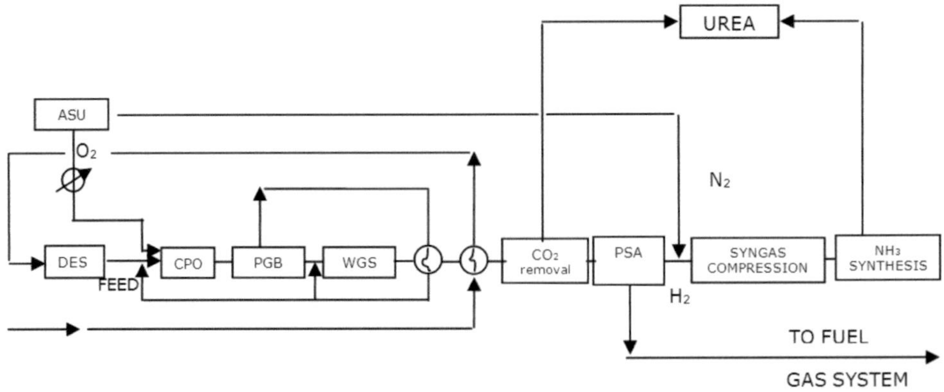

Figure 9. Ammonia and Urea production via O_2 Blown SCT-CPO

In the PFD schematised in Figure 8 the SCT-CPO reactor is fed with enriched Air, while the PFD shown in Figure 9 utilises an O_2-blown SCT-CPO reactor and a PSA unit, positioned after the CO_2 removal unit for increasing H_2 purity.

Both schemes consider that SCT-CPO allows producing a syngas mixture with an H_2/CO_2 ratio ≅ 3 v/v, very close to the most appropriate ratio for maximising the Urea production. In other words, the SCT-CPO technology would allow the increase of the urea production per unit volume of produced Ammonia. This point can be appreciated by considering equation (9) and assuming that the produced CO could be completely converted into CO_2 and H_2.

$$1.1\,CH_4 + 0.6\,O_2 + 0.3\,H_2O = 2.2\,H_2 + CO + 0.1\,CO_2 + 0.3\,H_2O \qquad (9)$$

Taking as basis an ammonia plant with a capacity of 3,000 MTPD (for an Urea production around 4,440 MTPD), we performed a preliminary technical economical comparison between the conventional process scheme, based on SR and ATR, and the two novel schemes based on SCT-CPO.

It is noted that an Urea production of 4,440 MTPD requires a syngas plant capacity corresponding to 352,000 Nm^3/h of syngas with a H_2/N_2 ratio equal to 2.78 v/v and CH_4 content lower than 1.0% v/v, delivered at a pressure around 30 bar_g. Table 1 includes a comparison between consumption and production features of the "state of the art" technological solutions and those achievable with SCT-CPO based on the process schemes of Figures 8 and 9.

	State of the Art Case	ENRICHED AIR (Figure 8)	OXYGEN (Figure 9)
FEED + FUEL (10^3 Nm^3/h) (1)	1	0.94	0.81
FEED + FUEL (MM Kcal/tonn of NH_3)	1	0.95	0.82
OXYGEN (tonn/tonn Urea)	0	0.26	0.41
EXTRA POWER (MWh) per Tonn of Urea	0	0.14	0.52(2)
IMPORT STEAM (tonn/h)	1	1.39	1.11
EXPORT STEAM (tonn/h)	1	0.86	0.65
CO_2 EMITTED (tonn/h)	1	0.00	0.00
CO_2 RECOVERED (tonn/h)	1	1.23	1.24
CO_2 TO UREA (tonn/h)	1	1.19	1.19
CO_2 EXPORT OR VENT(tonn/h)	1	2.02	2.24
AMMONIA PRODUCED (tonn/h)	1	1.00	1.00
AMMONIA TO UREA (tonn/h)	1	1.19	1.19
AMMONIA EXPORT (tonn/h)	1	0.00	0.00
UREA PRODUCED (tonn/h)	1	1.19	1.19

(1) 8650 $Kcal/Nm^3$
(2) Power for process air compressor was deducted

Table 1. Normalised values for comparing conventional technology (basic case) with the SCT-CPO process schemes of Figures 8 and 9.

It is quite evident that with the above mentioned assumptions, Urea productivity can be increased by ca. 15% utilizing the same Ammonia production capacity and avoiding any excess ammonia export. Moreover the produced CO_2 can be completely recovered achieving a 60% reduction of the CO_2 emissions. Feed and fuel consumptions would also be reduced while power consumption would increase. The specific consumption is lowered of about 18% compared with the "state of the art" case. However it is also noted that the HP Steam Export would be reduced and if considered in the energy balance, this point would strongly reduce the energy consumption advantage. In other words, it can be said that the SCT-CPO based processes would be more efficient, but would produce a lower amount of HP Steam export. It also appears that the SCT-CPO processes would be favored by high NG price and low electric energy price scenarios. The data of Table 1 have been transformed in cost and incomes and an overall economical balance was done to make a preliminary assessment. The conclusions indicated that SCT-CPO based processes become very convenient in situations in which the excess of ammonia has to be converted into Urea and/or when the variation of the Ammonia/Urea prices require a flexible process. A first qualitative comparison indicated that either operating and investment costs are reduced with both SCT-CPO based process schemes (Table 2).

	Operating Costs	Investment costs
Conventional	100	100
O_2 based	98	85-90
Enriched based	95	90-92

Table 2. Normalised values for comparing operating and investment costs of conventional technology (basic case) with the SCT-CPO process schemes of Figures 8 and 9.

3.4. Notes on SCT-CPO integration in methanol production processes

A typical large scale methanol production process (5,000 MTPD) requires a syngas flow (about 560.000 Nm^3/h) at 36 bar_g with the following characteristics:

- "Methanol Module" $M = (H_2\text{-}CO_2)/(CO_2\text{+}CO) = 2$ (v/v)
- ratio $CO_2/CO = 0.38$ corresponding to $CO/CO_2=2.6$ (v/v)
- methane residue in the syngas lower than 1.30% (v/v dry basis)

The conventional process scheme for producing such a syngas includes the following main steps:

1. hydro-desulphurisation (HDS)
2. pre-reforming
3. splitting of the feed into two separate streams (60/40%) one to be processed from a SR (60% of total feed) the other (40% of total feed) from an O_2 Blown ATR which combines the Steam Reformed Stream with the fresh feed
4. syngas cooling and high pressure steam generation

The utilisation of SCT-CPO integrated with MeOH synthesis loop can be envisaged accordingly to the scheme of Figure 10. Here in order to achieve a Methanol Module M = 2 v/v, 12% v/v of the synthesis gas has been shifted (WGS) and purified (PSA) for producing pure H_2, that was added to the main synthesis gas stream (directly produced from the SCT-CPO reactor).

The scheme based on SCT-CPO technology (Figure 10) includes the following main steps for synthesis gas production:

1. hydro-desulphurisation (HDS)
2. mixing the de-sulphurated feed with super-heated Steam and Oxygen
3. SCT-CPO reforming
4. splitting the reformed gas for obtaining a pure H_2 stream
5. mixing the pure H_2 stream with the syngas directly produced from the SCT-CPO reactor for reaching an M value of 2 v/v

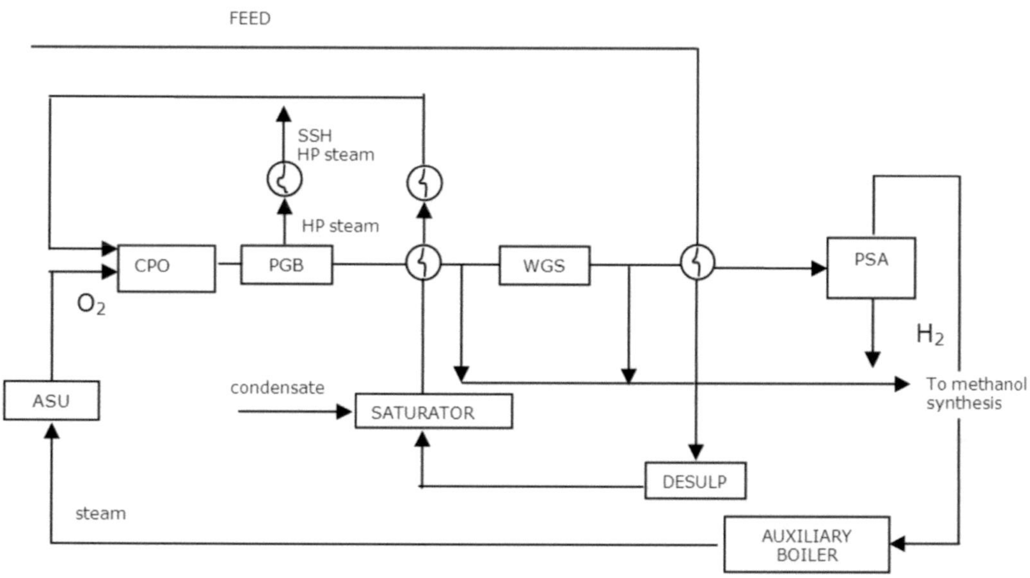

Figure 10. Block diagram for utilising the SCT-CPO technology in MeOH production

More in detail it is reported that desulfurized feed and steam flows are characterised by S/C = 0.6 v/v while in the "state of the art" process scheme this ratio corresponds to 1.8 v/v (this implies that extra heat is available from the synthesis loop). The feed + steam stream, at the outlet of saturator, is heated up to 480°C and mixed with Oxygen at 180°C for feeding the SCT-CPO reactor. The produced syngas at 1060°C is cooled down in the process gas boiler to raise HP steam, which is superheated in a separate furnace. The reformed gas at 320°C is split in two streams. One of these streams (ca. 42% v/v of total) is routed to a WGS section. The shifted gas is also split into two streams, the first, about 28% v/v of total is cooled and treated in a PSA unit to produce pure H_2, the second is mixed as such with the un-shifted gas. In this way the Methanol Module is raised by adding a proper amount of H_2 while CO/CO_2 ratio is controlled via the WGS section. The purge gas together with others streams from the methanol synthesis section, are fired in a SSH heater, in a feed preheater and in a boiler to raise HP steam to be used within ASU for making the extra O_2 required.

Ongoing technical and economical evaluations comparing the "State of the Art" and the SCT-CPO based technology figures are indicating that the new process scheme has the potential for reducing the CAPEX but slightly raises the OPEX. However it is clear that many of the possibilities accessible with this new "technological key", have still to be examined.

3.5. Notes on Technimont KT S.p.A. activities performed in the NEXT-GTL European project

Some Tecnimont KT S.p.A. R&D activities concerning the SCT-CPO have been developed within the framework of the European Project named: "Innovative Catalytic Technologies & Materials for Next Gas to Liquid Processes", Coordinated by INSTM, Consorzio InterUniversitario per la Scienza e Tecnologia dei Materiali, Messina (Italy) (website, http://www.next-gtl.eu). The project, started in November 2009, has the duration of 4 years and is pursuing a multi-disciplinary long term approach for developing catalysts and pre-competitive technologies. The project is examining novel (at a pre-competitive level) routes, particularly suited for remote and stranded areas, for transforming NG into liquid products either for chemical or for fuel production. In this latter case, a better balance between diesel vs. gasoline pools would be achieved. In particular, one of the three lines investigated, for which Tecnimont KT S.p.A. is workpackage leader concerns the definition of new routes for syngas production utilising membrane systems (Capoferri et al., 2011) coupled with steam reforming reactors integrated in process schemes also including SCT-CPO.

The basic concept is that the use of membrane reactors for steam reforming reactions, enables the shift of chemical equilibrium towards products by removing the produced hydrogen, thus enhancing feed conversion or achieving the same feed conversion at lower temperature (as an example, conventional plants usually require outlet temperatures around 850-880°C, that can be lowered to 600-650°C by adopting membrane reactors).

The integration of selective membranes in a chemical process can be made: (i) directly inside the reaction environment ("closed" architecture), (ii) after the reaction step, thus realizing a sequence of reactor/membrane modules ("open" architecture, Iaquaniello and Salladini, 2011). The integrated approach can be more efficient and allows higher system compactness, however the possibility to integrate the membrane module outside the reaction environment can i) assure higher system flexibility, ii) simplify the reactor mechanical design and membrane geometry, iii) make possible the optimization of the operating conditions in each stage separately.

Such a conceptual membrane reactor scheme derives from recent experience developed by Tecnimont KT/INSTM together with other partners in the framework of the Italian FISR Project "Pure hydrogen from natural gas reforming up to total conversion obtained by integrating chemical reaction and membrane separation". This consortium grouped Italian universities and the engineering company Tecnimont KT, cooperating in the development of this innovative technology, that individuated as critical points membrane manufacture and assembling and catalyst optimization. The project resulted in the design and operation of a demonstrative plant (capacity of 20 Nm^3/h oh hydrogen), placed in Chieti Scalo, based on the integration of a steam reforming reactor with Pd and Pd/Ag based membranes for hydrogen separation (Barba et al., 2008; Iaquaniello et al., 2008; De Falco et al., 2011a; De Falco et al., 2011b). This installation, the first of this type and size, made it possible to completely understand the potential of selective membrane application in industrial high-temperature chemical processes.

During the first stage of the Next-GTL project, an optimized process scheme was developed, characterized by the sequence of the following steps: (i) pre-reforming stage; (ii) SCT-CPO reactor; (iii) membrane for H_2 separation. The aim of the proposed architecture is to produce syngas and hydrogen with higher flexibility in terms of H_2/CO ratio and H_2 capacity at a more competitive production cost, exploiting all the advantages relevant to a prereformer unit, such as: (i) to increase the feedstock flexibility; (ii) to lower export steam production; (iii) to recover the maximum waste heat in the process (which in CPO architecture, downstream of the reactor, is largely available); (iv) to remove the bottleneck in an existing facility; (v) to protect downstream catalysts; (vi) to increase the capacity of an existing plant.

The experimental feasibility of the proposed architecture will be assessed at the pilot plant in Chieti Scalo, that will be operated at lower temperature (like a pre-reforming Unit).

The current plant configuration is reported in the block diagram in Figure 11.

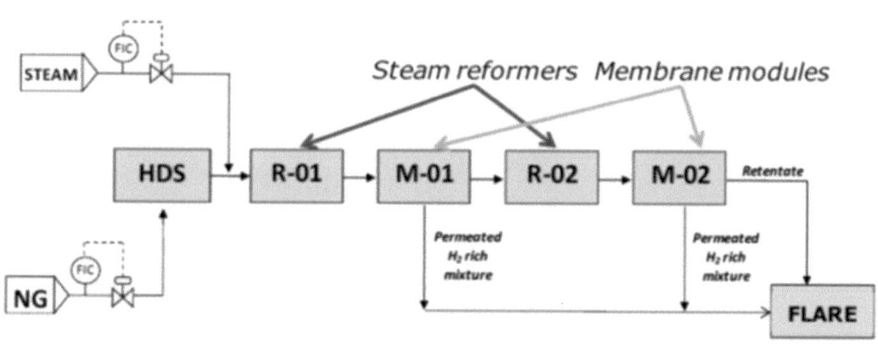

Figure 11. Block diagram of the current configuration of the existing pilot plant in Chieti Scalo

NG from battery limits is fed to the feed desulphurization (HDS) reactor for sulphur compounds removal. The desulphurised feed is mixed with steam, preheated in the convective section and fed to the first reforming reactor (R-01). Reforming temperatures can be set as required by the tests, adjusting the amount of fuel fed to two independent hot gas generators coupled to two steam reforming reactors. Typically the reactors operate at temperatures around 500-600°C.

The reformed gas produced in the first reactor is cooled down at 400-450°C and routed to the first separation module (M-01): the retentate is preheated and recycled to the second reformer stage (R-02), and a mixture of H_2 plus sweeping steam are produced. Reformed gas coming from the second reformer stage is cooled down and routed to the second separation module (M-02). H_2 from both modules are mixed together and sent to final cooling and condensate separation while retentate from the second stage is collected to the flare.

According to our considerations, the current configuration could be modified as reported in the following block diagram (Figure 12), where it can be observed that the retentate coming from the first membrane module is sent to a SCT-CPO reactor whereas the second steam reformer is eliminated.

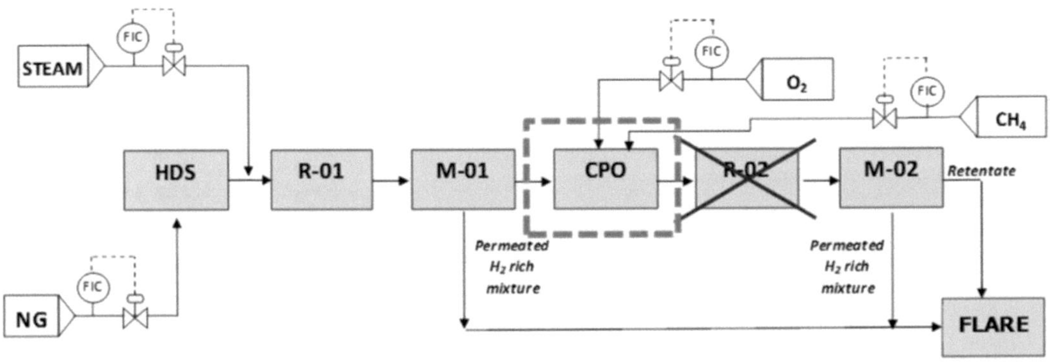

Figure 12. Block diagram of the modified configuration of the existing pilot plant in Chieti Scalo

The process scheme of the existing pilot plant after the rearrangement procedure is reported in Figure 13.

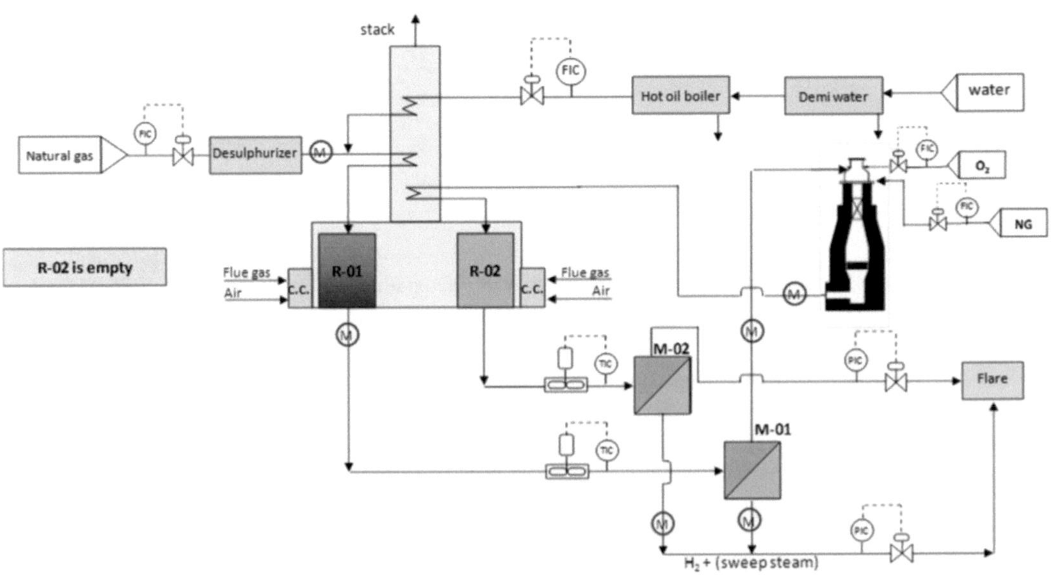

Figure 13. Process scheme of the pilot plant to be realized in the framework of "Next-GTL" project

It can be observed that the main difference with respect to the previous process scheme is represented by the fact that the retentate stream coming from the first separation module (M-01) is sent to the CPO reactor, which outlet is further routed to the second separation module (M-02). Before this passage, the exhaust stream coming from the CPO reactor is forced to flow through the second steam reformer reactor, which in this case is empty, so that any catalytic reaction occurs in it, thus simulating a reactor by-pass operation.

It is expected to have the first catalytic activity tests results within the end of 2012.

This system configuration could in principle be integrated also with membranes for O_2 and CO_2 separation since the former may contribute to a reduction of the costs for air separation, one of the more costly components of the overall GTL process, while the latter can be used to separate carbon dioxide to reduce greenhouse gas emissions and allow its reuse.

4. Conclusions

Syngas production is a key step for many industrial manufacturing schemes. The technological issues involved are quite relevant to determine the overall energy efficiency and the economic characteristics of syngas processes. Well established industrial solutions, namely the SR, the POx and the ATR will be soon challenged by a radical innovation; namely the SCT-CPO technology that has matured inside the ENI S.p.A., which has been extensively experimented in pilot scale plants and has reached the maturation for initiating the first industrial unit. Technimont KT S.p.A. has also been very active, among the other companies, in this field by studying in recent years, process scheme solution including SCT-CPO not only for H_2 production but also with NH_3/Urea and MeOH production units. This integration appears already feasible. Some more explorative studies are also ongoing for examining process scheme solutions including membranes separation units.

Author details

Gaetano Iaquaniello and Emma Palo
Tecnimont KT S.p.A., Roma, Italy

Alessandra Guarinoni, Andrea Lainati and Luca Basini
ENI S.p.A., Refining & Marketing Division, San Donato Milanese (MI), Italy

Elena Antonetti, Barbara Cucchiella and Annarita Salladini
Processi Innovativi s.r.l., Roma, Italy

Acknowledgement

This work received a financial support by the European Community throughout the Next GTL project- Contract NMP3-LA-2009-229183

5. References

Ahmed, S. & Krumpelt, M. (2001). Hydrogen from hydrocarbon fuels for fuel cells. *International Journal of Hydrogen Energy*, Vol. 26, pp. 291–301

Aasberg-Petersen, K. ; Bak Hansen, J.H. ; Christensen, T.S. ; Dybkjaer, I. ; Seier Christensen, P. ; Stub Nielsen, C. ; Winter Madsen, S.E.L. & Rostrup-Nielsen, J.R. (2001). Technologies for large-scale gas conversion. *Applied Catalysis A: General*, Vol. 221, pp. 379–387

Barba, D.; Giacobbe, F.; De Cesaris, A.; Farace, A.; Iaquaniello, G. & Pipino, A. (2008). Membrane reforming in converting natural gas to hydrogen (part one). *International Journal of Hydrogen Energy*, Vol. 33, pp. 3700-3709

Basini, L.; Guarnoni, A. & Aragno, A. (2000). Molecular and temperature aspects in catalytic partial oxidation of methane. *Journal of Catalysis*, Vol. 190, pp. 284-295

Basini, L; Aasberg-Petersen, K.; Guarinoni, A.; Ostberg, M. (2001). Catalytic partial oxidation of natural gas at elevated pressure and low residence time. *Catalysis Today*, Vol. 64, pp. 9–20

Basini, L. (2005). Issues in H_2 and synthesis gas technologies for refinery, GTL and small and distributed industrial needs. *Catalysis Today*, Vol. 106, pp. 34–40

Basini, L. (2006). Fuel rich catalytic combustion: Principles and technological developments in short contact time (SCT) catalytic processes. *Catalysis Today* Vol. 117, pp. 384–393

Basini, L. & Iaquaniello, G. (2011). Process for the production of hydrogen starting from liquid and gaseous hydrocarbons and/or oxygenated compounds also derived from biomasses. WO2011072877(A1)

Bizzi A; Basini L; Saracco G; Specchia, V. (2002). Short contact time catalytic partial oxidation of methane: analysis of transport phenomena effects. *Chemical Engineering Journal*, Vol. 90(1-2), pp. 97-106

Bizzi M; Basini L; Saracco G; Specchia, V. (2003). Modeling a transport phenomena limited reactivity in short contact time catalytic partial oxidation reactors. *Ind. & Eng. Chem. Research;* Vol. 42(1), pp. 62-71

Capoferri, D.; Cucchiella, B.; Iaquaniello, G.; Mangiapane, A. ; Abate, S. & Centi, G. (2011). Catalytic Partial Oxidation and Membrane Separation to Optimize the Conversion of Natural Gas to Syngas and Hydrogen. *ChemSusChem*, Vol. 4, pp. 1787-1795

Choudary, V.R.; Hamman, A.S.; Sansare, S.D. (1992). CPO process for NG conversion into synthesis gas. *Angew. Chem.* Int. Ed. Engl. Vol. 31, p. 1189

Christensen, T.S. (1996). Adiabatic pre-reforming of hydrocarbons-an important step in syngas production. *Applied Catalysis A–General*, Vol.138, pp. 285–309

Christensen, T.S.; Primdhal, I.I. (1994). Improve syngas production using autothermal reforming. *Hydrocarbon Processing* Vol. 73, pp. 39-46

De Falco, M. ; Iaquaniello, G. & Salladini, A. (2011a). Experimental tests on steam reforming of natural gas in a reformer and membrane modules (RMM) plant. *Journal of Membrane Science,* Vol. 368, pp. 264 – 274

De Falco, M. ; Marrelli, L. & Iaquaniello, G. (2011b). Membrane Reactors for Hydrogen Production Processes. Springer Ed., ISBN 978-0-85729-150-9

Faur Ghenciu, A. (2002). Review of fuel processing catalysts for hydrogen production in PEM fuel cell systems. *Current Opinion in Solid State & Materials Science,* Vol. 6, pp. 389-399

Grunwaldt, J.-D; Basini,L.; Clausen, B.S. (2001). In situ EXAFS study of Rh/Al_2O_3 catalysts for catalytic partial oxidation of methane. *Journal of Catalysis,* Vol. 200(2), pp. 321-329

Grunwaldt, J.-D.; Kappen, P.; Basini, L.; Clausen, B.S. (2002). Iridium clusters for catalytic partial oxidation of methane - an in situ transmission and fluorescence XAFS study. *Catalysis Letters* Vol. 78, pp. 13-21

Hickman, H.; Schmidt, L.D. (1992). Synthesis Gas formation by direct oxidation of methane over Pt monoliths. *Journal of Catalysis* Vol. 138, pp. 267-282

Hickman, H.; Schmidt, L.D. (1993). Production of Syngas by direct catalytic oxidation of methane. *Science* Vol. 259, pp. 343-346

Holladay, J.D. ; Hu, J. ; King, D.L. & Wang, Y. (2009). An overview of hydrogen production technologies. *Catalysis Today,* Vol. 139, pp. 244–260

Iaquaniello, G.; Giacobbe, F.; Morico, B.; Cosenza, S. & Farace, A. (2008). Membrane reforming in converting natural gas to hydrogen: Production costs, Part II. *International Journal of Hydrogen Energy,* Vol. 33, pp. 6595-6601

Iaquaniello, G. & Salladini, A. (2011). Method for hydrogen production. *European Patent Application* EP11150491

Joensen, F. & Rostrup-Nielsen, J.R. (2002). Conversion of hydrocarbons and alcohols for fuel cells. *Journal of Power Sources,* Vol.105, pp.195–201

Moulijn, J.A. ; Makkee, M. & van Diepen, A. (2001). *Chemical Process Technology,* John Wiley & Sons, Ltd, ISBN 978-0-471-63062-3, Chichester, England

Navarro, R.M.; Peña, M.A.; Fierro, J.L.G. (2007). Hydrogen Production Reactions from Carbon Feedstocks: Fossil Fuels and Biomass. *Chemical Reviews,* Vol. 107, pp. 3952-3991

Peña, M.A. ; Gómez, J.P. & Fierro, J.L.G. (1996). New catalytic routes for syngas production. *Applied Catalysis A–General,* Vol. 144, pp.7–57

Rostrup-Nielsen, J.R. (1993). Production of synthesis gas. *Catalysis Today,* Vol. 19, pp. 305-324

Rostrup-Nielsen, J.R. (2000). New aspects of syngas production and use. *Catalysis Today,* Vol 63, pp. 159–164

Rostrup-Nielsen, J.R. ; Dybkjaer, I. & Christensen, T.S. (1998). Steam reforming of liquid hydrocarbons. *Studies in Surface Science and Catalysis,* Vol. 113, p. 81

Song, X. & Guo, Z. (2006). Technologies for direct production of flexible H_2/CO synthesis gas. *Energy Conversion and Management,* Vol. 47, pp.560-569

Schwiedernoch, R.; Tischer, S.; Corea, C.; Deutschmann, O. (2003). Experimental and numerical study on the transient behavior of partial oxidation of methane in a catalytic monolith. *Chemical Engineering Science* Vol. 58, pp. 633–642

York, A.P.E. ; Xiao, T. & Green, M.L.H. (2003). Brief overview of the partial oxidation of methane to synthesis gas. *Topics in Catalysis*, Vol. 22, pp. 345-358

Chapter 13

Natural Gas: Moving to Chaos and Complexity in Financial Statements

Fernando Juárez

Additional information is available at the end of the chapter

http://dx.doi.org/10.5772/54131

1. Introduction

The study of complex systems is an interdisciplinary field [1 p. 4]; however, sciences of complexity remain under common and general principles [2], such as micro and macro level manifold [3, p. 41], emergency and chaos [4], interactions among a large number of agents [5], environment and aggregation [6], interdependencies [7], propagation, nonlinearity, feedback loops, open systems, memory and history, adaptation and self-regulation. Existing models are: a) autopoiesis (reproduction, replication, and ontogeny of structural change), b) dynamic systems (large-scale modeling, nonlinearity and unexpected behavior), c) dissipative systems (conditions far from equilibrium and triggering events), and d) dynamics of chaos (cumulative and chaotic nature of changes, moving away from equilibrium) [2]. Thus, complexity consists of a number of theories [8] and not a unified method and insight [9]; it is an intriguing concept with no characterization agreed on, and discussions are about intuitive notions [10]. Although metaphorical language is one of the existing approaches to complexity [11] and many models could have complex properties [4], concepts need to be properly understood, translating them into a useful model of reality [12], and caution is required, when trying to model a system as complex.

Sciences of complexity interact with other disciplines, such as cognitive and organizational science [6], in a natural interface [13], correlating with complex decision-making, strategies, processes, and emergent features of institutions and processes [8]. Complexity and uncertainty are evident [14] and organizational analysis should be performed within a conceptual-theoretical framework that may require of complexity [7], because the world itself is complex [15]. Organizations deal with complexity and complication [16], and, at some degree, any context involves both features [17]. Complicated contexts allow for prediction while complex contexts focus on interaction between systems [18]; besides, complexity is a state of the world while complication is a state of mind [19, p.2].

A dispute exists regarding the similarity between chaos and complexity. Although mathematical chaos shows that basic models acquire complex properties [13], no relationship exists between complex systems and chaos [20]. However, existence of chaos supports complex systems [2], and it locates within complexity [8], along with self-organization, nonlinear dynamics, dissipative structures [21], theory of catastrophes and co-evolution organizational models [22].

Management sciences have some interests in complexity [23, 11]; e.g., they explore the application of complex adaptive systems in supply chains [24], breeding program development [25], government processes [26], strategic management and system predictability [27]. Besides, strategies to absorb complexity [29], social project entrepreneurship [29], leadership in micro and mesosystem [30], culture creation [31], environmental complexity and transformational leadership [32] are also of interest. Other examples are complexity in bargaining games [33], markets with extensive-form games [34], competitive advantage and discontinuity [35], replacement of management proposals and feedback loops [36], development of new products [37] and landscape design [38], among others. In addition, chaotic behavior is a part of organizational change [2], co-evolution on the verge of chaos [6], and organizational behavior [39].

Finally, some issues, such as Financial Statement analysis, leadership, corporate finance, or small and medium enterprises and family businesses, have a normativism which needs to be overcome [12], as the hidden complexity underlying some problems [40] favors complexity in organizations [41]. According to this, organizations have to be prepared to deal with it.

The need for a different logic in management areas is clear [17, pp. 15-16] and, although the analysis of Financial Statements seems to be structured and based on fix rule, it is not true [12]. Items and transactions give raise to different systems, such as cost analysis, cash flow analysis, inventory control and asset management, among others. Besides, major indicators of Financial Statements such as balance sheet, cash flow and income statement, are in interaction with each other. Finally, the existence of multiple agents, such as analysts, accountants, managers, directors, supervisors, company stakeholder, investors, market analysts, etc., agree to a complex dynamic based on circumscription, belief and paraconsistent logics, resulting in quite a subjective and logic complexity. Moreover, the analytical approach to Financial Statements, which comprises Management Discussion & Analysis (MD&A) and Notes to Financial Statements sections, gives rise to various interpretations based on the analyst knowledge and subjectivity. The prescriptive orientation of Financial Statements does not coincide with their use, as it involves a multitude of accounting rules intertwined with interpretations and recommendations arisen from subjective knowledge.

Nevertheless, chaos and complexity theory can help in the analysis of these topics. For instance, they allow for examining sector incremental change and positive-negative results as a bifurcation with nonlinearity. This approach, along with models of health and epidemiology [45], applies to balance sheet (Stakeholder's Equity or Assets), cash flow (cash flow at end of year or increase in cash flow) and income statement (Profit and Loss) in crude oil and natural gas [42], health [43] and tourism sectors [44]. However, it has been pointed out that the application of chaos theory to natural gas industry, led to poor results [46].

2. Natural gas industry: Complex and chaotic characteristics

The future of natural gas has a chaotic nature, and market poses changing characteristics, what requires the analysis to integrate disciplines such as economics, politics, psychology and meteorology [46]. Besides, the needs for a nonlinear approach to balance the natural gas market [47] and complexity in opportunities for natural gas storage [48] have also been outlined. In recent years, market trends, in the natural gas industry, changed drastically, with a significant increase and volatility in prices [49]. In the United States, variations in the price of natural gas connect with consumer prices, and might be a leading indicator to guide the country's economic policies [50].

On the other hand, uncertainty prevails in the markets [51] and supply and prices [52], as well as contradictions between government policies [53]. There are substantial changes in natural gas market, due to deregulation, becoming more dynamic and unpredictable, with short-term contracts, which heightens the risks of investing in distribution networks [54] and liquefied natural gas market, but making expensive investments needed to ensure an adequate supply [55]. Construction and maintenance of distribution networks require lots of resources in the form of debt and equity [56], promoting companies to be cautious in high risk assets investment, and managing corporate finances. Moreover, prices tend to be volatile in response to underlying supply and demand factors, such as weather, availability of pipelines or consumption patterns [57] and short-term contracts [58].

All of this provides this sector with high volatility, uncertainty and a need for strong competition with other energy sources, allowing for models that include chaos and complexity. Although the industry has an exciting development opportunity, as indicated for the pipeline distribution networks, despite large turbulence in profits [59, 60, 61, 62], new models must be used to provide an explanation of such variations. The sector's future will depend on competitive forces able to create balance between supply and demand [63]; nonetheless just competition and a leader position do not guarantee stability, as turbulence, a sign of complexity, remains. The various interactions and changes in market conditions and business can make information always scarce and small and minor alterations result in acute transformations, as a suggestion of the existence of chaos.

Government regulation of prices can cause gas shortages in times in which consumption increases, but the regulations imposed by the market may also cause shortages by reducing production. It goes from one restriction to another, and avoidance of monopoly is a need, to keep prices below the monopolistic requirements [64]. Although monopoly could determine local stability of prices, the long-term estimate of thereof follow a chaotic model [65] and that there are a large variation in prices under uncertain demand [66].

However, due to price regulation, companies are forced to increase profits by managing operating costs and long-term forecast over the fluctuations in demand and competition from other sources of energy, what benefits from a defective, abductive and circumscription logic in the sector. These Logics go beyond linear comprehension, and they rely on interpretations probed to be true, at the same time that other interpretations are false.

Natural gas prices have increased since 2000, reaching high volatility, according to NYMEX and Henry Hub [49]. For years, these prices follow crude oil prices, a situation that changed [50] in recent years and, although the link could continue to a longer term, in the short term there are wide variations [67]. However, previous authors suggest that, for a long time, natural gas prices adjust to crude oil prices, as a thumb rule (i.e., natural gas is one-tenth the price of crude oil), and despite the creation of new methods, these mechanisms are still in use. Thus, these believes, or belief logic, instead of lineal mechanisms of error reduction, or cause-effect, is relevant, and we need to encourage other approaches.

Moreover, according to the above, reduction or stabilization of falling prices, lead to a decrease in the production of reserves, what it can affect even the most demanding times of the seasonal cycle of gas consumption. That may seem predictable, as consumption per month cycles constitutes regular wide variations, e.g., in the United States. While industrial consumption follows a fairly steady path, residential consumption is sometimes significantly increased and then decreased, in a nearly sinusoidal wave. Consumption in power generation plants is almost the opposite cycle, albeit less prominent; on the other hand, commercial and transportation consumption follows the residential one, with less strength [68]. All of these variations, for United States, are in Figure 1.

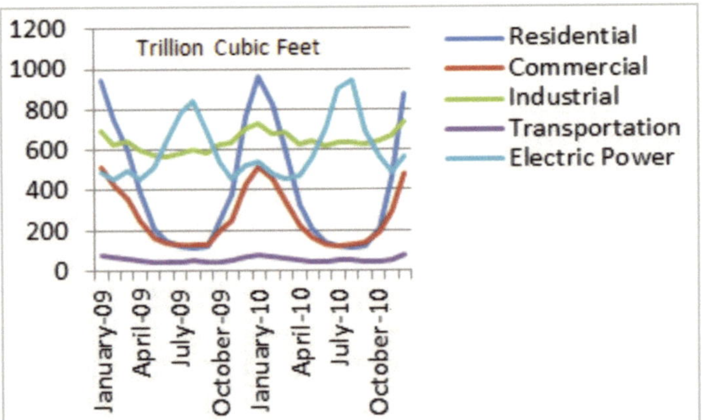

Figure 1. Consumption distribution of natural gas) in US 2009-2010 according to data published in Monthly Energy Review (2011).

Interestingly, although some of this consumption patterns model a sinusoidal oscillation, according to the spectral composition of time series, it consists of a sum of waves of similar or opposite effects, so that a change in one of them results in total consumption changes. Therefore, this repeats for all years and may be the result of a linear aggregation of variables; however, in annual consumption (Figure 2), there are many softer variations and a mismatch of types of consumption. It is difficult to see a periodic variation over the years, so it should be assumed that annual variations compose of other small ones producing this annual trend along the years. It depicts a growing trend in electric power consumption and a decrease in industrial consumption. Thus, aggregation over time causes different effects on consumption.

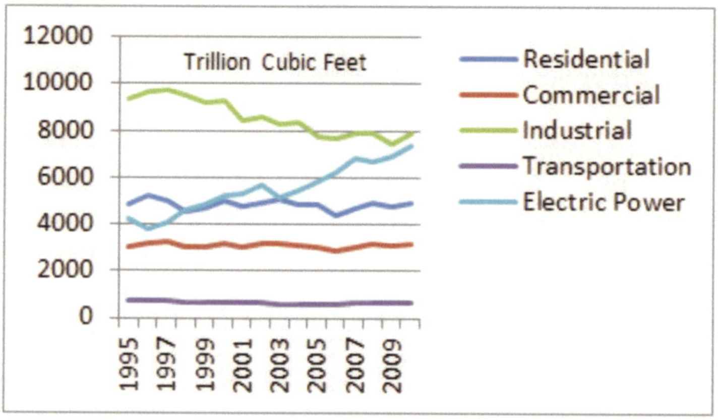

Figure 2. Annual consumption distribution of natural gas in US 1995-2009 according to data published in Monthly Energy Review (2011).

All of the foregoing might result in the dismissal of investment. However, this investment is a need, but it necessitates an environment of confidence in returns. In the short term, these returns become unstable and, in the long term, it turns into a safer fond but with low profitability. This is consistent with assuring sustainability in times of uncertainty, but it is in conflict with high-profitability anxious requirements that often exist in business operations. Stocks, in this industry, suit to conservative, income-oriented investors [56] so it is reasonable to make a bet on investments with long-term security in the low profitability of thereof. Moreover, when an environment of uncertainty and business dynamic exists, assets must be secured, and the payment of dividends needs to be postponed not to jeopardize the future of the company, which will depend on growth. In this environment, logical decisions to be made become more subjective and should take into account conflicting economic activities in the long term. Also, it requires control of resources, which, otherwise, would be handled by third parties in a monopolistic manner. The need for accurate predictions becomes true in situations where it seems impossible to make them.

Markets have become extremely sensitive to the volatility and price rises, missing, on occasions, a reassuring policy; further regulation by the market, without government intervention could produce strategic coalitions to gain control over resources or supplies, and impose prices. In this sense, government regulations apply to control price [49] and there is no doubt about the need for a minimal regulation [57]. However, FENOSA, the largest organization of gas and electricity in Spain and Latin America, served 5.6 million people in Latin America in 2010, but entering different markets did not give reasonably high values, so that regulatory policies could cause problems for the company, along with climatic conditions [69]. The same holds for other companies like Oil and Natural Gas Corporation Limited, for which, drilling and producing wells did not return investment [70].

In addition, the high volatility of market favors prediction-based positions depending on events or climate change; however, this is not a matter of predictions using hard data, but beliefs about what it will happen. Serious financial failures may occur [see 71], when circumscription, based on those beliefs, is not appropriate. In this sense, making restrictions could affect stock market behavior. Regulations improve transportation price efficiency, price structures and company pipeline transportation, by storage unbundling, release of capacity programs and change in rate design, allowing keeping storage to overcome fluctuations [72]. This results in substantial financial structure differences. Thus, all of that together with changes in demand due to seasonal variations, which become less predictable when looking at the microvariations and their aggregated effect, result in a financial system that must follow such changes or locate high, above them. There is a need to stabilize demand and find new markets, but also to maintain high fixed costs, in some companies.

However, deregulation has also risks. Thus, it can cause niches or disconnected networks from each other, limiting competition, initially proposed by deregulation, and causing a captive market for certain dealers, despite sophisticated transactions between networks [73]. Therefore, the complexity that emerges from open markets networks may overwhelm the capacity of businesses to coordinate actions [73]. This influences supply and may increase government intervention again, reducing industry efficiency.

The market volatility reflects in supply contracts, which are short term if less than one month, middle term between one and twelve months and long term if above one year [57]. In this context, the major sources of business risks are prices and volume [74]. Strategies to keep liquidity, storing gas, expand and use derivatives are essential, but require changes in accounting practices. These changes include reducing high dependence on liability, increasing investment so as to cover the periods of minimum demand or prices, or cost accounting adjusted to fit fluctuations in operation. On the other hand, while exploration costs and present value increase, and the number of tests grow, the possibility of being competitive, by maintaining demand, guarantees discount rates.

Initial predictions about opportunities and changes in natural gas industry, reveal times of a major reform in energy consumption and costs reductions with many countries maintaining natural gas reserves [75]. However, at the present time, in view of the above, there is a complex dynamic, which results in different markets and characteristics of contracts, independent distribution networks, management of stocks subject to wide fluctuations and changes in demand, finding new demands and intense competition with other energy sources; all of this, requires a different understanding of this industry.

According to the aforementioned, complexity and chaos theory must be in the Financial Statements of natural gas industry analysis. This approach can be based on the existence of different models [76].

3. Financial statements as a complex system in natural gas industry

Accounting transactions affect Assets, Liabilities or Stockholders' Equity, and it is necessary to maintain basic accounting equation in balance, that is: Assets = Liabilities + Stockholders′ Equity.

However, this equation is not a numerical calculation, but an intricated logical reasoning, which allows allocating an amount to an Assts item and to one of the Liabilities or Stockholders′ Equity items, or to another Assets item, but with a different sign. Nevertheless, this allocation composes of a real amount (for instance in Assets) and another identical amount, which constitutes an expectation, or belief, about some accounting events that will happen in the future (for instance, in Liabilities or Stockholders′ Equity).

Let′s put an example, while a bank loan is a real entry in Assets, acknowledging a debt in Liabilities is just a payment expectation. In the same way, a cash input in Assets can be recorded as unearned service revenue in Liabilities. In doing so, a future event includes in Liabilities, no matter whether it will be performed or not, and expectation is a part of Financial Statements. Moreover, when a service is provided, but payment is to be received on a later time, there is no reason to include the amount in Stockholders′ Equity. This is so, because it is merely the possibility of increasing the Stockholders′ Equity in the future. In the same manner, there is no reason to include it in accounts receivable, except to state that payment should be made in the future.

It is true that Assets are equal to Liabilities plus Stockholders′ Equity, but that is a logical truth under certain circumstances. Liabilities and Stockholders′ Equity do not imply the existence of Assets, but accountant must create it. Despite that there is no more money than what it is available or more property plant and equipment than those than there are, an asset is also a liability or a stockholder's equity. In doing so, the product of an accounting transaction is one thing, and another, simultaneously.

In natural gas industry, strong Assets investment may require substantial debt, with the expectation that payments must be secured by an adequate demand. However, on occasions, dependence on demand types, i.e., industrial demand [50], or the growing demand for electric power plants, makes it to be dependent on the consumption, an indicator of prices. Thus, the existence of these cycles and the uncertainty associated with the dynamics of change, in consumption market, produces a small expectation of compliance, and a lack of foresight and remedies for operation. In this sense, basic equation might not be true to some companies, considering expectations, and not conviction, in that they will meet their obligations in the future. This has an impact on contract prices and the consumer must be protected against significant variations in prices. According to this, there is an expectation about what it is going to be charged that could not be fulfilled. Cycles of consumer prices are extremely difficult to predict, even with the various available techniques [46] and the basic equation requires adjustments in expectations to show the existence of consumer protectionism regulations and price cycles.

Recognizing in the Financial Statements that a certain amount of money is owned (Asset) and, at the same time, it is owed (Liability) violates the classical logical principle (excluded third) that something cannot be its opposite at the same time. However, complexity agrees opposite forms, and, within it, another form of logic (dialogic) overcomes excluded third principle [4], allowing transactions in Financial Statements to follow a default reasoning. For instance, a transaction is assigned to an item, unless an example exists (or reasoning) that justify assigning it to a different item. Also, an abductive reasoning is used by following the ordinary standards, being possible to assign an amount to an item, whenever a reasonable explanation is available. Moreover, in the presence of a greater knowledge, or a different set of beliefs about Financial Statements, their organization would change, what it is an example of knowledge and beliefs argumentation. This, bases on arguments of circumscription logic (explanations not taken into account in the initial set of premises or assumptions), default reasoning (existence of examples that contradict the general law) and abductive reasoning (generalizations based on the need for economy and simplicity).

The principles of circumscription logic [77] indicate that a minimum must be obtained to meet the fact that the proposition is true only if it is necessary to be, or a proposition is false if it is possible to be. Therefore, if there is a possibility that an unearned service revenue amount is false, because the service could not be provided, then the statement "the transaction X is included in unearned service revenue" is false. Thus, $A \equiv L \cup S$ (A: Assets, L: Liabilities, S: Stockholders' Equity; ≡: Equivalence) is not confirmed, since deposit made in cash does not coincide with the service provided in the future, because that service may not be carried out.

The existence of a circumscription is evident in the natural gas markets, where the experience of participants determines the balance to be carried out between different contracts. In this way, it seeks to respond to changing market conditions, by the balance between supply and demand, both on short and medium term [57], what requires distributing assignments in the different items of Financial Statements, by proper expectations management. In this sense, allocation depending on the balance of the portfolio is a minimum circumscription based on the familiarity with the market. So a belief becomes a transaction in the Financial Statements; this transaction is true (accurate) unless it is likely to be false (not accurate). Not only that, but also financial trading of natural gas, causes many profits to be generated by investment operations, based on the expectation of reducing price risk, which also leads to a circumscription associated with beliefs and knowledge.

However, depending on the cost structure of the company this circumscription can be performed more or less widely. In this sense, the weight on the basic equation can be in the preservation of assets, and diversification of its use, or in increasing debt with a commitment to operations or short-term opportunity contracts, balanced with medium to long term contracts or more efficient investment to moderate cycles that happen and enhance the use of natural gas in other applications. In fact, the search for contracts that meet the requirements of the entities providing funds, ensuring the price with a moderate acceleration to allow for compliance with debt, is something that takes time [51]. This means that, in the Financial Statements, there are imbalances between the sets A (Assets), L (Liabilities) and S (Stockholders' Equity), promoting owning and use of assets, risk associated to a commercial and financial volatile market, or increase in debt and equity, searching for long-term contracts, operation diversification and use of fixed assets. However, these are different models in the accounting analysis, and they have a strong impact on the financial structure of firms.

According to this, circumscription on the market and Financial Statements will determine the future of organizations in this industry. Circumscription accepts a minimum truth in Financial Statements transactions. That is, accepting just what it is necessary to be truth and not what it has a chance to be false. This implies a logical circumscription on these Financial Statements. In general, we can say that circumscription on Financial Statements (FS) consists in a number of models, or minimum principles, so that:

C(FS) : (Tm |Tm is a minimum set of FS); C being the circumscription held on FS and T a set of statements which are necessary truth.

In this way, the basic equation of Financial Statements is openly interacting with the environment, and the logic of multiagents, multilevel and co-evolution with environment bases on actions that are true. This is recognition of the assumptions made.

Moreover, in the section of Management Discussion and Analysis (MD&A) in the report of Financial Statements, where, to a large extent, the framework is the scheme of management [78], a subjective interpretation of accounting data shows the position of business, investment and financial needs, results and other information, with an interpretive structure based on a different logic. This section, takes on a highly subjective narrative from different theoretical frameworks or models, and Critical Accounting Policy (CAP) makes financial management subjectively be based, but aimed at a satisfactory financial performance. Items discussed in CAP are the allowance for loan losses, loan sales, investment, intangible items, stock compensation, revenue recognition, capitalizing software and restructuring costs, among others [79]. Besides, Critical Accounting Estimates (CAE) presents an analysis of the variability or uncertainties to occur, along with estimation and accuracy [78].

The information about financial conditions, results of operations, liquidity and capital resources, market risks, other risk factors, and many others issues are included in MD&A, along with interpretations about financial data, in terms of cash flow, net product sales, expenses, stockholders' equity, earnings from operations, etc. On the other hand, Notes to Financial Statements, provides with interpretations based on nontraditional logic and models. They are another language and arguments explaining policies and solving doubts in Financial Statements. According to this, company policies aim to provide a structure for Assets management, reserve management, and debt or portfolio schema design, and they are the basis of interpretation of this section, where the company future is designed. In MD&A, the following suggestions can be made:

The assets increased, due to property acquisition and extension of loans, based on a reasonable price prediction. It is possible to purchase reserves and sell them in long-term contracts with limited price escalation. That might result in long-term sustained revenues.

Implicit in these assertions, several assumptions guide the operations performed. However, the acceptance of these assumptions within a monotonic logic of reduction of uncertainty, causes the final conclusion seems to be the result of a linear accumulation of explanatory factors. This, however, is not true because different scenarios may occur. For example, while it is true that, in contracts, procedures to protect profits are included, it is also possible that price dynamics relativize them, thus losing a certain value. On the other hand, industry dynamics may lead to trade more contracts in the short term, in the spot market; so finding an escalated sell, along the time, does not provide the same benefits and, therefore, opportunities are lost. Finally, although reserves might have been sustained, it is also possible that market changes cause a lost in the volume of drilling, increasing stock prices and compromising long-term benefits.

Thus, predictions made in MD&A bases on a set of beliefs, which are not a cumulative process of reducing uncertainty, but they move within uncertainty, framing suggestions on assumptions that are believed most successful. Yet, in a changing market and business dynamics, other assumptions are also successful, i.e., the fact of making some recommendations does not mean that other recommendations that could be provided and are contrary to the formers, are not appropriate. This cannot be solved by linear predictions or logic trying to find out what the best option is, but by logic capable of incorporating contradictory thesis.

The aforementioned problem can be solved by basic principles of paraconsistent logic. This logic supports contradictions. Paraconsistent logic requires caution in the use of contradictions, but takes into account the fact that if an inconsistency exits, it does not imply that something is wrong. It can be assigned V to the truth value, 0 (False) or 1 (True); determining, accurately, the decision to make, and having in mind that every assumption (B) is true or false, imply that we assume:

$V(B, 1) \leftrightarrow V(\neg B, 0)$; i.e. B is true if and only if no-B is false

And also

$V(B, 0) \leftrightarrow V(\neg B, 1)$; i.e. B is false if and only if no-B is true

It means that only one of the propositions B and no-B (the opposite of B) is true, if and only if the other is false. Thus, at least one of the propositions is accomplished, but not both of them. This is different to determine that one of them is false if the other is true. The way of solving contradictions and paradoxes is by creating relationships between terms and not by making inferences from a term to another, as making inferences assumes that one term is true and, consequently, the other is false.

According to this, it means that following initial recommendations is a right choice whenever the other ones are false, so determining the false value of the other scenarios is a prerequisite to opt for the former. This logic poses a high standard of cautiousness, for reaching conclusions it must follow a process of revising every proposition, and determining the relationship among them and their truth value. This complex logic, expressed in MD&A, is not trivial, but it leads to different conclusions than those obtained by classical logic, where an inference guarantees that something is true or false. The key is that the company does not know what it will happen in the future, and expected or unexpected results could be obtained anyway, so a mean to guarantee results is to know that predictions are right because no possibility exists they are not.

Moreover, in Notes to Financial Statements, clarifications about profits from sales, supplies and promotion costs which are recognized when payment is somehow guaranteed, etc., are included. This indicates that the final structure of the Financial Statements links to changing expectations about the behavior of agents in the system, who according to unexpected events, might change their behavior, leading to new emerging systems resulting from interactions between actors. Thus, Notes to Financial Statements participates of belief logic, e.g., development of a number of assumptions which are valid only until one of them ceases to be. Therefore, a set of beliefs B exists, which is not empty, and consists of beliefs b1, b2, b3, ..., bn:

$$B(b1, b2, b3, \ldots, bn)$$

These beliefs can presume the market behavior, the commitment of the buyer, the performance of commodity prices, the maintenance of goodwill, etc. According to this, depending on the reasoner [80] that analysts are, several options exist. One of them is to equal the analyst to the best accurate reasoners, who never believes a proposition that is false; what can be expressed as:

$$\forall p\ (Bp \rightarrow p)$$

Thus, the belief B in a proposition p signs that the proposition must be true, i.e., they believe that market dynamic will raise prices, and it happens. Thus, they only believe in things that will occur.

In Notes to Financial Statements, managing certain items reflects the complexity, for instance, trading of liabilities on stocks markets or anticipating contract conditions. It deals, in a broad sense, with the uncertainty of not knowing how the price will behave. Strongly grounded on fixed assets, some companies in this industry need to calculate depreciation according to expectations about demand and reserves, and these expectations constitute a set of beliefs. Moreover, Assets follow an explosive growth once they have reached a certain threshold value. This growth has already been shown in other financial sectors; surpassing a certain amount of assets originates many more assets to be acquired in an exponential manner. However, it must go through a period of turbulence, as shown in Figure 3. This turbulence causes uncertainty in the performance of companies, which may choose to go back and maintain their Assets whenever turbulence is not fully overcome; once this turbulence is overcome, turning back is more difficult, because it is to reverse a process and give up virtually to what it has been achieved.

As it is shown in Figure 3, once the turbulence is overcome, a rapid increase might resemble a catastrophic phenomenon, due to a sudden and irreversible change in a parameter. It must be noted that there is no possibility to return to initial conditions once this phenomenon has taken place. This occurs due to the incremental increase in Asset differences. Decreasing the differences in Assets would reduce business operations, number of employees, properties, plant and equipments, and so on, with serious consequences for business sustainability.

What could a company do, after a substantial investment in fixed assets? Company cannot go back, and the market does not forgive that distribution or production networks are not functioning properly; demand cannot wait, and the cost of doing nothing is even higher [53]. Accordingly, the company's future would be jeopardized.

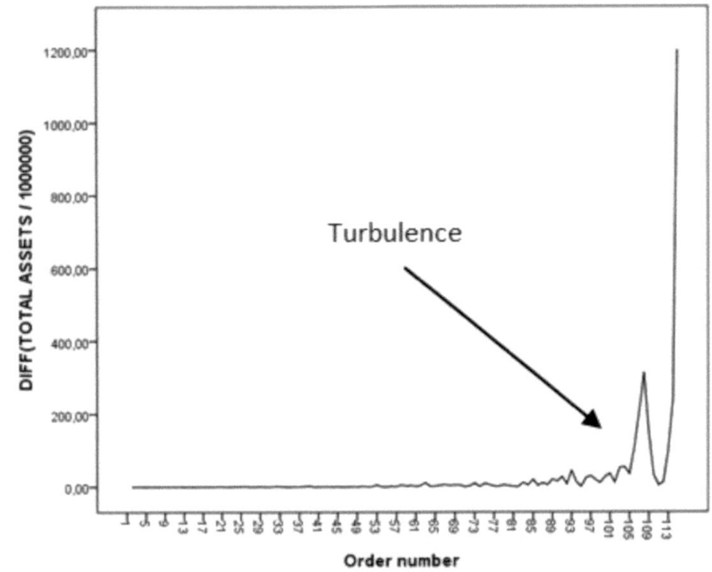

Figure 3. Differences in Total Assets, in orderly values: Total Assetsn+1 – Total Assetsn (Data from Colombia, Energy Industry)

Operation revenues of pipelines companies reached record figures in 2010 [61] but expected future earnings heavily rely upon at a cycle's peak [81]. Interest rates need to be competitive for investors, because a lot of money is necessitated in the form of debt and equity, leading to a financial structure more weighted towards debt; this is cheaper than the cost of stockholder's equity [56].

Financial ratios are open to different explanations depending on the tacit knowledge that people have of the company, and the reasons to compute them [82]. In the Gas Natural Industry, common ratios are those related to debt. For instance, companies with modest Long-Term Debt Ratios are more tolerant to grow, and a high debt ratio shows operating problems, other ratios are Return on Common Equity, Net Profit divided by Common Equity, and Cash Flow, which is a more stable measure than profits [56].

However, relationship among these ratios gives an idea of liquidity, profitability and sustainability, and it is vital to analyze the joint behavior of several of them. Examples of these indicators are:

Liquidity:

$$Working\,capital = Current\ Assets - Current\,liabilitites$$

When working capital is negative, the company is at risk of not paying sort-term obligations and could fall under bankruptcy.

Sustainability:

$$Free\ Cash\ Flow = Cash\ Provided\ by\ Operations - Capital\ Expenditures - Cash\ Dividends$$

It is the ability of the company to generate cash.

Profitability:

$$Gross\,Profit\,Rate = \frac{Gross\,Profit}{Net\,Sales}$$

It provides an insight about how the company is doing regarding sales in relation to gross profits, which can be obtained by other means than sales.

Relationship among these indicators depicts in Figure 4.

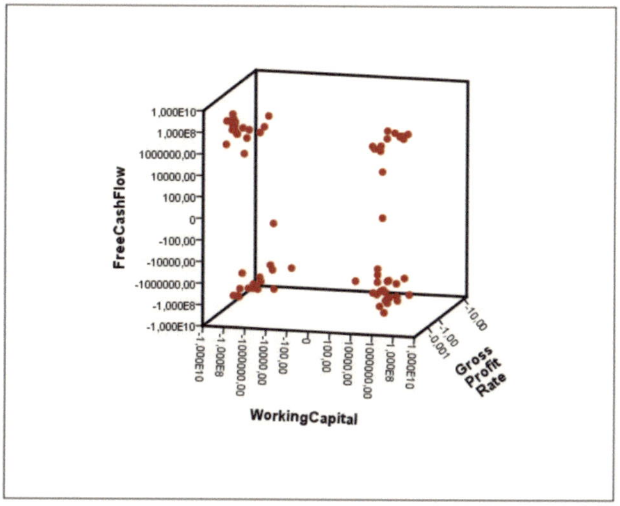

Figure 4. Scatter plot of log Free Cash Flow, log of Working capital, and log of Gross Profit Rate

In this figure, the distribution of the ratio values, resemble a grouping in four areas. It is crucial to note that sustainability (Free Cash Flow) does not associate to one liquidity (Working Capital) figure, but to opposite ones; i.e., positive and negative liquidity figures lead to the same sustainability ratio. It is the same with profitability, where Gross Profit Rate results in different states of liquidity (Working Capital). According to this, there is a complex relationship among these ratios that can be modeled with chaos theory [see 12, 42, 43]. In this figure, several attraction points exist, which give rise to areas where profitability, sustainability and liquidity satisfy several options of performance, and not a fixed rule associates to a better performance. Having enough liquidity, or profitability do not guarantee the sustainability of the company, but it is the relationship among these indicators what determines the future of the company. Small oscillations in Free Cash Flow, accompanied by small oscillations in Working Capital, can lead to large changes in Gross Profit Ratio, impacting stockholders' rights to payment, and sustainability of the company can be reduced by small changes in Working Capital producing a decrease in Free Cash Flow.

4. Conclusion

The approach to Financial Statements from a complex perspective and chaos theory in Natural Gas Industry is a need. This is true not only because of the existing complexity in the industry, but also because they reflect a form of interpretation and analysis that might help to understand some of the phenomena that occur in this field. Thus, complex logic and the relationships between the indicators of Financial Statements, can contribute to new developments and ideas about what it is occurring in Natural Gas Industry.

Author details

Fernando Juárez
Universidad del Rosario, Bogotá, Colombia

5. References

[1] Mitchell M (2009) Complexity: A Guided Tour. Cary, NC, USA: Oxford University Press.

[2] Dooley, K.J. (1997) A Complex Adaptive Systems Model of Organization Change. Nonlinear Dyn. Psych. L. Sci., 1: 69-97.

[3] Pribam KH (1996) Interfacing complexity at a boundary between the natural and the social sciences. In: Khalil EL, Boulding KE, editors. Evolution, Order and Complexity. London, GBR: Routledge. pp. 40-60.

[4] Morin E (2007) Complejidad restringida y Complejidad generalizada o las complejidades de la Complejidad. Ut. Prax. Latinoam., 12: 107-119.

[5] Maubussini MJ, Sullivan T (2011) Embracing complexity. Hvd. Bus. Rev., 1271: 89-92.

[6] Anderson Ph (1999) Complexity Theory and Organization Science. Organiz. Sci., 10(Special Issue: Application of Complexity Theory to Organization Science), 216-232.

[7] Elsner W, Hocker G, Schwardt H (2010) Simplistic vs. Complex Organization: Markets, Hierarchies, and Networks in an Organizational Triangle — A Simple Heuristic to Analyze Real-World Organizational Forms —. J. Econ. Issues, XLIV: 1-29.

[8] Klijn E-H (2008) Theory and public administration: What's new? Key concepts in complexity theory compared to their counterparts in public administration research. Publ. Manage. Rev., 10: 299–317.

[9] Bousquet A, Geyer R (2011) Introduction: complexity and the international arena. Cambridge Rev Int. Affairs, 24: 1-3.
[10] Ekstig B (2010) Complexity and Evolution: A Study of the Growth of Complexity in Organic and Cultural Evolution. Found Sci., 15: 263–278.
[11] Richardson KA (2008) Managing Complex Organizations: Complexity Thinking and the Science and Art of Management. E:CO, 10: 13-26.
[12] Juárez F (2012) Chaos and Complexity in Financial Statements. In: Barnejee S, editor. Chaos and Complexity Theory for Management: Nonlinear Dynamics. IGI Global.(In press).
[13] Morel B, Ramanujam R (1999) Through the Looking Glass of Complexity: The Dynamics of Organizations as Adaptive and Evolving Systems. Organiz. Sci., 10(Special Issue: Application of Complexity Theory to Organization Science): 278-293.
[14] Magellan D (2011) The antidote to complexity. MWorld, 10: 27-29.
[15] Urry J (2011) ¿Complejidades y futuros?. Revista CIDOB d'afers internacionals, 95: 11-20.
[16] Vasconcelos FC, Ramirez R (2011) Complexity in business environments. J. Bus. Res. 64: 236–241.
[17] Juárez F, Contreras F (2012) Liderazgo y complejidad: Conceptualizaciones e implicaciones para la organización actual. Madrid, España: EAE
[18] Sargut G, McGrath RG (2011) Learning To Live with Complexity How to make sense of the unpredictable and the undefinable in today's hyperconnected business world. Hvd. Bus. Rev., 1271: 69-76.
[19] Norman DA (2010) Living with Complexity. Cambridge, MA, USA: MIT Press.
[20] Cilliers P (2000) What Can We Learn From a Theory of Complexity? EMERGENCE, 2: 23-33.
[21] Alhadeff-Jones M (2008) Three Generations of Complexity Theories: Nuances and Ambiguities. Educ. Philos. Theor., 40: 66-82.
[22] McKelvey B (1999) Avoiding Complexity Catastrophe in Coevolutionary Pockets: Strategies for Rugged Landscapes. Organiz. Sci., 10(Special Issue: Application of Complexity Theory to Organization Science): 294-321.
[23] Lewin AY (1999) Application of Complexity Theory to Organization Science. Organiz. Sci., 10(Special Issue: Application of Complexity Theory to Organization Science): 215.
[24] Choi ThY, Dooley KJ, Rungtusanatham M (2001) Supply networks and complex adaptive systems: control versus emergence. J. Oper. Manage., 19: 351–366.
[25] Teisman GR (2008) Complexity and management of improvement programmes. An evolutionary approach. Publ. Manage. Rev., 10: 341–359.
[26] Teisman GR, Klijn E-H (2008) Complexity theory and public management. An introduction. Publ. Manage. Rev., 10: 287–297.
[27] Bovaird T (2008). Emergent strategic management and planning mechanisms in complex adaptive systems. The case of the UK Best Value initiative. Publ. Manage. Rev., 10: 319 –340.

[28] Boisot M, & Child J (1999) Organizations as Adaptive Systems in Complex Environments: The Case of China. Organiz. Sci., 10(Special Issue: Application of Complexity Theory to Organization Science): 237-252.

[29] Swanson LA, Zhang DD (2011) Complexity Theory And The Social Entrepreneurship Zone. E:CO Issue, 13: 39-56.

[30] Hannah ST, Lester PB (2009) A multilevel approach to building and leading learning organizations. Leadersh. Q., 20: 34–48.

[31] Frank KA, Fahrbach K (1999) Organization Culture as a Complex System: Balance and Information in Models of Influence and Selection. Organiz. Sci., 10(Special Issue: Application of Complexity Theory to Organization Science): 253-277.

[32] Ussahawanitchakit P (2011) Effects of learning capability, technology change, globalization force, business experience, and environmental complexity on transformational leadership: Evidence from Thailand. J. Int. Bus. Econom., 11: 84-97.

[33] Chatterjee K, Sabourian H (2000) Multiperson Bargaining and Strategic Complexity. Econometrica, 68: 1491-1509.

[34] Gale D, Sabourian H (2005) Complexity and Competition. Econometrica, 73: 739-769.

[35] Stefanović I, Prokić S, Vukosavljević D (2011) The response to the changing landscape of tomorrow: Reconfigurable organizations. Afr. J. Bus. Manage., 5: 13344-13351.

[36] Sterman JD, Wittenberg J (1999) Path Dependence, Competition, and Succession in the Dynamics of Scientific Revolution. Organiz. Sci., 10(Special Issue: Application of Complexity Theory to Organization Science): 322-341.

[37] Kim J, Wilemon D (2007) The Learning Organization as Facilitator of Complex NPD Projects. Creativity Innov. Managem., 16: 176-191.

[38] Levinthal DA, Warglien M (1999) Landscape Design: Designing for Local Action in Complex Worlds. Organiz. Sci., 10(Special Issue: Application of Complexity Theory to Organization Science): 342-357.

[39] Dooley KJ, Van de Ven AH (1999) Explaining Complex Organizational Dynamics. Organiz. Sci., 10(Special Issue: Application of Complexity Theory to Organization Science): 358-372.

[40] Green DG, Leishman TG, Sadedin S, Leishman GD (2010) Of Ants and Men-the Role of Complexity in Social Change. Evol. Inst. Economy Rev., 6: 259–275.

[41] Siggelkow N, Rivkin JW (2005) Speed and Search: Designing Organizations for Turbulence and Complexity. Organiz. Sci., 16: 101-122.

[42] Juárez F (2010) Applying the theory of chaos and a complex model of health to establish relations among financial indicators. Procedia Comput. Sci., 3: 982-986.

[43] Juárez F (2010) Caos y salud en el sector económico de la salud en Colombia. Int. J. Psych. Res., 3: 29-33.

[44] Juárez F (2011) Financial health and risk in the tourism sector in Colombia. Int. J. Math. Models Methods Appl Sci., 4: 747-754.

[45] Juárez F, Farfán Y (2011) A Statistical, Epidemiological and Financial Health Approach to the Retail Trade Sector in Colombia. Afr. J. Bus. Managem., 6: 5606-5614.

[46] Chwee V (1998) Chaos in natural gas futures? Energy J., 19: 149-164.

[47] Gabriel SA, Supat K, Zhuang J (2005) A Mixed Complementarity-Based Equilibrium Model of Natural Gas Markets. Oper. Res., 53: 799-888.
[48] Schumacher D (2010) Natural Gas Storage Offers Big Opportunities And Big Complexities. Pipeline & Gas J., 237: 58-59.
[49] Costello K, Huntington HG, Wilson JF (2005) After the Natural Gas Bubble: An Economic Evaluation of the Recent U.S. National Petroleum Council Study. Energy J., 26: 89-109.
[50] Serletis A, Shahmoradi A (2005) Business cycles and natural gas prices, EPOC Rev., 29: 75-84.
[51] Clements JR, Graeber FD (1991) Dealing with Natural Gas Uncertainties. Power Eng., 95: 37-39.
[52] Blankinship S (2003) Natural gas uncertainty prevailing concern at COAL-GEN '03. Power Eng., 107: 9.
[53] Inhofe JM, Fannon F (2005) Energy and the environment: the future of natural gas in America. Energy Law J., 26: 349-388.
[54] Petrash JM (2006) Long-term natural gas contracts: Dead, dying, or merely resting? Energy Law J., 27: 545-582.
[55] Knowles GL (2003) Liquefied natural gas: Regulation in a competitive natural gas market. Energy Law J., 24: 293-319.
[56] Gallagher R (2011) Industry Analysis: Natural Gas Utility. Retrieved at December, 19, 2011 from http://iiiprxy.library.miami.edu:3134/Stocks/Industries/Industry_Analysis__Natural_Gas_Utility.aspx
[57] Juris A (1998) Competition in the Natural Gas Industry. The emergence of spot, financial, and pipeline capacity markets. Publ. Policy Private Sect., 137: 1-7.
[58] Linn SC, Zhu Z (2004) Natural gas prices and the gas storage report: Public news and volatility. J. Fut. Markets, 24: 283-313.
[59] Freedenthal C (2001) Natural gas prices bring rich profits, lower growth. Pipeline & Gas J., 228: 80.
[60] Smith ChE (2008) Natural gas pipeline profits surge; oil flat. Oil & Gas J., 106: 50-69.
[61] Smith ChE (2010) Natural gas pipelines continue growth despite lower earnings; oil profits grow. Oil & Gas J., 108: 102-123.
[62] Smith ChE (2011) Natural gas pipeline operators' 2010 profits reach record levels. Oil & Gas J., 109: 92-113.
[63] Barcella ML (1996). Natural gas in the twenty-first century. Bus. Econom., 31: 19-24.
[64] Méra X (2010) Factor Prices under Monopoly. Q. J. Austrian Econom., 13: 48–70.
[65] Jablanovic VD (2011) The Chaotic Monopoly Price Growth Model. Chinese Bus. Rev., 10: 985-989.
[66] Humphreysa BR, Soebbing BP (2012) A test of monopoly price dispersion under demand uncertainty. Econom. Lett., 114: 304–307.
[67] Brown SPA, Yûcel MK (2008) What Drives Natural Gas Prices? Energy J., 29: 45-60.
[68] U.S. Energy Information Administration (2011, october) Natural gas. Mon. Energy Rev., 4: 68-74.

[69] Datamonitor (2011) Fenosa. Company Profile. Retrieved on April 27, 2012 from www.datamonitor.com.

[70] Datamonitor (2011) Oil and Natural Gas Corporation Limited. Company Profile. Retrieved on April 27, 2012 from www.datamonitor.com.

[71] Marthinsen JE, Gai Y (2010) Did Amaranth's absolute, relative and extreme positions affect natural gas futures prices, spreads and volatilities? J. Derivatives & Hedge F., 16: 9–21.

[72] Finnoff D, Cramer C, Shaffer Sh (2004) The Financial and Operation Impacts of FERC Order 636 on the Interstate Natural Gas Pipeline Industry. J. Regul. Econom., 25: 243-270.

[73] De Vany A, Walls WD (1994) Open access and the emergence of a competitive natural gas market. Contemp. Econom. Policy, 12: 77-96.

[74] Géczy ChC, Minton BA, Schrand C (2006) The use of multiple risk management strategies: evidence from the natural gas industry. J. Risk, 8: 19-54.

[75] Flavin Ch, Lenssen N (1995) The unexpected rise of natural gas. The Futurist, 29: 34-37.

[76] Alhadeff-Jones M (2008) Three Generations of Complexity Theories: Nuances and ambiguities. Educ. Philos. Theor., 40: 66-82.

[77] McCarthy J (1980) Circumscription – A form of non-monotonic reasoning. J. Artif. Intell., 13: 27–39.

[78] O'Shaughnessy J, Rashty J (2005) Critical Accounting Policy Disclosures of Selected NASDAQ Companies. J. Acc. Finance Res., 13: 19-40.

[79] Henry TF, Holtzman MP (2006) Critical Accounting Policy Disclosures for Financial Institutions. Bank Acc. & Finance, 19: 14-18.

[80] Smullyan RM (1986) Logicians who reason about themselves. In J.Y. Halpern (Ed.), Proceedings of the 1986 conference on Theoretical aspects of reasoning about knowledge. San Francisco, USA: Morgan Kaufmann Publishers. pp. 341-352.

[81] Mitkowski JrR (2011) Equipment Industry Analysis: Natural Gas Utility. Retrieved at December, 19, 2011 from http://iiiprxy.library.miami.edu:3134/Stocks/Industries/Industry_Analysis__Oilfield_Services_and_Equipment.aspx

[82] Bjurklo M (2008) Narrative Accounting. A New Form of Management Accounting? Int. Stud. Manage. & Organiz., 38: 25–43.